COMBINED ROOF-BOLTING SYSTEMS OF MINE WORKINGS

V. Bondarenko
I. Kovalevska
H. Symanovych
M. Barabash
V. Chervatiuk
O. Husiev
V. Snihur

Monograph

CRC Press
Taylor & Francis Group
Boca Raton London New York Leiden

2020

CRC Press is an imprint of the
Taylor & Francis Group, an informa business
A BALKEMA BOOK

CRC Press/Balkema is an imprint of the Taylor & Francis Group, an informa business

Typeset by Roman Lysenko & Olha Malova, Department of Underground Mining,
Dnipro University of Technology, Dnipro, Ukraine
Printed and bound by LizunoffPress Ltd, Dnipro, Ukraine

Published by: CRC Press/Balkema
P.O. Box 11320, 2301 EH Leiden, The Netherlands
e-mail: Pub.NL@taylorandfrancis.com
www.crcpress.com – www.taylorandfrancis.co.uk – www.balkema.nl

ISBN: 978-0-367-53330-4 (Hbk)
ISBN: 978-1-003-08143-2 (eBook)

INTRODUCTION

At the current level of stope works intensification in coal mines, the task of timeous preparation of new production sites is extremely topical, one of the directions of which is to reuse the preparatory mine workings. This technology is widely used in the mines, but it is not always possible to obtain the desired results in terms of a combination of resource savings and ensuring the required operating conditions for the reuse of mine workings. The successful implementation of this combination in technical solutions for the maintenance of reusable mine workings, in our view, is one of the fundamentals for intensification of coal mining in the Western Donbas.

Two main perspective components of the technology for mine workings reuse have been considered.

The first is to reduce the costs for maintaining mine working by involving in the work not only of border rocks, but also of more distant rocks in order to withstand the rock pressure forces. This direction has its own history already in terms of the wide application of load-bearing roof-bolting based on the resin-grouted roof bolt. The border rocks of the roof and side rocks are subjected to strengthening for a limited distance into massif, mainly up to 2.4 m, due to the moderate length of the roof-bolts. In recent years, the so-called "deep" strengthening of adjacent massif with rope bolts up to 6.0 m in length (mainly) are actively implemented. These bolts already allow forming a certain load-bearing structure not only from the rocks of immediate roof in the coal stratum, but also involving the underlying seams of rocks in the main roof. The load-bearing armored-rock structure (plate), formed in this way, takes up a part of the rock pressure which makes possible to unload the side rocks and the rocks of mine working bottom, frame support, as well as the elements of security system. Consequently, when using a combination of resin-grouted roof bolts and rope bolts (combined roof-bolting systems) to strengthen the roof rocks of mine workings, there is a real possibility to reduce the material consumption of traditional fastening and security systems, as well as the labor intensity of their setting.

The second component of the effective maintenance of reused preparatory mine workings is caused by the choice of rational parameters for their fastening (first of all) and then by subsequent protection (secondarily). The substantiation of the fastening parameters of mine workings is inextricably linked with the peculiarities of rock pressure manifestations, both in the primary operation of mine working as a belt entry, so in its reuse as a ventilation drift.

Previously, from the experience of maintaining mine workings with the traditional "frame-support-load-bearing roof-bolting" fastening system, it is necessary to note one of the main features – the predominant loading on the prop stays of the frame support with the relatively unloaded state of its cap board. In the opinion of experts, this behavior is due to the formation of high oblique and lateral loads on the frame while reducing the vertical rock pressure on the central part of the mine working arch by application of load-bearing roof-bolting. Against the background of negative phenomena (for example, rock swelling of the mine working bottom), like plastic deformation of the prop stays of the frames with bending into the mine working, to eliminate it with the preservation of the required horizontal dimensions of the drift is an extremely difficult and labor-consuming process. If ensuring the height of mine working has been technologically worked out and realized through the periodic bottom breaking and setting the prop stays of strengthening support, then preserving its acceptable width (according to operational standards and rules) is a difficult task. Its successful solution is based, in our opinion, on the study of the displacement processes of coal-overlaying formation (in the area of mine working location) by performing large-scale visual and instrumental observations of the drifts condition, revealing the main patterns of the rock pressure manifestations and the causes leading to their unsatisfactory operational state. Further, on the basis of generalization of experimental data, the concepts are formed about the features of the displacement mechanism of an adjacent massif of laminal rocks with a low hardness, and geomechanical models are developed to describe this process for a long period of time for primary and repeated use of mine workings.

At the final stage, the analysis of the results of multivariate computational experiments creates a scientific and technical basis for the development of calculation methods for the rational parameters of reusable mine workings fastening based on established and mathematically described patterns of the connection between the fastening means parameters and the geomechanical factors for maintaining mine working.

The need for such a sequence of studies is conditioned not only by the tendency for a more objective and adequate reflection of the geomechanical processes of loading the fastening system elements in mine working, but also by unsuccessful attempts (in some mining-geological conditions) to preserve the belt entry section for its reuse. In this regard, it is necessary to investigate the effectiveness of the application of certain structural and technical solutions for strengthening the adjacent roof rocks with combined roof-bolting systems.

In addition, despite the almost century-long history of the studying the phenomenon of bottom rocks swelling and the accumulated production experience in dealing with swelling, a common understanding of this process, as well as the effective ways and means for limitation or complete elimination of the negative effects of bottom rocks swelling, has not yet been formed. At the same time, it should be taken into account the considerable volumes of repair and restoration works with a low degree of their mechanization, which requires timely planning of measures to resist swelling, the manifestations of which are varied and ambiguous in the changing mining and geological situation of in-seam workings maintenance not only in a specific area of the mine field, but even within the length of a single mine working. Therefore, in the current, actively developing direction of continuous monitoring of coal mining technological processes, the task of assessing the bottom condition in the in-seam workings on a real-time basis and predicting the swelling for proved making technical decisions is timeous and very important. Thus, the substantiation and development of the method for predicting the bottom rocks swelling, which would reflect the features of the mining and geological conditions, is an important scientific and technical task for in-seam workings maintenance at different depths under conditions of intensive mining of thin and very thin flat-lying coal seams.

Chapter I was written in collaboration with R. Lysenko and K. Prokopenko.

The authors express their sincere appreciation to O. Malova for preparing the manuscript for publication and O. Stoliarska for examination the English text.

1. ANALYSIS OF ROCK PRESSURE MANIFESTATION IN THE PREPARATORY MINE WORKINGS AT THE STRUCTURAL TRANSFORMATION OF ENCLOSING ROCKS BY THE ROOF-BOLT STRENGTHENING

1.1. EXPERIENCE OF COMBINED ROOF-BOLTING SYSTEMS APPLICATION FOR STRENGTHENING THE ROOF ROCKS OF MINE WORKINGS

Analysis of the world and domestic trends in the use of roof-bolts in the structure of fastening systems in mine workings has shown the expansion in the area and in the volume of application of traditional resin-grouted roof bolts and rope bolts combination for strengthening the roof rocks as an independent support, as well as in a combination with other types of standing supports. Such technical solutions are termed differently, for example, "two-level roof-bolting", but the term "combined roof-bolting systems" is more accurate, since, in fact, it is used a combination of two types of roof-bolts that differ in their parameters and in tasks for strengthening the roof rocks in mine working.

In this regard, a wide range of computational experiments has been conducted on the calculation of the stress-strain state (SSS) of the massif around the preparatory mine workings (supported behind the stope face) and their fastening elements which integral components are combined roof-bolting systems in the roof of the drift. The designs of these schemes differ from each other mainly in the number of resin-grouted roof bolts in the roof and the parameters of their arrangement in combination with the rope bolts.

The research with the application of rope bolts in the fastening system in the mining and geological conditions of the "Stepova" Mine, DTEK "Pavlohradvuhillia" deserves close attention, where the influence of the rate of stope face advance on the intensity of rock pressure manifestations in the drifts is considered [1]. The positive effect of the rope bolts application has been obtained to limit the displacements of rocks in the coal-overlaying formation. The refusal from the prop stays for the strengthening support in the front bearing pressure zone and behind the longwall face is substantiated, due to the fact that the armored and rock plate in the roof of the drift is quite capable to limit the rock pressure manifestations and preserve the mine working section, sufficient for its reuse.

In the conditions of Kuzbas mines, a study [2] has been conducted to assess the effectiveness of the "two-level roof-bolting". An attempt was made here to calculate the parameters of the roof-bolts setting depending on the structure and mechanical properties of the roof rocks. However, the problem has been solved at the level of schematic concepts about the formation of rock prisms of sliding and the hypotheses about the arch of ultimate equilibrium, which can be considered as an initial stage to the description of the geomechanical processes of displacement in the inhomogeneous coal-overlaying formation, broken by natural and technogenic fractures into rock blocks and layers. The mining and geological conditions of the Kuzbas Basin are fundamentally different from those in Donbas. The reliability and durability of the “two-level roof-bolting” is noted, with a significant reduction in the material consumption and labor intensity of setting the fastening structures, which leads to a sharp decrease in the cost of maintaining mine workings.

The combination of frame support with roof-bolting systems is used in most European coal mines. However, in the work [3] it is asserted that there are no unified technical solutions for such fastening schemes, since the mining-geological and mining-technical conditions of coal mining are very diverse. These conclusions are based on instrumental observations of the mine workings state with various fastening schemes, including rope bolts.

A number of empirical methods used in Poland for designing the schemes for the combination of frame support and roof-bolting systems is analyzed. The results of mining studies indicate a sharp decrease in displacements of the rock contour of mine workings and their quite satisfactory condition when using the rope bolts systems. As a result of a three-year measurement period, the high efficiency of using the combined roof-bolting systems for maintaining mine workings in various mining and geological conditions for mining the coal deposits has been confirmed.

The roof-bolt support is widely used in the coal mines of China [4]: annually, about 8 000 km of mine workings are developed, of which 80% are fastened with the roof-bolts. The choice of setting schemes for the roof-bolts is based on accumulated experience and analogies in identical mining and geological conditions. For more substantiated recommendations on the parameters setting schemes for the combined roof-bolting systems, a complex of underground measurements and computational simulation using the FLAC 3D has been carried out to study the evolution of stresses in the process of drifting by the longwall face of the experimental site in the preparatory mine working. It is indicated the general similarity of the results in analytical and mining research, which has made it possible to offer several variants for combined roof-bolting systems application. Thus, a positive experience has been described of a computational experiment usage for the selection of rational parameters of setting the resin-grouted roof bolts and rope bolts in the preparatory mine workings in the zone of stope works influence. It is difficult to use these recommendations for the mines of other coal deposits, but the principle in itself of the combined roof-bolting systems operation is assessed as quite effective.

At present, a computational experiment on the study of the SSS of a rock massif around mine workings fastened by combined roof-bolting systems, is becoming more and more developed [5 – 8]. This scientific and practical direction should be considered very promising in view of the wide possibilities of numerical methods for describing the complex geomechanical processes.

As for taking into account the structural peculiarities of the surrounding massif (stratification, fracturing), then the mining instrumental observations are used with some hypothetical concepts [9 – 12] and the accumulated experience of fastening the mine workings with resin-grouted roof bolts and rope bolts [11 – 14].

1.2. A MECHANISM FOR INCREASING THE STABILITY OF MINE WORKINGS BY COMBINED ROOF-BOLTING SYSTEMS

The edge area of the coal seam in the preparatory mine workings is the most complex multifactorial object of geomechanical processes. Therefore, it seems expedient to assess the degree of influence of the roof-bolt strengthening on the mechanism of structural transformation of the roof rocks in the preparatory mine working.

The classical theory of bending the girders and plate [15, 16] asserts that the maximum horizontal stresses of tension (compression) σ_z arise in the surface areas of the girder (plate), and their value is inversely proportional to the square of the girder thickness (thickness of the rock layer). For example, if two rock layers of the same thickness are connected by roof-bolts, then the resistance to bending of the combined layers increases fourfold, while their total resistance increases only twice in the case of separate bending. Such a mechanism for increasing the stability of the bolted roof was studied in the works [17, 18].

Based on the data presented in Fig. 1.1 the mechanism has been illustrated of increasing the resistance to bending of two rock layers of the immediate roof, strengthened with the roof-bolts.

In case of separate deformation of rock layers, the tensile stresses have an increased value (see Fig. 1.1, *a*), and with a significant bending of the layer, they exceed the resistance of the rock to tensile stress – the development of the tension cracks begins on thickness of the layer. At the joint deformation of rock layers connected by roof-bolts, the moment of resistance of their section increases, causing a decrease in the stresses σ_z maxima, including tensile stresses. The process of formation and development of tension crack is being slows down and not every alternating bend of immediate roof strengthened with roof-bolts leads to the partitioning of doubled rock layers into blocks. In addition, with the increased thickness of the load-bearing girder (plate), its rigidity increases, which limits the amount of deflection of the layers and reduces the probability of destruction of this rock volume and the decomposition of the thrust system into separately descending blocks (see Fig. 1.1, *b*). The resistance to bending of strengthened rock layers is provided by the so-called restoring moment from the action of a pair of forces F as a result of stresses σ_z in the holistic area of the cross section. Undoubtedly, the process of partitioning into blocks also occurs in the immediate roof strengthened with roof-bolts, but with an increased length and thickness of the blocks.

Consequently, the number of interacting elements in the thrust system of the roof rocks decreases, and this increases the stability of the roof rocks – the reaction of their resistance to the rock pressure increases and the value of lowering into the mine working cavity decreases [19].

We note the general patterns of the anomalous zones formation of the state of coal-overlaying strata near the longwall face, which have been mined over many decades of studying this phenomenon and have created the basis for geomechanical research, regardless of the mining-geological and mining-technical conditions of coal seam mining.

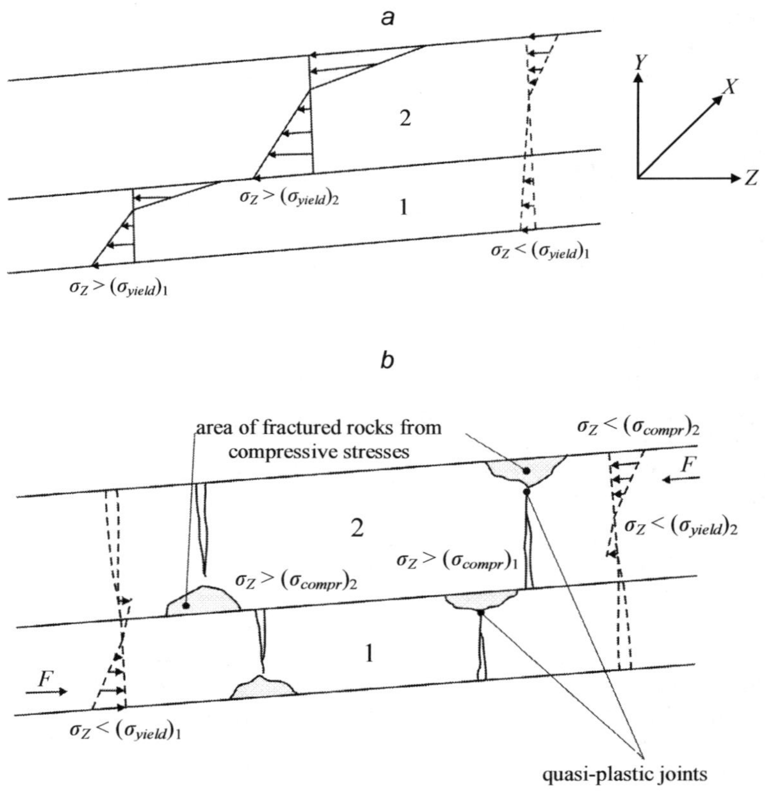

Fig. 1.1. The curves of horizontal stresses σ_z distribution on thickness of the roof rock layers (*a*) and the scheme of stability assessment (*b*) at separate (——) and joint (– –) their deformation

Ahead of the longwall face, a bearing pressure zone is formed, the length of which along mine working in the Western Donbas mines is assessed in the range of up to 20 – 40 m [20]. As the longwall approaches, the concentration of vertical stresses σ_y in the rocks massif increases, reaching a maximum at a distance of 2 – 15 m from the face. The value of the variable depends on the ratio of the rigidities of the coal seam and the roof rock layers.

The change in the initial state of the massif at the approach of the longwall face induces the weakening of soft coal-bearing rocks, especially of the immediate roof and bottom, and, as a consequence, a sharp increase in their deformation. At a short distance from the face, the vast majority of the vertical displacements U_y is realized by lowering the easily deformable roof layers. The reduced coal seam deformations contribute the approximation of the rock pressure maximum to the stope face. The traditional for the Donbas rocks, the so-called "non-rigid pinching" the rock cantilevers above the longwall face in the Western Donbas acquires properties approaching the concept of "rigid pinching" (used in structural mechanics) with a concentration of the bending moment in the rock cantilever in the area of the face. This induces an intensive gradient of displacements growth of the rock contour of mine working in a limited area ahead of the longwall face with length of up to 20 – 40 m [21 – 32]. Here, the rate of the roof and bottom rocks convergence, as well as the sides of mine working, sharply increases from a fraction of mm/day to one and even a few tens of mm/day, according to various studies.

The traditional technology of maintaining the preparatory mine workings in the bearing pressure zone ahead of the longwall face is connected with setting the central prop stays of the strengthening support, mainly in the form of wooden (rigid) prop stays. The central prop stays of the strengthening support resist to the vertical rock pressure (in combination with the frame and the bearing roof-bolt fastening), but do not increase the resistance of the frame to lateral loads.

A number of anomalous areas are formed around the preparatory mine working, associated with the redistribution of rock pressure and the processes of rock weakening. This has been established by numerous studies, including the modeling of the state of the border rocks and support in the zone of the stope works influence when using the bearing roof-bolt fastening. The peculiarities of the enclosing rocks state are as follows.

On the one hand, in the border rocks of the roof an area of stratification is formed from the action of vertical tensile stresses σ_y, as well as horizontal stresses σ_z and σ_x, which activate the process of weakening. It is known that the weak rocks, for example, of the Western Donbas have very low resistance to tensile stress (usually up to 1.5 – 3.5 MPa), moreover, due to the action of factors weakening the rock, such resistance is virtually absent. Nevertheless, because of the dense reinforcement grid of the roof with roof-bolts, the area of stratification in the form of an arch is not so significant, and the weight of rocks inside the arch is many times less than the resistance of the frame support to vertical loads. In addition, if the length of the roof-bolts usually exceeds the arch dimensions, then they themselves (due to the crown runners and the metal grid of the interframe fence) are quite capable of retaining the volume of the loose rocks.

Another feature of the coal-bearing thin-bedded strata of the soft rocks is the very weak adhesion over the contact surfaces of the constituent rock layers; in many areas such an adhesion is practically absent. It should be also taken into account the action of geostatic anomalies of rock pressure near mine workings, which disturb the already weak contacts. The rock layers near mine working are partially deformed independently of each other [33 – 42].

It has been determined by means of analytical, numerical and experimental research methods that in the sides of the mine working, the increased rock pressure weakens a certain volume of soft rocks, which increases their mobility. The descending overlying rock layers load slightly the underlying volumes of rocks, including the weakened ones in the immediate roof. The coal seam restricts the vertical displacements of the lateral rocks and their displacement vector changes the direction into the cavity of mine working, loading the frame prop stays. This process in the bearing pressure zone is developed more intensively due to the increased vertical component of the rock pressure. The intensification of the "stamp" effect also occurs in the rocks of immediate bottom. From the side of the longwall face, similar processes of pressing-out the rocks from the sides and the bottom of mine working are developed, but they are intensified by the deformation of the rock blocks. Therefore, there is an asymmetry of the load along the entire perimeter of mine working, which should be taken into account when constructing geomechanical models and calculating the parameters of the combined fastening system.

A schematic representation of the displacement mechanism of the coal-overlaying strata in the cross section of the preparatory mine working is supplemented with the consideration in its longitudinal direction. There is a generally accepted pattern of smooth bending of rock layers of the roof ahead of the stope face and their layer-by-layer collapse into the mined-out space with the formation of rock cantilevers of variable length [25 – 34, 43 – 49]. For Western Donbas, there is a characteristic disturbance of contacts between layers of coal-bearing strata and a very low resistance to tensile forces with developed fracturing (usually two or three basic systems of fractures) condition some specificity of the process of deforming the coal-overlaying formation ahead of the stope face. The increase in the bearing pressure as the longwall face approaches, intensifies the relatively independent deformation of the inhomogeneous layers of the coal-overlaying formation, at bending of which horizontal tensile stresses σ_x occur (in the upper area of each layer thickness) even on the approach to the maximum of the bearing pressure. In weak fractured rocks, the extremely small tensile σ_x are enough for the nucleation and growth of fractures of technogenic origin. In the section of the layer where the fracture was formed, the resistance moment decreases sharply and further development of fractures is most likely here (as the longwall face approaches). These fractures eventually join with the area of broken rocks (due to the concentration of compressive σ_x) in the lower parts of each layer and their partitioning into rock blocks occur.

From the viewpoint of the preparatory mine working stability, the formation of fractures in the adjacent rock layers ahead of the longwall face increases their deformation and reduces the resistance to the bearing pressure, which inevitably leads to an intensification of the process of pressing-out the rocks in sides and bottom of mine working. Therefore, to improve the reliability of prediction of rock pressure manifestations in the geomechanical models being developed, it should be taken into account the discrete disturbance of rock layers, which is intensified as the stope face approaches. Here, one of the main questions is how to take into account this discrete disturbance, for which the fact has been considered of deformation of a partially disturbed rock plate with curves of horizontal stresses σ_x and σ_z distribution, and due to the action of which a reaction of resistance to rock pressure occurs (see Fig. 1.1). In this view, between the partially disturbed rock plates in the bearing pressure zone, a similarity of the thrust system occurs, which, on the one hand, has increased deformation, and, on the other hand, a certain load-bearing capacity due to the action of only compressive stresses σ_x and σ_z on the thickness of the layer. In the absence of tensile stresses σ_x and σ_z in the section of the rock plate, its load-bearing capacity is realized due to the restoring moment from the shoulder of the equivalent forces on opposite transverse planes YX and YZ of the plate as integral values of the compressive σ_x and σ_z, over the areas of their distribution on the end planes. Such a mechanism of deformation and interaction of adjacent rock layers with discrete spatial disturbances is more reliable than the traditional concepts about the continuity of the massif (in the zone of bearing pressure ahead of the longwall face) and the occurrence of some areas with disturbed rocks.

Thus, the conclusions and recommendations have been formulated on the principles of creating a geomechanical model for the displacement of the massif and loading of the support in the preparatory mine working in the zone of the stope works influence.

First, the partial weakening of the rock layers of the coal-overlaying formation in the form of discrete disturbances in the areas of tensile stresses (mainly horizontal σ_x and σ_z) and high concentrations of compressive stresses. This leads to an increase in the volumes of rocks exposed to active deformations, and their separation into a system of interacting plates intensifies the process of pressing-out the partially weakened border massif in the sides and bottom of mine working.

Secondly, the creation and improvement (toward the way of increasing the reliability and objectivity) of the geomechanical model should go through a series of successive stages, reflecting the real process of the origin and development of discrete disturbances in the thin-bedded coal-overlaying formation of weak rocks.

1.3. THE INTERACTION OF SUPPORT IN MINE WORKINGS WITH BORDER ROCKS UNDER THE CONDITIONS OF AREAS FORMATION OF THEIR LIMITING STATE

The development of views and the formation of hypotheses about the interaction of the support of mine workings with the enclosing rock massif has more than a century of history. By now, there are definite ideas about the parameters of the loading process of various types of supports with account of their operating mode [50]. The term "operating mode of mine working" means [51] the ratio of the repulse reaction q of support with its displacement u under the action of the load (in principle, the pliability of support). This link is called the deformation-strength characteristic of the support and is represented as a function $q(u)$. This function was previously considered in [50] in a two-parameter formulation (parameters q and u), and later, in the course of the concepts development about the interaction of rock massif and support, more general ones were proposed: three-parameter formulation (the third parameter is the perimeter of the cross section of mine working) [51], and the four-parameter scheme [52], where the longitudinal coordinate of mine working is also used. The conditions were investigated of maintaining mine working outside the zone of stope works influence, in which the intensity of the rock pressure manifestations is significantly lower than in the immediate vicinity of the longwall face. In addition, the areas of disturbed rocks around a single mine working are quite local in comparison with the zones of active displacements of the coal-bearing massif in the course of performing the stope works. Nevertheless, in our opinion, the nature of the occurrence and course of geomechanical processes has a common basis, regardless of the conditions for maintaining mine workings, for example:

– formation of unloading zones and increased rock pressure in the vicinity of mine working;

– formation of areas of weakened and broken rocks, their interaction with the support and holistic massif;

– development of stratification in the planes of weakening, in the thickness of the lithological variety, and in the planes of laminations of adjacent lithotypes;

– partitioning of the rock layer into blocks by fractures perpendicular to the planes of weakening and lamination, as well as the interaction of these blocks during mutual displacement relative to each other;

– an increase in the volume of loosening rocks in constrained conditions of deformation (generates increased rock pressure with the need for its redistribution);

– zonality of changes in the mechanical properties of the rock massif and its structural transformations in the vicinity of mine working.

This is by no means all general positions in terms of the mechanism of formation of the rock pressure on support, but they are enough to use the patterns obtained for single mine workings in the process of studying the tendencies of connection between the operating modes of support and the stability of preparatory mine workings maintained in the zone of stope works influence.

This approach is conditioned by two reasons: firstly, the patterns of the rock pressure manifestations and the process of interaction between the rock massif and the support for single mine workings have been studied more thoroughly; secondly, the processes of interaction of the massif with the support for single workings are exposed to the influence of fewer factors, and the pathway of knowledge from a simple phenomenon to a more complex one seems to be the most promising in accordance with the classical canons of philosophical thought.

In the light of the foregoing, a scheme is considered (for simplicity, a two-parameter one), of the interaction between the support of the mine working and the surrounding rock massif under the conditions of areas formation of its limiting state. The emphasis on these areas is made because of the increased interest precisely to the conditions of intensive rock pressure manifestation.

There are two known tendencies [50 – 52] in the formation of the load on the support, which are illustrated in Fig. 1.2 by two dependences:

- *the process of the rock massif deformation into the cavity of mine working* (Fig. 1.2, curve *1*), which has a tendency to reduce the load P (and, accordingly, the support reaction q) with increasing its pliability u. In a situation with a certain value of pliability $u_{\max}$, the load on the supports disappears; the pattern *1* reflects the process of SSS redistribution around mine working, when an excess volume of rocks, which in the superlimiting state moves in the direction of least resistance, and, if it is not "interfered" with the help of support reaction, then there is possible an occurrence of a "virtual" equilibrium state of $P = 0$ at $u = u_{\max}$;
- *the process of stability loss by a part of the rock volume in the limiting or superlimiting state* (see Fig. 1.2, curve *2*). To restore their stability, a support reaction is necessary, which tends to grow with an increase in pliability, since in conditions of quasi-free deformation of the rock massif, unstable volumes of rocks are distributed along its depth.

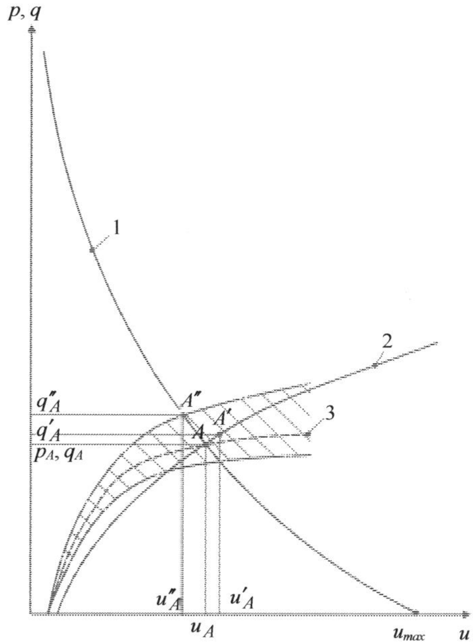

Fig. 1.2. Scheme of interaction between the support of a single mine working and the rock massif

Two mutually opposite tendencies predetermine the existence of a minimum load on the support P_A (in Fig. 1.2 it is shown as a point A at the intersection of curves *1* and *2*), for the determination of which it is necessary that the deformation-strength characteristic of the support $q(u)$, which is represented by curve *3* in Fig. 1.2, was passing through the point A. According to the canons of mathematical analysis, the point A with pliability parameters u_A of the support at its reaction q_A, reflects the only optimal solution to the problem of minimizing the load on the support. It is impossible to achieve such an optimal solution in practice for different structural and technological reasons, which give some variation in the values $q(u)$. Therefore, it is more correct to speak about a certain range of fluctuations of the deformation-strength characteristic of the support, shown in Fig. 1.2 with the shaded area of the spread of the function $q(u)$ values.

Within the interval of the function fluctuation, the points A' and A'' are singled out, which are approximated to the point A of optimal value; they outline the areas of the rational values variations of the support reaction $q'_A - q''_A$ at its pliability $u''_A - u'_A$ (see Fig. 1.2). The point A' is determined by the intersection of the vertical at an increased pliability u'_A (the vertical passes through the point of intersection of the lower boundary $q(u)$ with curve *1*) of curve *2*, since an increased pliability of the support (in comparison with the optimal one) leads to an increase in load in accordance with the second tendency to its reduce according to the first one, in this case, not dominant. With limited pliability u''_A (point A''), the degree of influence of both tendencies is changed diametrically opposite: the increased load q''_A is shown by curve *1*.

Thus, the above scheme of the interaction of the support with the surrounding rock massif makes it possible to select rational intervals for variation of the support reaction $q'_A - q''_A$ and its pliability $u''_A - u'_A$ with the known dependences (curves *1* and *2* in Fig. 1.2) of the behavior of the weakening adjacent massif and the formation of the unstable rocks area. It should also be clarified that at the time of the geomechanical processes development (in terms of parameters q and u), proceeding from the mine working stability, it is expedient that the deformation-strength characteristic of support (curve *3*) was located above the curve *2*; then, at any time, the support is capable of bearing the stable rocks.

The presented data are reflected (to a certain extent) in the normative and technical documentation [53], where the methods for calculating the load on the support are given, depending on its pliability; i.e., the influence of the deformation-strength characteristic of the support on the loading process in specific mining and geological conditions is taken into account. Of particular interest is the assessment of the influence degree of the structural pliability of the support on the change in the so-called normative load, according to which its required parameters are subsequently calculated.

When considering this interaction, the four values of the ratio of structural pliability of conditionally rigid u_r and conditionally pliable u_p supports are taken as an example: $u_r / u_p = 0.1$; 0.25; 0.5 and 0.75. For these values, the graphs have been constructed of the ratio P_r / P_p of the normative load on more rigid P_r and more pliable P_p supports depending on the geomechanical conditions for maintaining the mine working, which in the normative documents are characterized by depth H of the mine working location and by the average calculated compressive resistance R of adjacent rock massif. The results of the calculations are shown in Figs. 1.3 and 1.4. They are interpreted as follows: the patterns of ratios P_r / P_p and u_r / u_p relation are sufficiently stable regardless of the H / R parameter, and characterize the general tendency of reducing the load on support with an increase in its pliability. Thus, with a high rigidity of the support ($u_r / u_p = 0.1$) a load P_r is formed on it, exceeding the load P_p on a pliable support by 7.1 – 7.3 times (like a frame support from the special SCP profile); in the process of increasing the pliability (till $u_r / u_p = 0.75$) the ratio of these loads decreases to 2 – 2.95. The lower values of the given intervals correspond to more favorable conditions for maintaining mine working ($H / R = 21$ m/MPa), and the upper ones to the intense rock pressure manifestation ($H / R = 54$ m/MPa). However, the relative deviation of the P_r / P_p parameter is 3 – 32% for such a wide range of changes in mining and geological conditions for maintaining mine working. Moreover, there is a tendency to a decrease in deviations with an increase in the difference in pliability of the compared supports. The constancy of patterns, probably, indicates their sufficient objectivity in predicting the development of the load on the support, with account of pliability.

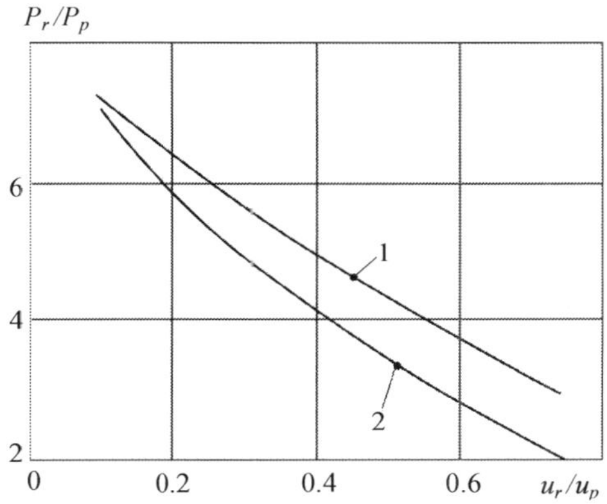

Fig. 1.3. The dependence of the load ratio P_r/P_p onto a rigid and pliable supports on the ratio u_r/u_p of their pliability: 1 – $H / R = 21$ m/MPa; 2 – $H / R = 54$ m/MPa

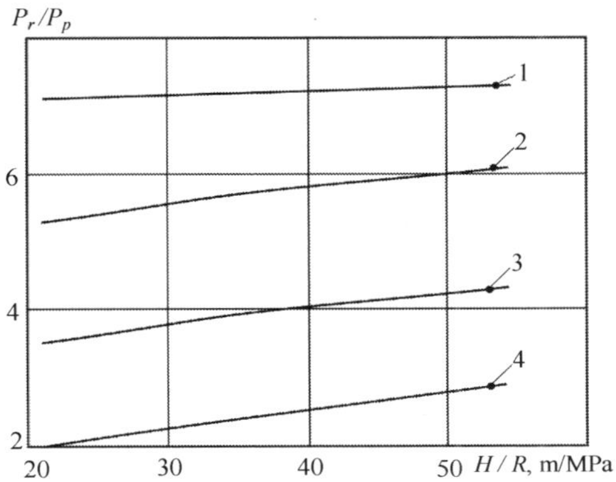

Fig. 1.4. The Influence of the geomechanical parameter H/R on the load ratio P_r/P_p onto a rigid and pliable supports: 1 – $u_r/u_p = 0.1$; 2 – $u_r/u_p = 0.25$; 3 – $u_r/u_p = 0.5$; 4 – $u_r/u_p = 0.75$

Thus, in accordance with the performed analysis of the formation of the load on the support of mine working, outside the zone of stope works influence, it is possible to state that there is a significant effect of the operating mode of support on this process. The current ideas about the interaction of the support with the rock massif (in the conditions of the unstable equilibrium areas formation) reveal sufficiently the mechanism of the phenomenon and generalize both the production experience and the results of analytical studies.

The opinion about the existence of a certain rational mode of support operation will be very useful when studying the geomechanical processes around the preparatory mine working, located in the zone of stope works influence. Therefore, further studies should be done with account of the assessment of the link between the deformation-strength characteristic of the fastening system of mine working and the degree of its stability during drifting the stope face in the area from the bearing pressure zone till the zone of stabilization of the displacement processes in the coal-overlaying formation behind the longwall face.

Thus, based on the analysis and study of the theoretical basis for the formation of the main provisions for the occurrence and course of geomechanical processes around mine workings, the patterns have been obtained of changing the load on rigid and pliable supports on the basis of solving the problem of support interaction with border rock massif in the conditions of formation of areas of its limiting state. The influence dependence has been studied of the ratio of the depth for maintaining mine working and the average calculated compressive resistance of an adjacent massif, which makes it possible to predict the development of the load on the support with account of its pliability.

CONCLUSIONS

The analysis of modern works in the field of maintaining the preparatory mine workings in the zone of stope works influence by combined roof-bolting systems testifies that there is a perspective of their wide-scale application:

– firstly, it is very important to take into account the peculiarities of mining-geological and mining-technical conditions for maintaining mine workings as a determining factor in the choice of parameters for roof-bolt fastening, as indicated by most researchers;

– secondly, the most promising method for studying the data of complex geomechanical systems is a computational experiment in combination with the mining instrumental observations;

– thirdly, when performing the computational experiments, the structure of an adjacent massif, its natural and technogenic disturbances should to be taken into account obligatory, which will make it possible to increase the adequacy of research and the accuracy of the recommendations being developed for parameters calculation of the combined roof-bolting systems.

2. RESEARCH AND ANALYSIS OF THE EXPERIENCE OF THE PREPARATORY MINE WORKINGS REUSE AT YUBILEINA MINE

2.1. MINING RESEARCH METHODOLOGY

The 590 belt entries and the 592 upper entries of the C_6 seam mine field have been selected as an object of mining research performing and being the variants for various basic schemes for the roof-bolts setting in the roof. The research was carried out in the area of the eastern wing of the inclined part of the C_6 seam mine field, where three panels of 590, 592 and 594 longwall faces were mined. Mining and geological conditions are approximately the same both in depth of development and in structure, as well as the properties of the coal-bearing strata.

At the baseline of research, the 590 belt entry has been driven with a internal section of $S_{int} = 15.0$ m^2, with setting the TSYS-15.0 support, with a step of 0.8 m. The resin-grouted roof bolts according to the scheme in Fig. 2.1 were set in the interframe space with a step of 0.8 m. When mining the 590 longwall face, the mine working was used as a belt entry, and in the course of stope works in the adjacent 592 longwall face, the mine working was re-used as an upper entry. The mine working state was assessed according to indicators of the rock pressure manifestations in the form of movements of the drift contours in key areas of the cross section: general convergence towards the roof – the bottom and sides of mine working, the change in the height of the drift in the zone of people's passage, the diagonal from the angle in the bottom of mine working to the upper end of the frame prop stay in the opposite flank of the drift, the height from the lower edge of the coal seam to the bottom of mine working, the degree of the yielding joists actuation; in addition, a visual assessment was made in the geometry change of the frame cap board (its flattening or bending in the area of setting the central prop stay of the strengthening support) and its prop stays in the form of converging their bearings, the SCP profile twisting or rectilinear bending of prop stay part into the cavity of mine working.

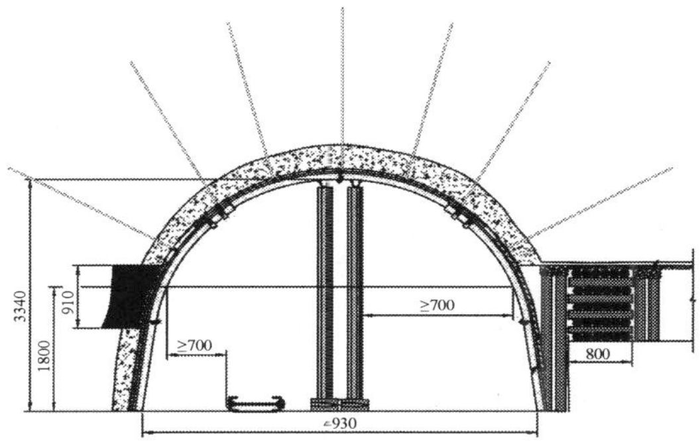

Fig. 2.1. Scheme for maintaining the 590 belt entry of the C_6 seam

To specify the measurements of the current dimensions of the drift, the corresponding scheme is shown in Fig. 2.2. In time and space, the fixation of the mine working dimensions was made on a case by case basis. In the area outside the zone of the stope works influence (not less than 60 – 80 m to the stope face), the measurements were made once every 3 – 4 days; in the area of the front bearing pressure action ahead of the longwall face, in the area of its "window" and behind the longwall face, at a distance of at least 60 to 80 m, the readings were fixed once a day; then, in the area of stabilization of the rock pressure manifestations, the measurements were made once for every 3 to 4 days. Such a technique made it possible with the use of metering station to reflect more accurately the different intensity of movements development of the mine working contour during the stope face passage of the drift section.

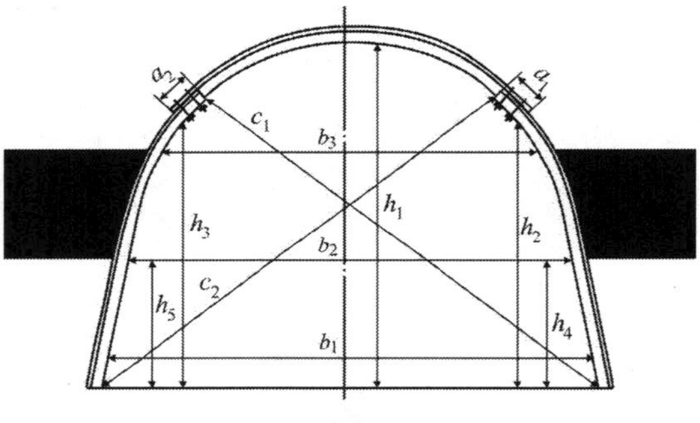

Fig. 2.2. The scheme of measuring the current dimensions of the preparatory mine workings as the stope face advances

For the completeness of the representation about the curve of the rock contour movements in mine working and the deformation features of the frame support, it is placed a sufficient number of measurement points (see Fig. 2.2). Thus, the height h_1 of mine working from the arch keystone to the bottom and its width $в_1$ (at nearby of the bottom) gives the general data on the section loss when compared with the design dimensions of the drifts. The parameters b_2 and b_3 are not less important to assess the section loss of the drift, but also for accounting the convergence of its sides in the zone of people's passage. In addition, a comparison of the values of the parameters $b_{1,2,3}$ change makes it possible to identify the areas of the greatest deformations of the frame prop stays that correspond to the areas of increased rock pressure in the mine working sides. The measurement of the drift height $h_{2,3}$ in the working and non-mining flanks gives an opportunity to assess two positions at once: the convergence of the roof and bottom in the zone of people's passage, refining the remaining height of mine workings in this place; the degree of asymmetry of the frame deformation in different sides of the drift. The two more parameters $c_{1,2}$ are tending to refine the level of frame unevenness, and they measure the diagonal from the drift angle to the upper end of the frame prop stays in the opposite flank of mine working. If initially the same values c_1 and c_2 vary in different ways, this indicates the non-uniform loading of the frame. In addition to the specified parameters for assessing the load asymmetry, the overlap value $a_{1,2}$ of a cap board and a prop stay was fixed in the yielding joists of the frame; at a $a_1 \neq a_2$, there is a concentration of the load from the side of a larger overlap value. In addition, by the values $a_{1,2}$, the performance of the frame support in the yielding mode is assessed: how far the joist was actuated, and whether its constructive pliability was exhausted. At the same time, the quality of the connection is visually assessed between the cap board and the frame, as well as possible disturbances of the yielding joist itself. The two dimensions $h_{4,5}$ have still remained, which determine the distance from the lower edge of the seam to the bottom of mine working. These dimensions indirectly mark the value of swelling the drift bottom rocks with a certain error, which is conditioned by deformation of the edge part of the seam under the influence of the bearing pressure.

In general, in our opinion, the described measurements make it possible to fully reflect the real rock pressure manifestations in the drifts that ensures the objectivity of a comparative analysis of various schemes for setting the roof bolts in the roof of mine workings. The measurement results are presented in the form of graphs of a particular dimension changes depending on the distance z to the stope face. Thus, the dynamics of the increase in movement of mine working contour is observed, covering all the most characteristic areas of its maintenance.

2.2. ANALYSIS OF MINING RESEARCH RESULTS

The consideration of visual and instrumental observations results of the drifts state was carried out sequentially, in the process of mining the panels within the above area of the C_6 seam mine field at the "Yuvileina" mine. However, to reduce the volume of the book, the results are presented of the analysis of the 590 belt entry state, when mining the 590 and 592 longwall faces of the C_6 seam. Therefore, two series of fixations of the mine working contour movements were made at the metering stations – during the drifting of the 590 longwall face, and then, after a long period of time, when approaching and removing the 592 longwall face from the metering station. Thus, it was possible to assess the state of reusable mine working during the whole period of its operation.

The metering stations were located at different sites lengthwise the 590 and 592 panels of the longwall faces:

– PK47 was located in the area of the face entry of the 592 longwall face, and the rock contour movements were fixed here when the longwall face recedes, that is, during the primary roof caving, and its secondary roof caving happened already at a sufficient removal from the PK47 and had insignificant influence; therefore, the period of mine working deformation was estimated at the PK47 behind the stope faces from the active phase not far from the longwall face to the zone of rock pressure stabilization at its removal for more than 60 – 80 m;

– PK31 was located at a distance of about 290 – 320 m from the face entry and during the periodic readings fixation, all the characteristic zones for maintaining the 590 belt entry were considered: outside the zone of stope works influence, in the bearing pressure zone, the "window" of the longwall face, the zone of the unstable rock pressure in the immediate vicinity behind the longwall face and in the area of stabilization of the displacement processes of the coal-overlaying formation;

– the metering stations at PK9 – PK11 were located at the end of the 590 and 592 longwall face areas at a distance of about 60 – 100 m from the stopping lines of the stope faces and also made it possible to assess most of the above areas for maintaining the 590 belt entry with some restrictions on the end zones – there is no stope works influence and stabilization of the rock pressure manifestations.

It is necessary to point out one more feature of making measurements, related to the technological difficulties of taking some readings due to the equipment placed in mine working. Therefore, at separate metering stations, it was made only a partial fixation of the movements available for the measurement. Nevertheless, when analyzing the patterns of changes in the rock contour movements of the 590 belt entry, depending on the distance z to the stope faces, all measured distances are present, obtained by averaging the readings on individual pickets at the same equal distance z to the longwall face.

Fig. 2.3. Fragment of the 590 belt entry in the area from PK0 to PK4

The overall assessment of the 590 belt entry state has been made, starting from the top of the extraction panel of the 590 longwall face, that is, during the initial use of this mine working. A survey of the 590 belt entry, starting from the incline from the West haulage gate No. 3 to PK8, has shown its stable state without any special rock pressure manifestations, which could draw any attention. It is evidenced by the fragment of mine working in the area of PK0 to PK4 (Fig. 2.3). This situation is quite expected, since this area was located in the protecting pillar of the main mine workings outside the zone of stope works influence – even at their end in the 590 panel, the stopping lines of the 590 longwall face was at a distance of not less than 35 – 40 m from the indicated section of the drift.

At PK9 metering station, the effects of the rock pressure manifestations have been already fixed, when the mine working fell into the zone of stope works influence. A few meters before the “window” of the longwall face, the following measurement results were obtained:

– the height of mine working was growing smaller by 690 mm, despite the fact that ahead of the longwall face the bottom ripping was made by an average of 980 mm; that is, a significant development of vertical rock pressure has been fixed;

– the yielding joists of the frame have actuated on average for 160 mm from the value of the initial overlap, and therefore, to a certain extent, thay have fulfilled their functions of protecting the frame from excessive rock pressure; the assessment of efficiency of their work is ambiguous: on the one hand, after bottom ripping, its swelling was actively developing due to a very weak compressive resistance ($\sigma_{compr} = 4.2 - 7.5$ MPa) of water-saturated siltstone and the above-mentioned loss of the drift height was partially compensated by bottom heaving, the value of which could not be fixed (in the area of the longwall face "window" for technological reasons); on the other hand, the areas of their plastic bending with loss of the initial form were observed in the cap boards of the frames, and, therefore, the actuation of yielding joists for 160 mm was insufficient; in addition, the wooden central prop stays of the strengthening support obstructed the normal operation mode of the yielding joists;

– other dimensions according to the scheme in Fig. 2.2 could not be fixed (at a given distance to the longwall face) due to the density of technological equipment in the mine working area.

In general, despite the proximity of the longwall face "window", the drift was in satisfactory condition according to the main dimensions and gaps necessary for performing the technological operations. Of course, there were disturbances of the initial geometry of the cap board and prop stays of the frame because of their plastic deformations, the gusts of the interframe fence grid with partial pressing-out the side rocks from the side of the 590 longwall face (Fig. 2.4), but these disturbances did not have a significant impact on the rate of stope works operation.

Fig. 2.4. The state of the 590 belt entry in the "window" area of the 590 longwall face

Simultaneously with taking of readings at PK9, the movements of the mine working contour were fixed at the metering station PK10, located (at that time) at a distance of about 11 – 13 m behind the stope face, that is, in the zone of unstable displacement of the coal-overlaying formation. Fig. 2.5 shows the fragments of the mine working state in the area of PK10, in which the following features are observed:

– the frame cap board experienced plastic deformations in the form of its bending from the side of the mined-out space of the 590 longwall face, which was induced by two rows of rigid wooden prop stays of the strengthening support; therefore, there was no proper actuation of the yielding joists in the working flank of mine working;

– from the side of the non-mining (at that time) flank, some flattening of the frame cap board was observed, but these deformations did not interfere with the actuation of the yielding joists, – an average overlap is 700 mm, which is by 300 mm larger than its initial value;

– visible plastic deformations of the frame prop stays were not observed near the non-mining flank, and from the side of the 590 longwall face there was a certain bending of the prop stays into the mine working cavity.

In this section of the drift, only vertical convergences of the 450 mm level have been fixed, which is 240 mm less than at PK9 near the “window” of the long wall face. This can be explained by the fact that the PK10 was already partially located in the unloading area over the mined-out space of the 590 longwall face in the absence of a front bearing pressure, which loads the cap board of the frame and causes an intensive bottom rocks swelling.

a

b

Fig. 2.5. Fragments of the state of the 590 belt entry behind the 590 longwall face (at a distance of 11 – 13 m, PK10) in the area of unstable displacement of the coal-overlaying formation from the side of non-mining (*a*) and working (*b*) flanks of mine working

The metering station PK11 was already located at a distance of more than 30 m behind the longwall face; here, the complete stabilization of displacement in the coal-overlaying formation has not yet begun, but rheological phenomena have already begun to develop in the form of creeping the deformations of mine rocks. The measurement results indicate the following changes in the frame support geometry:

– the height of mine working has decreased by 790 mm, and its width by 630 mm; these movements affected seriously the value of the section loss, but were not an obstacle to the performance of technological operations thanks to the increased initial dimensions of the 590 belt entry for the TSYS-15.0 support;

– the yielding joists worked relatively reliably with an increase in the overlap of the cap board and the frame prop stay up to 700 mm, that is, to a certain extent they provided “deviation” of the frame from excessive rock pressure;

– the asymmetry of the support form was weakly manifested: the difference between the diagonals c_1 and c_2 was only 100 mm; the diagonals themselves decreased more significantly: c_1 – by 1070 mm, c_2 – by 970 mm; thus, with a relatively small asymmetry in the frame deformation from the side of mined-out space of the 590 longwall face, quite significant movements were observed throughout the entire contour of the drift.

The specified features of the rock pressure manifestations are confirmed by the fragments of mine working at PK11, shown in Fig. 2.6 Indeed, the asymmetry of the frame form is poorly visible, the condition of the yielding joists is quite working, because of which the plastic deformations of the cap board and prop stays are low-observable. In general, the state of the 590 belt entry at a distance of at least 30 m behind the 590 longwall face can be assessed as satisfactory.

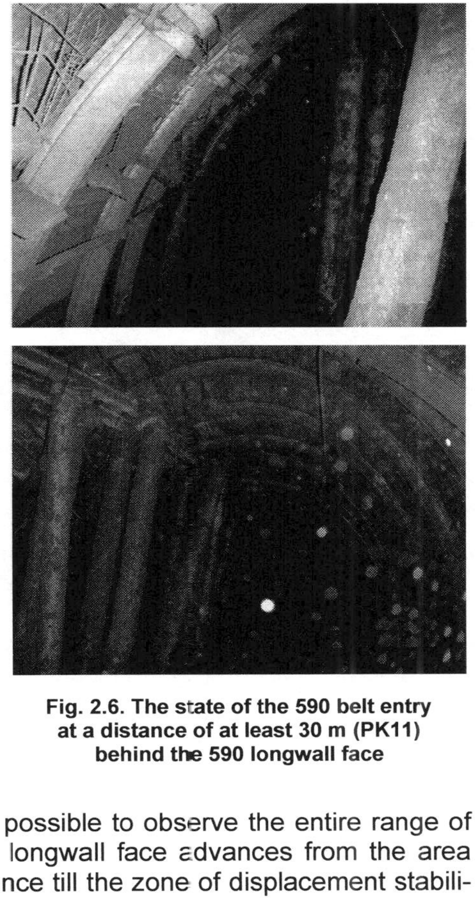

Fig. 2.6. The state of the 590 belt entry at a distance of at least 30 m (PK11) behind the 590 longwall face

The metering station at PK47 was located in the area of the 592 face entry and in the whole coal in relation to the face entry of the 590 longwall face. Therefore, the stope works influence in the 590 longwall face is not so significant. The mine working was in a satisfactory state with some deviations (from the initial section) conditioned by the action of rheological factors:

– the height of the drift decreased by 490 mm, but here the main contribution was made by the swelling of soft bottom rocks;

– the width of mine working has decreased by 700 mm, which is caused by the convergence of the frame prop stays in the area of their bearings during the horizontal shifts of bottom rocks into the mine working cavity;

– the overlap of the cap board and the frame prop stay was 350 mm, that is, the yielding nodes have not actuated yet, which once again indicated a loss in the height of mine working due to its bottom swelling.

The metering station at PK31 made it possible to observe the entire range of rock pressure manifestations as the 590 longwall face advances from the area outside the zone of the stope works influence till the zone of displacement stabilization in the coal-overlaying formation.

The change in the drift dimensions at the PK31 before the beginning of the stope works is practically the same as compared to the PK47, which is quite expected due to the similar geological and mining conditions for maintaining the 590 belt entry in these areas. Such a stability in the state of mine working over its length of more than 300 m, made it possible to take it as the initial one when assessing the stope works influence only, but the impact of geomechanical processes in the massif before mining the 590 longwall face was not taken into account. Thus, to objectively analyze the development of the contour movement of the 590 belt entry depending on the distance z to the longwall face, a separation was made of directly the stope works influence and the impact of geomechanical processes that occurred before the beginning of mining the 590 longwall face.

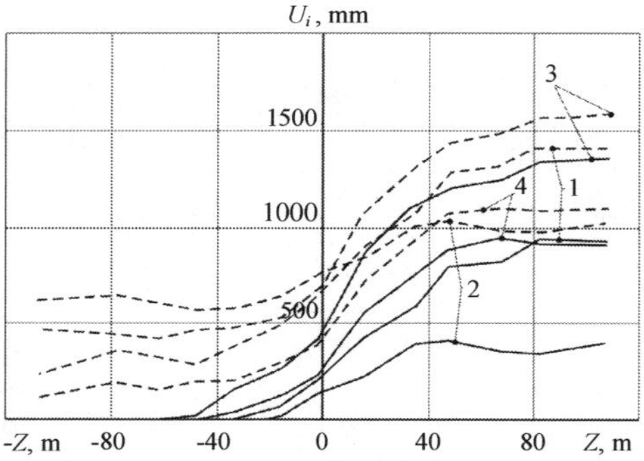

Fig. 2.7. Dependences of U_i movements of the 590 belt entry contour on the coordinate Z of the 590 longwall face position till the measured section: 1 – U_{h_1}; 2 – U_{b_1}; 3 – U_{c_1}; 4 – U_{c_2}; general movement of the i-th measurement parameter; share of movements from stope works influence

The results are presented in the form of graphs of changes (Fig. 2.7) in any fixed dimension depending on the distance z of the specified section from the 590 stope face. Since the measurements were made at several pickets, the results obtained were averaged.

For each fixed dimension of mine working, two graphs are given:

– the dotted line marks the contour movements in relation to the initial dimensions of the TSYS-15.0 support, according to the passport of conducting the 590 belt entry. These data give an idea of the general situation of maintaining mine working, including the need for its periodic bottom ripping;

– a solid line shows the contribution to the convergence of the rock contour of the stope works influence, which was determined as the difference between the general displacements and movements that occurred during the maintenance of the mine working section outside the zone of the stope works influence; such an indicator makes possible to assess more objectively the effectiveness of counteraction of the drift fastening system to the rock pressure manifestations in the zone of active displacement in the coal-overlaying formation.

The analysis of vertical convergences U_{h_1} of the roof and bottom allowed to establish the following. Outside the zone of the stope works influence, a decrease has been fixed in the mine working height in the interval of 420 – 470 mm; most of U_{h_1} is conditioned by the development of bottom rocks swelling, but according to the value of the yielding joists actuation and the state of the cap board and frame prop stays, the lowering of the roof rocks is assessed at the level of 120 – 180 mm. At the same time, there is some change in the initial form of the frame due to the plastic deformations. Hence, it follows that even outside the zone of stope works influence, the fastening system of the 590 belt entry (with the 7 resin-grouted roof bolts set within the arch) does not fully cope with the formed load in the roof of mine working.

The reason for this situation, in our opinion, is not only in the liability of adjacent massif rocks to the phenomenon of creeping the deformations. According to the information of the mining and geological prediction, the argillites and siltstones of the main and immediate roof have a developed fracturing, with alternation of lithotypes and a high probability of moisture saturation. Such conditions are favorable for the active stratification of the roof rocks and the formation of a vast arch of weakened rocks. Presumably, the arch's dimensions exceed the length of resin-grouted roof bolts, and this prevented them to resist effectively to vertical rock pressure, because the thickness of the adjacent roof being strengthened is rather limited. If the roof rocks, occurring above the armored and rock structure, change to unstable state, the load-bearing capacity of the strengthened layer is not sufficient to protect against excessive loadings of the frame support.

The influence of the approaching stope face of the 590 longwall face was marked at a distance of $z=-34$ m; the displacements U_{h_1} begin to increase sharply with a fairly constant growth gradient in the area from $z=-16$ m till $z=47$ m; this area of the drift is characterized as a zone of active rock displacements in the coal-overlaying formation. Further on, the growth gradient of movements U_{h_1} decreases, and at $z>82$ m their growth is not fixed. This mark is the beginning of the zone of displacement processes stabilization in the coal-overlaying formation, at which the decrease in the height of mine working is poorly expressed and is conditioned mainly by the creeping of rocks deformations of the adjacent massif. In the area of stabilization of the rock pressure manifestations, the stope works influence is 60 – 70% of the total value of the roof and bottom rocks convergence, which reaches 1400 mm. Such a decrease in the height of the drift constitutes a danger from the point of view of norms and rules for its operation. But, here it should be noted that the bottom rocks heaving exceeds, as a rule, the lowering of the roof rocks; therefore, when performing the bottom rocks ripping before and after the stope face passing, the mine working height is retained at the level of $h_1=2850-3000$ mm, which is quite enough for reliable performance of technological operations.

The lowering of the roof rocks can be approximately estimated at up to 300 – 400 mm (when performing the stope works in the 590 longwall face). On the one hand, this value slightly exceeds the vertical structural pliability of the frame, but, on the other hand, there is a plastic bending of the cap board and the prop stays with the arising of some asymmetry in their deformations. Therefore, it should be noted that according to the vertical displacement factor U_{h_1}, there is a moderate disturbance of the initial form of the frame support.

The tendencies are also considered of change in the mine working width $в_1$ according to its bottom, and represented by graphs 2 in Fig. 2.7. Outside the zone of stope works influence, the decrease in the width of mine working in relation to the initial structural dimensions of the TSYS-15.0 support varies in the interval of

$U_{b_1} = 580 - 600$ mm. Such a significant convergence of the sides is conditioned by the plastic bending of the frame prop stays, which is induced, first of all, by the rock layer of the immediate bottom moving horizontally in the area of the prop stays; this phenomenon is caused by the active swelling of the bottom rocks, extending beyond the width of mine working, where the differently vectored movement of the border rocks has, as well, a horizontal component. The second factor, which intensifies the prop stays bending, is a significant lateral load as a result of the bearing pressure impact in the sides of mine working. The third factor – the vertical load on the frame contributes to the plastic bending of the prop stays, which have already lost their initial stable form.

Thus, initially, still outside the zone of the stope works influence, there is a significant convergence of the mine working sides with a changeover of the stable form of the prop stays to a less stable one.

The further increase in the convergence of the drift sides is fixed starting from the mark $z = -(20 - 25$ m) ahead of the longwall face and is conditioned by the influence of its front bearing pressure. The growth gradient of movements $U_{в_1}$ is stable enough till the mark $z = 35$ m behind the longwall face and represents the active action of the geomechanical processes of displacement in the coal-overlaying formation towards the mined-out space of the 590 longwall face. Then, as the stope face recedes, the value of convergence of the drift sides (under the influence of stope works) is stabilized in the range of 330 – 410 mm; the total value of the convergence of the sides for the entire period of maintenance of the 590 belt entry drift was $U_{b_1} = 970 - 1030$ mm.

According to the above results, the following should be noted. The value of convergence of the mine working sides approximately of 1 m in relation to the initial structural dimensions of the TSYS-15.0 support is quite significant in terms of the section loss of the drift at its initial use as a belt entry. If the mine working height can be partly restored by means of the bottom ripping, the increase in width is caused by the drawing out the rock, retimbering, high additional costs, and is not advisable in the current situation of applying a larger standard dimension of the mine working section. At the same time, it is necessary to take into account the prospect of reusing the 590 belt entry and the accompanying intensification of the rock pressure manifestations. In addition, it is necessary to pay attention to the discrepancy of the lateral structural pliability of the prop stay (100 mm) to the fixed convergence value of its prop stays, which at an average has five-fold exceeding. This ratio, of course, causes not only a significant plastic bending of the frame prop stays with the loss of their initial form, but also disturbs the normal operation mode of the yielding joists with the possible destruction of some of them.

One of the reasons for this situation is the combination of increased lateral bearing pressure and very soft rocks of the immediate bottom, which causes both the formation of a significant lateral load on the frame and the intensification of the processes of bottom rocks swelling. If it is technically difficult to change the mechanical properties of the rocks of the immediate bottom, then the control of the lateral bearing pressure is real enough; here it is meant the following. In the applied fastening scheme of the 590 belt entry, seven resin-grouted roof bolts strengthen mainly the rocks of the central part of the arch along the length of the cap board, creating the armored and rock structure with certain load-bearing capacity. But, this structure rests mainly upon the frame support and disturbed border rocks in the sides of mine working (above the coal seam and above the security system). All of the above quasi-bearings have some rate of pliability, especially the disturbed border rocks of the immediate roof. When lowering (on pliable bearings), the armored and rock structure induces (due to low repulse reaction) stratification in the rocks of the main roof and, as a result, forms a vast area of unstable rocks in the roof, which creates not only an increased load on the frame, but also a higher lateral bearing pressure. If to concentrate the strengthening on the rock volumes that perform the functions of bearings in the sides of mine working, the resistance of the armored and rock structure in the roof will be harder with the restriction in the stratification of the overlying massif. And, if to add to this a more "deep" strengthening of the roof rocks with the expansion in the area of the armored and rock structure bearing, then obviously the intensity of the lateral bearing pressure decreases many times, which will result in a sharp restriction in the impact of the negative factors of increased lateral load and bottom rocks swelling. Therefore, it is advisable, in our opinion, to change both the scheme of setting the roof-bolts in the arch of mine working, and the principle of strengthening of adjacent roof rocks.

One of the phenomena that have a negative effect on the stability of the frame support and the mine working in general, is the asymmetric deformation of the frame, caused by an increased load from the side of the 590 longwall face. A quantitative estimation of the asymmetry degree is represented by measuring the diagonals c_1 and c_2 (see Fig. 2.2) and revealing the patterns of their changes as the 590 longwall face advances (graphs *3* and *4* in Fig. 2.7).

In the area outside the zone of stope works influence, the diagonal c_1 decreases in the range $U_{c_1} = 230 - 330$ mm, but starting from a distance $z = -(40 - 45$ m), ahead of the longwall face, the growth gradient U_{c_1} begins to increase steadily and after the stope face passage, the displacement value along the diagonal c_1 is partially stabilized; nevertheless, without taking into account the bottom rocks ripping, the displacements along the diagonal c_1 reach 1430 – 1580 mm in the zone of stabilization of the rock pressure manifestations ($z \geq 40 - 50$ m). At the same time, the share of the stope works influence is also decisive in the area of the displacement processes stabilization of the coal-overlaying formation and is about 85%. The displacement value U_{c_1} indicates the formation in the mine working roof behind the 590 longwall face of significant oblique loads (from the side of the mined-out space).

The patterns of change in the diagonal c_2 are as follows. Outside the zone of the stope works influence on the extended area up to $z = -30$ m, the displacement value varies insignificantly in the interval of 120 – 190 mm, which is much less than the value U_{c_1}. Thus, already outside the zone of the stope works influence, some asymmetry of the frame deformation is manifested. In the zone of the front bearing pressure of the 590 longwall face, the intensity of the build-up in the displacements U_{c_2} increases, and a decrease in the growth rate of U_{c_2} occurs behind the longwall face, starting from a distance $z = 40 - 50$ m. By absolute value, the decrease in the diagonal c_2 reaches 1050 – 1100 mm (excluding the bottom ripping). As can be seen from the graphs, the asymmetry of the frame deformation is intensified as the stope face recedes, and reaches a value of 370 – 480 mm in the zone of the displacement processes stabilization in the coal-overlaying formation. The main share of the asymmetry is conditioned by uneven loading of the frame in the zone of the stope works influence and amounts to 60 – 70% of the total difference in the current dimensions of diagonals c_1 and c_2.

The phenomenon of asymmetry of frame support deformation changes substantially its initial form and reduces the load-bearing capacity. At the same time, the normal operation mode of the yielding joists is disturbed, which entails an increase in the load on the fastening system, a more intensive deformation of the frame and a consequent section loss of mine working. Therefore, even taking into account the periodic ripping of bottom rocks, the section loss of the 590 belt entry is quite significant already at the first stage of its maintenance while mining the 590 longwall face; this leads to the conclusion about insufficient effectiveness of strengthening the frame support by a system of 7 resin-grouted roof bolts set in accordance with the technical documentation by the scheme shown in Fig. 2.1.

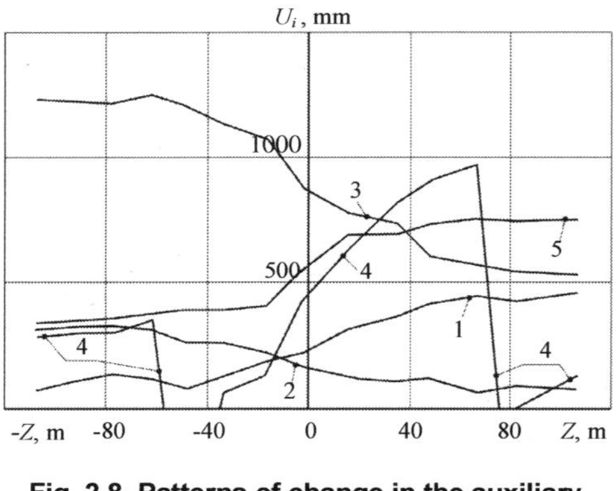

Fig. 2.8. Patterns of change in the auxiliary parameters when assessing the state of the 590 belt entry during the period of mining the 590 mining site:
1 – $h_2 - h_3$; 2 – $b_1 - b_2$; 3 – $b_1 - b_3$; 4 – U_{h_4}; 5 – a_1

Let us consider the change in the auxiliary parameters of the frame support movement as the stope face advances; the corresponding graphs are shown in Fig. 2.8. The first is the difference $h_2 - h_3$ in heights h_2 and h_3 from the side of the non-mining and working flanks of the drift, measured along the verticals from the yielding joists of the frame to the bottom. As mentioned above, this parameter allows to estimate the degree of asymmetry of the frame deformation in the vertical direction, but it should be beard in mind some ambiguity of indications conditioned by uneven bottom rocks heaving from the side of working and non-mining flanks of mine working. Outside the zone of the stope works influence, there is a slight asymmetry of the frame deformation in the vertical direction in the range of 80 – 140 mm, but there is probably an error from the uneven heaving (in the width of mine working) of the bottom rocks. Starting from an area $z = -(30 - 40$ m) ahead of the longwall face, there is a steady increase in the difference $h_2 - h_3$ at approximately constant growth gradient up to a mark $z = 45 - 50$ m behind the longwall face; in the zone of stabilization of the rock pressure manifestations, the difference in the analyzed dimensions varies in a relatively narrow interval of 420 – 460 mm. Here, it is impossible to explain this difference only by the swelling unevenness; obviously, an increased oblique load is being formed from the side of mined-out space of the 590 longwall face, which significantly changes the initial form of the frame and, to a large extent, of its cap board.

Next, we refer to such auxiliary parameters as $b_1 - b_2$ and $b_1 - b_3$, which allow to assess the deformation degree of the frame prop stays and changes in their initial form (graphs *2* and *3* in Fig. 2.8). The difference in dimensions $b_1 - b_2$ in the area of the drift outside the zone of stope works influence is changed in a narrow interval of 270 – 340 mm; starting from the distance to the longwall face $z = -(30 - 35$ m), this parameter gradually decreases to a value of 110 mm at $z = 35$ m; in the future (in the zone of stabilization of the rock pressure manifestations), the fluctuations of the value $b_1 - b_2$ are in the range of 70 – 120 mm. Thus, the initial difference in the structural dimensions b_1 and b_2, as the longwall face approaches and recedes, is reduced by 200 – 220 mm; such a value of the prop stays movement exceeds the limits of elastic deformations, which is confirmed by the development of the plastic state areas in the frame prop stays. In the process of their bending, the approach to the vertical position of the initially inclined prop stays of the TSYS-15.0 support happens, which reduces their resistance reaction to the side loads impact.

The difference in dimensions $b_1 - b_3$ is even more changed in the course of the 590 longwall face advance. This is quite expected, because the dimension b_3 is at a higher altitude in relation to b_2 and an arm of the lateral pressure forces action on the prop stays increases, and with it the bending moment increases, as well as the corresponding plastic deformations. In the area of the drift, outside the zone of the stope works influence, the difference in dimensions b_1 and b_3 is close to the initial one, and the range of fluctuations does not exceed 4%. The decrease in this difference is due to the anticipatory movement of the prop stay in the area of its bearing above the movement of the prop stay in the area of the upper edge of the coal seam; this is observed at a distance of about 60 – 65 m ahead of the stope face. Further, there is a stabilization of the analyzed parameter within 530 – 610 mm, that is, the convergence of the prop stays in the area of their bearings exceeded the convergence of the prop stays in the area of the upper edge of the coal seam by 630 – 710 mm, which once again confirms the conclusions of visual observations of significant plastic deformations of the frame prop stays.

In terms of the obtained results of instrumental observations, it is necessary to emphasize once again the opinion on the relevance of reducing the intensity of the lateral bearing pressure by choosing the rational schemes for fastening the roof rocks in mine working by means of roof-bolting systems.

The problem of intensive bottom rocks swelling of the 590 belt entry is solved by its periodic ripping (at least, before the approach of the longwall face and behind the stope face in the zone of stabilization of rock pressure manifestations). It was noted previously that the fixation of dimensions $h_{4,5}$ (see Fig. 2.2) has a certain error in estimating the value of swelling. Nevertheless, with a general assessment of the geomechanical situation of maintaining the 590 belt entry, the information characterizing the behavior of its bottom is also of interest (graph 4 in Fig. 2.8). Outside the zone of the stope works influence, the value of swelling is quite stable at a level of 290 – 350 mm. In order to prevent the expected activation of the swelling processes, a ripping has been made till the zone of the front bearing pressure of the 590 longwall face. However, the swelling phenomenon (already after the ripping of the bottom) is not stabilized, but continues to develop intensively: thus, in a relatively short area of the 35-meter-long drift, up to the stope face, the bottom heaving was already 440 mm, and it continued to grow behind the longwall face (with some slowing down of the rate) until the second ripping; after that it was fixed the bottom heaving by 120 mm at the level of 107 m behind the stope face.

The last of the fixed parameters, the overlap value $a_{1,2}$ (graph *5* in Fig. 2.8), assesses the degree of compliance of the yielding joists work with the normal mode. Here it is necessary to clarify that from the side of the working flank of the drift, in the area of the stope works influence, there are significant frame plastic deformations are observed in the area of the yielding nodes location, and their operation mode should be considered unsatisfactory. From the side of the non-mining flank of the drift, the plastic deformations also occur in the frame, but they proceed without significant bending of the SCP profile, which causes the yielding joists actuation. It is for this group of yielding nodes the change in the overlap value a_1 is fixed, as the stope face advances.

In the area of the drift, outside the zone of the stope works influence and at the beginning of the zone of the front bearing pressure from the 590 longwall face, the overlap value is 350 – 420 mm; that is, within the limits of the design standards of the TSYS series support. Starting from the mark $z = -16$ m, at first a sharp increase in the overlap with length to 700 mm occurs in the area up to 16 m behind the stope face, and then slowing down of the growth a_1 with stabilization at the level of 740 – 750 mm with $z \geq 48$ m.

Summarizing the analysis of the patterns change in the auxiliary parameters (depending on the distance to the 590 longwall face), it is necessary to note their conformity and a useful addition to the similar patterns for the basic displacement parameters (see Fig. 2.7). All the main conclusions stated above are confirmed, including the general assessment of the insufficient effectiveness of the applied roof-bolting schemes for strengthening the roof rocks of the drift, according to the main criteria for the maximum possible preservation of its section and the load-bearing capacity of the fastening system with the subsequent reuse of the 590 belt entry.

When mining the 592 mining site, the 590 belt entry was reused as an upper entry. The metering stations remained the same, as the technique for the readings fixation of the displacement of the mine working circuit. The presentation of the measurements results requires, in our opinion, to be adjusted based on the following considerations. During the initial use of the 590 belt entry in the course of stope works in the 590 longwall face, the mine working state has already been fixed by a set of parameters that quite fully characterize the movements of its contour. It is quite expected that the stope works in the adjacent 592 longwall face (reuse of the 590 belt entry) will activate the rock pressure manifestations, and the fixed parameters will vary within the range of their value. The simple summation of readings of the movements will make it difficult, in our opinion, to objectively interpret the results of assessing the degree of influence (on the state of the drift) of mining of each of the 590 and 592 extraction panels. The quantitative indicators of the influence of the 590 longwall face drifting are shown in the graphs of Figs. 2.7 and 2.8 and have already been analyzed. Therefore, it is visually more clear and methodologically more expedient to assess the degree of the stope works influence in the 592 longwall face separately, which will allow to systematize patterns, conclusions and recommendations during the initial and repeated use of the drift. Therefore, when constructing and analyzing the patterns of change in the geometric parameters of the 590 belt entry depending on the distance to the stope face of the 592 longwall face, we consider it logical to reflect only the influence of the last one, taking the established dimensions of the drift as a basis after its initial use. The patterns of changes in the movements $U_i(z)$ of the contour of the 590 belt entry are shown in the graphs of Fig. 2.9.

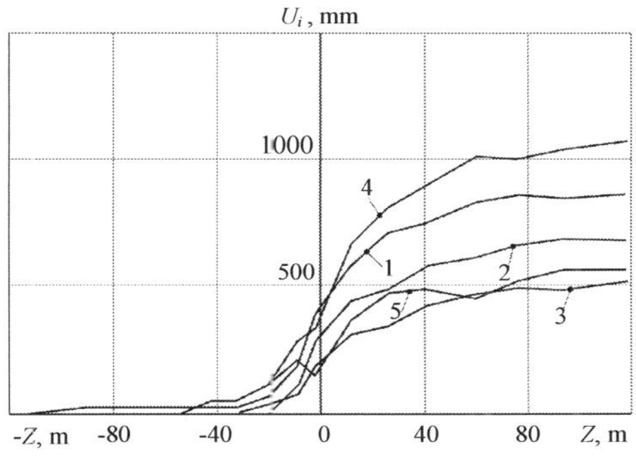

Fig. 2.9. Dependences of the contour movements of the 590 belt entry on the distance z to the 592 longwall face (reuse of the drift): 1 – U_{h_1}; 2 – U_{b_1}; 3 – U_{c_1}; 4 – U_{c_2}; 5 – $U_{c_2} - U_{c_1}$

The tendencies of U_i movements development are similar among themselves in the qualitative terms, but differ in quantitative indicators. The value of the convergence h_1 of the roof and bottom in the area outside the zone of the stope works influence is insignificant (up to 25 – 30 mm) and is determined mainly by the bottom rocks heaving. The length of the area of the front bearing pressure zone ahead of the 592 longwall face is 30 – 35 m, judging by the beginning of the more intensive growth of U_{h_1} ; this length of the area is almost the same as that for the 590 longwall face. When it approaches a distance of $z = -20$ m, the growth gradient U_{h_1} sharply increases and behind the longwall face at a distance of $z = 25 - 30$ m, the convergence of the roof and the bottom is already 720 mm. When, further on, the longwall face recedes, the growth U_{h_1} intensity decreases and at $z = 70 - 80$ m, there is a stabilization of the convergence value of the roof and the bottom at the level of 850 – 860 mm.

Compared to the degree of the 590 longwall face influence, the value U_{h_1} has decreased by only 70 – 80 mm, that, with a certain measurement error from the swollen bottom rocks, makes it possible to consider the influence of the 590 and 592 longwall faces to be equal. If to sum up the general convergence of the roof and the bottom rocks over the entire period of the 590 belt entry operation, then it reaches a value of 2250 – 2280 mm, that is, reduces the height of mine working by 60 – 65%. However, due to periodic ripping of the bottom rocks in mine working, its height is maintained at a technologically acceptable level of $U_{h_1} = 2600 - 2900$ mm, and a decrease in the height of the drift is determined only by lowering the roof of approximately 600 – 900 mm with the corresponding deformation of the frame support.

The process of reducing the mine working width is activated in the zone of the front bearing pressure ahead of the 592 longwall face only 20 m before the stope face. Such a "delay" in the manifestation of the prop stays movements along the drift bottom can be explained by the presence of the mined-out space from the side of mined 590 panel; here, the broken rocks have an increased deformability and "absorb" partially (due to compaction) the vertical displacements of the coal-overlaying formation (from the influence of the 592 longwall face), and this reduces the intensity of the lateral bearing pressure and the level of deformation of the frame prop stays. Nevertheless, the growth gradient U_{b_1} is extremely high near the stope face in the area from $z = -20$ m till $z = 12$ m, so that at the end of this area the convergence of the prop stays reaches 440 mm. Then, the intensity of the U_{b_1} growth gradually decreases, but the stabilization of displacements at the level of 660 – 680 mm is observed only at a distance of 75 – 80 m behind the stope face. A more extended area of the stabilization of the rock pressure manifestations can be explained by the fact that the stope works in the 592 longwall face expand the area of rocks displacement in the coal-overlaying formation with the activation of a part of the massif around the already mined-out 590 longwall face. The processes of "secondary" stratification and lowering of the massif affect the higher-lying rock layers, the deformation of which takes place over a longer period of time, during which the 592 longwall face has enough time to recede further away from the studied section of mine working. Such considerations refer not only to the patterns of convergence of the drift sides at the level of its bottom, but also to other fixed parameters.

The total convergences of the frame prop stays (at the bottom level) for the entire period of the 590 belt entry operation reach 1630 – 1710 mm, which amounts to 31 – 33% of its initial width. Loss of up to one third of the mine working width is a serious problem for the reliable and stable stope works operation in the 592 longwall face, and in this respect it is necessary to emphasize once again the importance of limiting the convergence of the drift sides by reducing the intensity of the lateral bearing pressure, as it was described earlier in studies during the 590 longwall face mining. Then, it was concluded that the applied scheme for setting the resin-grouted roof bolts is insufficient to create in the roof an armored and rock structure with a high load-bearing capacity, which would protect the frame support from excessive loads.

The tendencies of decreasing the diagonal c_1 and c_2 dimensions have been radically changed in the zone of the stope works influence of the 592 longwall face in comparison with the period of the 590 longwall face mining. The movements of U_{c_1} in a diagonal c_1 direction developed to a much lesser extent: their noticeable growth began at a distance of 20 – 25 m from the stope face of the 592 longwall face and the value U_{c_1} steadily increased with a variable gradient up to the area at a distance of 75 – 80 m behind the stope face, where the maximum value of U_{c_1} =480 – 510 mm was fixed. The value of displacements U_{c_2} in a diagonal c_2 direction, on the contrary, developed more intensively, beginning with the mark z =–(30 – 40 m) ahead of the 592 longwall face up to the mark z = 60 m behind the stope face; though, and in more remote areas, there was a slight increase of U_{c_2} up to a value of 1070 mm.

The noted patterns are quite logical, in our opinion, since numerous studies of the state of preparatory mine workings in various mining and geological conditions indicate the predominant movements of their contour from the side of the active longwall face. This fact is also fixed in our studies: the displacement U_{c_2} exceeds the U_{c_1} by 560 mm, which quite significantly distorts the form of the frame, which has already been formed when conducting the stope works in the 590 longwall face.

When estimating the total displacement in diagonals c_1 and c_2 direction (see Fig. 2.2) for the entire period of the 590 belt entry operation, the following results have been obtained: U_{c_1} =2030 – 2090 mm, U_{c_2} =2140 – 2170 mm. In relative values, the dimensions of the diagonals are reduced by 40 – 43% of the initial value, which is a critical indicator of the section loss of the drift. However, due to the periodic bottom rocks ripping, the diagonal dimensions increase and the displacement U_{c_1} and U_{c_2} are limited to no less than by half, which makes it possible to achieve a satisfactory operational state of mine working by a factor of its dimensions along the diagonals.

It should also be noted that as a result of the 592 longwall face mining, the final difference in displacement has been decreased to 80 – 110 mm. But, this does not mean that the form of the frame has returned to its initial position: the plastic deformations of the cap board and frame prop stays, that have occurred, are irreversible and lead to a significant loss of the load-bearing capacity of the frame support, despite the approximate equality of the diagonals c_1 and c_2. In terms of the direction for limiting the plastic deformation of the frame, it is necessary to reduce the asymmetry of its loading, first from the side of the 590 longwall face, and then from the side of the 592 longwall face when reusing the drift. This can be achieved by forming a powerful armored and rock structure in the mine working roof, which could take up most of the rock pressure.

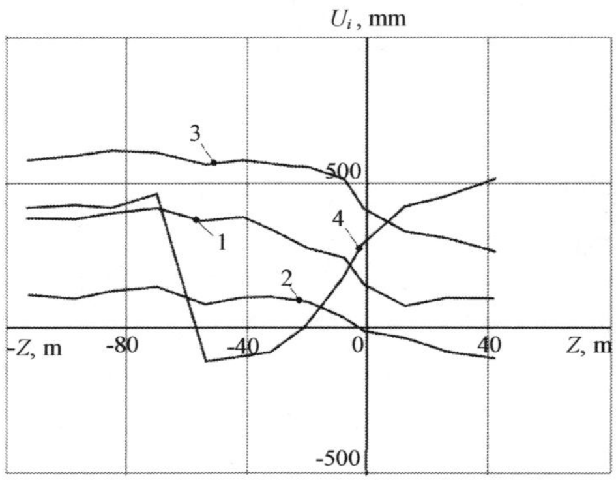

Fig. 2.10. Patterns of change in the auxiliary parameters when assessing the state of the 590 belt entry during the period of the 592 longwall face mining: 1 – $h_2 - h_3$; 2 – $b_1 - b_2$; 3 – $b_1 - b_3$; 4 – U_{h_4}

The patterns of change in the auxiliary parameters (Fig. 2.10) complement the general idea of the 590 belt entry state when it is reused during the period of stope works operation in the 592 longwall face. In qualitative terms, the presented patterns are partially similar to those established for the 590 longwall face, but in quantitative terms there are some differences.

The difference in vertical dimensions $h_2 - h_3$ over the expanded area to $z = -42$ m remains slightly (40 – 80 mm) less than during the drifting of the 590 longwall face, which, in fact, has induced this asymmetry of the frame form. But, the approach of the 592 longwall face contributed to the change in the curve of loading with the concentration from the side of its working flank. Therefore, the dimension h_2 decreased more rapidly than the dimension h_3 (see Fig. 2.2) and their difference $h_2 - h_3$ decreased in the zone of the stope works influence of the 592 longwall face, starting from the mark $z = -42$ m till 12 m behind the stope face; further, the dimension $h_2 - h_3$ difference was stabilized at a level of 80 – 110 mm, repeating completely the asymmetry values when measuring the diagonals c_1 and c_2. Thus, the asymmetry of the frame form has been reduced, but with a much smaller residual section and significant plastic deformations of the cap board and the prop stays.

The difference $b_1 - b_2$ in horizontal dimensions, when reusing the 590 belt entry to the mark $z = -20$ m, has changed in the range of 90 – 150 mm and slightly exceeded that (70 – 90 mm), fixed in mining the 590 mining site; this discrepancy is conditioned by a long time interval (between the previous measurements and the last), during which the error is introduced in the rheological process of creeping the deformations of adjacent rocks. When the 592 longwall face approaches to a distance of less than 20 m, the dimension difference $b_1 - b_2$ has decreased to such an extent that at $z = -3$ m it is equal to zero, and has already taken negative values approximately of 90 – 110 mm behind the stope face. This indicates that the frame prop stays in the bearings convergence each other by such a value that the residual distance b_1 between them becomes smaller than the residual distance b_2 at the level of the seam lower edge. Such a curvilinear form of the lower part of the prop stays sharply reduces their load-bearing capacity, which again indicates the need to limit lateral shifts of the border rocks.

The difference in horizontal dimensions $b_1 - b_3$ to the mark $z = -32$ m remained practically constant (560 – 610 mm), slightly exceeding the residual difference of 440 mm during the period of the 590 longwall face mining. The approach of the 592 longwall face has led to a further plastic bending of the prop stays, as a result of which the difference $b_1 - b_3$ decreased to 270 mm with $z = 42$ m behind the stope face.

If to summarize the results of measurements of horizontal convergences of the frame prop stays, then it is necessary to point out their significant plastic deformations, which reduce sharply the load-bearing capacity of the frame support and the stability of mine working in general. Therefore, special attention should be paid to limiting the convergence of the mine working sides by reducing the intensity of the lateral bearing pressure.

The bottom heaving value U_{h_4}, after the previous ripping, has stabilized at the level of 420 – 460 mm outside the zone of the stope works influence. Before the zone of the front bearing pressure from the 592 longwall face, the bottom rocks ripping has been made to a depth of about 0.6 m, after which a gradual increase in the bottom heaving occurred with the highest intensity from the mark $z = -32$ m ahead of the longwall face to the mark $z = 12$ m behind the stope face; the maximum fixed value of the bottom swelling was 520 mm.

CONCLUSIONS

The main conclusions from the results of observations of the 590 belt entry state during its repeated use are as follows. In the studied mining and geological conditions, the significant displacements of the mine working contour in all fixed directions are developed and it is necessary to have a reserve section of the drift in order to preserve it for repeated use. Therefore, the technical decision on the 590 belt entry mining, with a section inside 15.0 m^2 (for the TSYS-15.0 support) is fully justified. Nevertheless, it is critically important to limit the convergence of the mine working sides; one of the ways to solve this problem is in reducing the intensity of the lateral bearing pressure impact by forming above the mine working an armored and rock structure with high load-bearing capacity. This will protect the frame support from excessive rock pressure and limit the negative impact of the asymmetry of the frame deformation to preserve its load-bearing capacity at the highest possible level. It is advisable to implement such a direction by choosing the rational parameters for resin-grouted roof bolts setting, but also “deep” strengthening of the roof rocks with the help of rope bolts.

3. THE COMPUTATIONAL EXPERIMENT SUBSTANTIATION AND STRESS-STRAIN STATE ANALYSIS OF BORDER ROCKS, AS WELL AS OF THE BASIC FASTENING SYSTEM OF PREPARATORY MINE WORKINGS IN THE ZONE OF STOPE WORKS INFLUENCE

3.1. CONSTRUCTION OF GEOMECHANICAL MODELS IN CONDUCTING THE COMPUTATIONAL EXPERIMENTS

The computational experiments based on the finite element method (FEM) were carried out with the use of the recommendations [20, 54 – 58] of modeling the geomechanical processes of the coal-bearing massif displacement in the vicinity of the preparatory mine workings. The modeling technology assumes the synthesis of a common geomechanical system of two components: a model of the coal-bearing massif and the scheme for mine working fastening. With the purpose of objective assessment, the last one is performed in two variants: the existing fastening system according to the technical documentation for mine working development, which is called the "basic" scheme and the improved fastening scheme, which is substantiated on the basis of studies of the rock massif state and the experience of maintaining the reusable mine workings.

In the methodical plan, the components of the study are represented as follows:

– generalization of ideas about the mechanism of displacement in the coal-overlaying formation and loading the fastening system in the preparatory mine workings;

– analysis of SSS of the rock massif in the vicinity of mine working;

– assessment of SSS of the load-bearing elements in the basic fastening system;

– development and substantiation of the improved fastening scheme;

– calculation and analysis of SSS of the geomechanical system "massif – frame – combined roof – bolt support";

– establishing the patterns of the main geomechanical factors influence on the SSS of the load-bearing elements of the combined roof-bolting system when strengthening the mine working arch and in combination with the TSYS support;

– development of a methodology for calculating the parameters of the roof-bolting system in the preparatory mine workings;

– carrying out and analysis of underground tests of the developed fastening system of reusable mine workings.

To implement this research scope, a general geomechanical model and its components have been created using the following main provisions.

The first is the substantiation of the general model dimensions and the conditions on its boundaries. For this, a test calculation has been made of a general geomechanical model, differing in its increased dimensions by coordinates: y – height, x – width and z – thickness along the axial coordinate of mine working. Here, the SSS perturbations are considered, which are conditioned by mine working developed in the mine field. To establish the boundaries of anomalies in the stress components distribution, it is necessary to compare them with those of the virgin massif. In this regard, the substantiation of the model dimensions is inextricably linked with the conditions of its boundaries loading, which are conditioned by the choice of the initial state model of the virgin massif.

In the existing mining and technical literature [49 – 61], there is no common understanding of the initial state of the virgin massif: hydrostatic ($\sigma_y = \sigma_x = \sigma_z$) or non-hydrostatic ($\sigma_y \neq \sigma_x = \sigma_z$). In the solved geomechanical problems, the non-hydrostatic initial state has been chosen as prevailing in the description of the coal-bearing massif behavior. The authors of the work use the model of a non-hydrostatic initial state of a virgin massif, characterized by the following initial conditions:

$$\sigma_y = \gamma H; \tag{3.1}$$

$$\sigma_x = \sigma_z = \frac{\mu}{1-\mu}\gamma H, \tag{3.2}$$

where H – the depth of mine working's location; γ – the weight-average unit specific gravity of the rocks in the coal-overlaying formation till the daylight area; μ – the Poisson's ratio (lateral deformation) of mine rock.

According to the data of the geological service of the mine, the maximum depth of the mine working development area was chosen and its weight-average unit specific gravity was calculated in accordance with the coal-overlaying formation structure, and then, using the formula (3.1), the initial vertical stresses of the virgin massif have been determined.

The horizontal stresses $\sigma_x = \sigma_z$ will differ for each lithological variety due to the different values of the Poisson's ratio. This circumstance was taken into account with reference to the method developed in the FEM of introducing the 'symmetry' condition along the vertical boundaries of the model. With that, the horizontal stresses $\sigma_x = \sigma_z$ calculation is performed automatically according to the values μ assigned for each lithotype. The information on the coefficient μ is taken from the data of the geological service of the mine, and the missing information is obtained from the results of studies of the physical and mechanical properties of the Western Donbas rocks, described in the works [62, 63].

Thus, on the upper horizontal boundary of the model, a uniform calculated geostatic pressure σ_y is applied, and on the vertical boundaries there are stresses $\sigma_x = \sigma_z$ in accordance with the formula (3.2). The lower horizontal surface of the model rests on a rigid base in accordance with the developed recommendations [64 – 67].

Based on substantiated boundary conditions, a test calculation has been made of the massif model with a hole in the form of mine working for the TSYS support without its setting, since the influence of the support extends only to the adjacent border rocks. The test model dimensions were intentionally outsized, in order all the SSS anomalies, caused by mining operations, are guaranteed to be placed in it. In addition, the experience of performing the computational experiments demonstrates the occurrence of SSS anomalies in the area of the model boundaries, which are caused by the application of the load and the action of other borderline conditions. These edge effects disappear in the movement to mine working and, if the model dimensions are chosen appropriately, then the influence of the edge effects and mine working is not imposed upon each other.

Based on the above conditions, the test calculations were performed on the model with dimensions: height $y = 60$ m, width $x = 70$ m, thickness $z = 8$ m, based on the calculation of modeling in the future of ten sets of frame support. The results of test calculations have shown that it is quite sufficient for the model height $y = 40$ m, widths $x = 40$ m, but as for thickness z of the model, the edge effects distribute for less than 1.0 m into the massif. Therefore, at the first stage of test calculations, $z = 6$ m is accepted with a certain margin.

The structure of the rock massif model fully corresponds to the structure of the coal-bearing strata in the area of mine working location and its orientation relative to the seam C_6. To increase the reliability of calculations, the real processes of contacts disturbance along the planes of bedding the adjacent lithological varieties are simulated in the model [55, 56, 67, 68].

The listed provisions are the content of the second stage of test studies. As for the coal-bearing massif structure, there is no need for any substantiation, and the model was constructed in accordance with the data of the prediction geological sheet of mine workings.

A more important structural feature was the determination of the degree of contacts cohesiveness of adjacent lithological varieties, since it is known from the work [55] that the disturbance of contacts between the rock layers drastically changes the SSS of the massif surrounding the mine working.

The research method consists in comparing the value of shearing stresses τ_{yx} and τ_{yz}, which act in the plane of contact of any pair of adjacent layers with the value of the maximum permissible shearing stresses τ_p, which, according to the Mohr-Coulomb failure theory, are determined by the formula

$$\tau_p = C + \sigma_y \cdot f_{fr}, \tag{3.3}$$

where C – the adhesion value along the plane of bedding the layers; f_{fr} – the friction coefficient of adjacent lithotypes in the contact plane.

From the data of the mining and geological prediction, it follows that the structure of the coal-bearing strata, adjacent to mine working, is characterized by a regular alternation of the rock layer (argillite and siltstone) with coal seams C_6 and C_6^1 or coal interlayers with a thickness of 0.1 m. That is, a coal seam or interlayer occur in any of the contacts. It is known from works [62, 63] that in the conditions of Western Donbas there is practically no adhesion of coal with other rocks. Therefore, with some margin in the "minus" of the massif stability, a value of adhesion can be take as equal to zero ($c=0$) and the formula (3.3) will be extremely simplified. The value of the friction coefficient f_{fr} of coal against the argillite (siltstone) is taken from the studies [69, 70] and reduced by 50% in accordance with the recommendations [71], by reason of wetness or water content of the contacts.

With the use of above-mentioned methodology, the contacts extension has been revealed, where the conditions of their integrity are not satisfied

$$\left.\begin{aligned} \tau_{yx} < \tau_p; \\ \tau_{yz} < \tau_p. \end{aligned}\right\}. \tag{3.4}$$

By the area of these disturbed contacts, the condition "frictionless media" is adopted, which simulates the slip regime of adjacent lithotypes relative to each other at their deformation. As a result, a model of the massif structure in the vicinity of mine working has been constructed, which is as much as possible adequate to the real conditions of its maintenance.

The physical and mechanical properties of each lithological variety of the model correspond to the data of the mining-geological prediction and experimental studies of the lithotypes characteristics of Western Donbas [62, 63]. The strength properties of rocks of the immediate and main roof and bottom are introduced into the model with account of the moisture saturation according to the IGTM test data. For other lithological varieties, the water content of the rocks was taken into account in accordance with normative documents [71, 72], and fracturing and rheology also from documents [71, 72] and studies represented in [62, 63].

The third stage of geomechanical model preparing for conducting a computational experiment was to substantiate the physical and mechanical properties of each lithotype of the coal-bearing massif based on the data obtained from the above-mentioned sources. Here, in the first place, it was necessary to determine the choice of the model of the mine rock state: elastic, elastic-plastic, visco-plastic, with or without account of weakening and loosening at the superlimiting stage of deformation. The simplest and the most reliable, when performing calculations (the minimum number of so-called "failures" of the computational process), is an elastic model of mine rock behavior in which the inelastic deformations are taken into account by using the deformation modulus instead of the elasticity modulus in the linear diagram of stresses and relative deformations correlation [67]. However, the elastic model does not reflect the process of formation and the influence of areas of the limiting and superlimiting state of the rock on the change in the SSS of the surrounding massif. The model of a complete mine rock deformation diagram and the rheological model are very complex [73], and are characterized by a low degree of reliability due to constant "failures" of the computational process for reasons of software imperfections or a lack of computing resources.

The way out of this situation was found in a compromise approach, taking into account the specific mining and geological conditions of the computational experiment performance. Sufficiently hard roof rocks contribute to the weakening areas localization in small volumes of border rocks. In proportion to these volumes, there is an impact on the SSS change in the coal-bearing massif. The degree of influence of local weakening areas can be assessed as insignificant in the adjacent relatively hard roof rocks and the coal seam rocks. Nevertheless, the state of these small volumes of disturbed rocks is modeled by the condition of plastic flow, which is very typical for clay rocks [74]. Siltstone and argillite, occurring in the immediate and main bottom, have extremely low strength characteristics, both in naturally-wet and water-saturated states. Therefore, these lithotypes are very liable to plastic flow. The displaying of the plastic flow state is provided by a bilinear diagram of mine rock deformation, which is used in computational experiments. The first linear section models the prelimiting stage of loading with the use of the value of deformation modulus of each lithotype as deformation characteristic. The limiting and superlimiting stages are simulated by the plasticity condition when the elementary volume of mine rock is constant [74]. As a result, this model of the coal-bearing massif behavior adequately reflects the real conditions of its deformation and is relatively stable in the technological process of the computational experiment performance.

The principle of displaying the mechanical characteristics of the rock is as follows. In the model of each lithotype, the following physical and mechanical indicators are used: the uniaxial compressive strength σ_c, uniaxial tensile strength σ_t, the deformation modulus E, Poisson's ratio μ and the unit specific gravity γ. This information is taken from the data of the geological service of the mine and is supplemented with the results of studies of the mechanical properties of the Western Donbas mine rocks [62, 63].

When performing the calculations, the effect of factors weakening the rock was taken into account: water content, fracturing and rheology by introducing the correction factors in accordance with the effective normative document [71] and with works [53, 58, 59, 64].

The calculated compressive resistance R of the rock and the calculated deformation modulus E_{cal} were determined by the formulas [71]

$$R = \sigma_c \cdot K_c \cdot K_w \cdot K_t; \tag{3.5}$$

$$E_{cal} = E \cdot K_t^2, \tag{3.6}$$

where K_c – the coefficient, which takes into account the "additional disturbance of the rock massif by the weakening surfaces without adhesion and by low coherence", is determined according to the Table 3.1; K_w – the coefficient, which takes into account the "weakening of the water-flooded rocks as a result of filtration along the water column of aquifers", is determined according to the Table 3.2; K_t – the coefficient, which takes into account the decrease in the mechanical characteristics of the rock under long-term loading; is calculated by the formula

$$K_t = \sqrt{1 - \frac{x}{\beta}}, \tag{3.7}$$

here $\frac{x}{\beta}$ – the rheological index, which is determined in accordance with the works [62, 63] by the formula

$$\frac{x}{\beta} = 0.8 - 0.326 \lg \sigma_c .$$

Table 3.1

COEFFICIENT OF STRUCTURAL WEAKENING OF ROCKS

Average distance between surfaces of rocks weakening, m	K_c
More than 1.5	0.9
1.5 – 1.0	0.8
1.0 – 0.5	0.6
0.5 – 0.1	0.4
Less than 0.1	0.2

Table 3.2

COEFFICIENT OF WATER-FLOODED ROCKS WEAKENING

Type of rocks	K_w
Sandstones and siliceous shales	0.8
Limestone sandstone and limestone	0.7
Clayey sandstones and siltstone	0.6
Argillites and marls	0.5
Clays	0.4 – 0.5

The third stage is the final in the complex of test calculations for the creation of a geomechanical model of the massif; based on their results, the model of the coal-bearing strata in the vicinity of the studied mine working is fully substantiated and prepared for the computational experiment.

The next (fourth) compulsory stage in the preparation of a computational experiment is the substantiation and testing of the second component of the general geomechanical model – fastening system of mine working and the conditions for its interaction with the surrounding massif.

The basic fastening scheme (Fig. 3.1) has been modelled fully in accordance with the passport of mine working development. The mechanical characteristics of the fastening materials are taken from works [16, 19].

To most accurately reflect the work of the fastening system, the methodical developments [57, 58, 75, 76] were used.

Studies of the fourth stage to substantiate the geomechanical model included the following scope of works.

The orientation of mine working relative to the seam C_6 was made in full accordance with the technical documentation for its development. The cross section of mine working is oriented along the strike of the seam, the longitudinal section is to the dip. A combined ripping of the immediate roof and bottom of the seam C_6 have been fulfilled. Along the contour of the frame support with the interframe fencing, as well as the rock contour with the interframe fencing, the normal and tangential forces act.

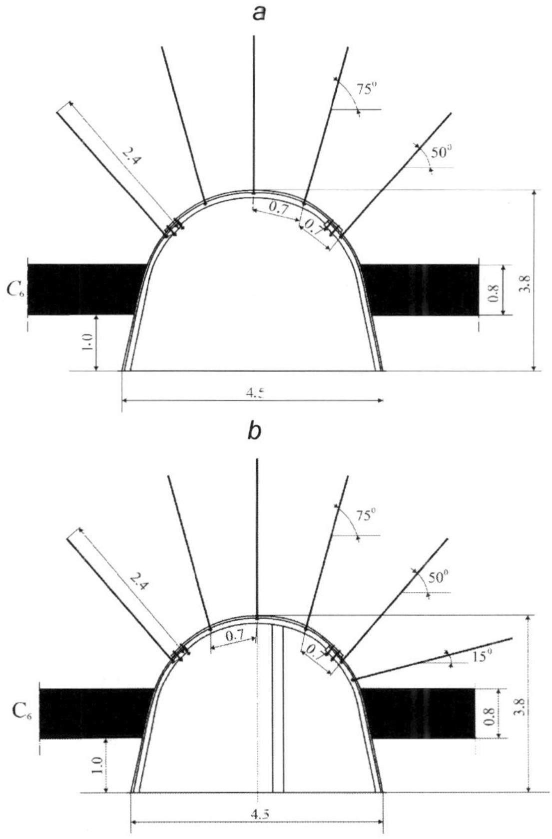

Fig. 3.1. The models of the "basic" variant of the scheme for mine working fastening: ahead of the longwall face outside the zone of (*a*) the bearing pressure and in it (*b*)

Thus, the real geometric parameters are completely reflected, including, the cross section of the SCP profile is reproduced. For this purpose, the grid of the SCP coating with finite elements had a scale that was different by two or more orders of magnitude from the scale of coating the rock massif. This ensures high accuracy in determining the stress components in the frame support. The modeling of the yielding nodes of the frame was carried out according to the recommendations [75], the essence of which is as follows. The accurate reflection of the yielding joists construction in the model of the TSYS support does not cause any fundamental difficulties. However, because of structural imperfections in terms of geometric, the cap board of the frame has local contact areas with the prop stay, especially in the yielding mode of support, when the area of these contacts is changed in arbitrary way. Thus, some static imbalance of the structure arises with an increase in the degree of freedom of movement of the prop stay coupling with the cap board in any direction. Due to the inconsistency of these movements with the incremental step of stresses and deformations in the finite-element grid of the frame, the "failures" of the computing system arise steadily. In order to eliminate these drawbacks in this model of the TSYS frame support, a simulator of the yielding node in the form of a SCP part with a constant cross section geometry, but with a different yield limit of material has been used, compared to the St.5 steel, from which the SCP special profile is usually manufactured. By the length of the yielding joist, the mechanical properties of the SCP steel remain unchanged except for its yield limit. Thus, the complete compliance of the constructive pliability of the real frame and the model has been achieved without distorting the nature of the process.

As for the mechanical characteristics of the frame support material, the SCP profile is made of steel St.5 and its mechanical characteristics are implemented into the model according to the data [19]. Since plastic bending of the cap board or frame prop stays is often observed, an elastic-plastic model of steel behavior was used to display it, the deformation diagram of which, when changing to the yield segment, is displayed quite reliably through the condition of perfect plasticity, used in the computational experiment performance.

To model the resin-grouted roof bolts and rope bolts, an even finer finite element grid was used than for the frame support. This is conditioned by the structural features of resin-grouted roof bolts and rope bolts, which are fully reflected in the model. A bearing rod with mechanical characteristics according to data [16], but a quick-hardening polymeric composition with mechanical characteristics according to the manufacturer's technical documentation with additions from the data of the work [74]; back plate with tension nut with the properties of steel St.3.

The parameters of roof-bolts setting are used from the technical documentation for the maintenance of mine working, and at the new (experimental) site – according to the improved fastening system. In both variants, the roof-bolts are set in the middle of the interframe gap length.

From the point of view of displaying the real operation mode of resin-grouted roof bolts and rope bolts, the analysis of mining and technical literature, as well as the experience of operating mine workings with support fastened by the roof-bolts, indicates its rather rigid deformation-strength characteristics. This means that the roof-bolts do not permit any significant lengthening or sliding relative to the rock walls of the bore hole because of the high adhesion of the polymer to the rock or metal.

Therefore, the modeling of the rigid contact of the roof-bolt reinforcement with the rock walls of the bore hole is the most objective, which has been performed in the computational experiment.

According to the data, the work of the fastening system has been modeled as a whole and a test calculation has been performed, pursuing two goals.

The first is the identification of any fundamental contradictions in comparison with the results of modern studies of geomechanical processes in the vicinity of in-seam working. The test results confirm the existing ideas about the displacement processes of the coal-bearing massif around mine working outside the zone of the stope works influence.

The second is to substantiate the thickness of the model (according to the coordinate Z) with account of the frame support and roof-bolts setting. Therewith, it was found that the distortion of the model's SSS near its vertical boundaries (along the planes YX) occurs no deeper as till the second frame, starting from the boundary plane. Then, all subsequent frames will be sufficiently objective in displaying the nature of their SSS. It was decided to model five sets of frames and six rows of roof-bolt support in the longitudinal section of mine working. The central frame and the adjacent rows of roof-bolts were outside the zone of influence of the SSS model anomalies adjacent to a border. Thus, all the provisions of the research plan for the fastening system have been implemented.

In addition, performing of the assigned tasks required a change in the parameters of the geomechanical model compared to the area of maintaining the preparatory mine working outside the zone of the stope works influence. On the one hand, the stress state of the surrounding rock massif is changed as compared to that in the area outside the zone of the stope works influence – the anomalies of the rock pressure occur around mine working caused by the stope works. The accounting for the impact of stope works (in the area of the longwall face approach) in substantiation of geomechanical and constructed models of preparatory mine working has been made on the basis of the following provisions.

First, the support zone ahead of the longwall face is characterized by increased vertical stresses σ_y in the coal-bearing massif, the concentration of which is enhanced as it approaches the coal seam [51, 77 – 85]. With that, the question arises: what concentration value σ_y should be set as a borderline condition when constructing and calculating the geomechanical models.

Secondly, the areas of bearing pressure on the sides of the preparatory mine working are also characterized by concentration of σ_y, which increases as it approaches to mine working and to coal seams, but the value σ_y is different for the left and right flanks of mine working and is different in the zone of bearing pressure ahead of the longwall face. Here, the same question arises concerning the concentration σ_y, used in the model.

Therefore, it is necessary to match these anomalous areas for the most adequate and reliable reflection of the stope works influence in the model. This matching has been made on the basis of the analysis of the field σ_y in the work [55], under the following conditions:

– *firstly*, the areas with the same concentration criterion $k = \dfrac{\sigma_y}{\gamma H}$ should be distributed into the roof of the coal seam at approximately equal height in the bearing pressure zone ahead of the longwall face and in the areas of bearing pressure in the sides of mine working;

– *secondly*, the height (according to the coordinate Y) of the distribution of the concentration criterion should exceed the total thickness of the main roof layers, which form the load both on the end parts of the powered support and on the fastening system of the preparatory mine working, and this does not change their SSS.

Simultaneous satisfying the specified conditions is achieved in the basic model at a height of 18.5 – 18.9 m from the coal seam, where the concentration criterion is $k = 1.8 - 2.1$. According to these parameters, the borderline conditions of the model loading are substantiated: the initial vertical stresses $\sigma_y = 2\gamma H$ act on the upper boundary of the model, passing through the siltstone and sandstone contact at a height of 18.95 m; the other borderline conditions remain unchanged.

The other side in the geomechanical model change concerns the introduction of additional fastening elements:

– ahead of the stope face at a distance of not less than 20 m, the central wooden prop stays of the strengthening support are set under the cap board of the frame with a slight shift towards the working flank of mine working;

– in the area of the yielding joist of the frame, an additional roof-bolt 2.4 m long is set at an angle of 10 – 20° to the horizontal. This roof-bolt is intended to maintain the roof directly above the drift berm after passing the longwall face.

3.2. ANALYSIS OF STRESSES INTENSITY IN THE BORDER ROCKS OF THE PREPARATORY MINE WORKING IN THE BEARING PRESSURE ZONE AHEAD OF THE LONGWALL FACE

The stress intensity analysis was performed separately for the rock massif surrounding the preparatory mine working and for its fastening system, which contributes to a more detailed and accurate study of the stresses fields.

The considerable changes in the stresses fields were taken into account as the longwall face approaches the studied mine working section. The part of the mine working length, where the bearing pressure from the approaching longwall face acts, is relatively small, and the measures to strengthen the fastening system are carried out in the area – not less than 20 m. With a stable longwall face operation, this distance is passed in 4 – 7 days. During this time, the rheological processes are in the initial stage. But, considering the increased stress intensity of the massif, three states are modeled at different distances to the longwall face: at the beginning of the bearing pressure zone (concentration criterion 1.1); in the middle of the bearing pressure zone ($k = 1.65$); near the stope face ($k = 2.0$).

The main component of the resulting assessment of the state of the rock massif adjacent to mine working is the stresses intensity σ.

At the beginning of the bearing pressure zone, the curve σ is characterized by a slight asymmetry of distribution (relative to the vertical axis of mine working) towards the approaching longwall face. The unloading zones affect the first layers of the main roof and bottom with a level of acting stresses $\sigma = 1 - 3$ MPa. In the sides of mine working, the concentration of $\sigma = 14 - 17$ MPa has a very limited distribution up to 0.4 m, and the width of the $\sigma = 10 - 13$ MPa action does not exceed 1.9 m in the upper part of the immediate bottom. Also, a low concentration of $\sigma = 10 - 14$ MPa acts under the bearings of the frame support prop stays. The specified data indicate a weak influence of the bearing pressure zone in the initial period of its impact, which is quite consistent with experimental studies on the measurement of the rock contour displacements of the preparatory mine working.

As the longwall face approaches to the studied mine working section, its influence increases:

– the degree of unloading σ in the roof and bottom is reduced and localized in small areas, and in the remaining volumes of rocks the value $\sigma = 5 - 7$ MPa already slightly exceeds the initial state of the massif;

– in the sides of mine working, the asymmetry of σ distribution increases: if from the side of the adjacent area the stresses value $\sigma = 11 - 12$ MPa extends to a height of up to 13 – 16 m and a width of up to 7 – 8 m, then from the side of the approached longwall face such stresses reach the model boundaries;

– in the border lateral rocks, the destructive concentrations $\sigma = 20 - 70$ MPa, spreading across the width up to 2.1 m, act on the entire height of mine working and under the bearings of the frame prop stays.

Thus, there is a rather unstable state of the border rocks in the sides of the preparatory mine working, which occupy a significant volume and for this reason are capable of creating an increased lateral load on the fastening system. It takes the notice the active border rocks stratification of the bottom to a depth of 0.8 m, which induces its swelling.

3.3. RESEARCH AND ANALYSIS OF THE STRESSES INTENSITY OF THE BASIC FASTENING SYSTEM IN THE BEARING PRESSURE ZONE AHEAD OF THE LONGWALL FACE

The purpose of this chapter is to assess the SSS of the fastening system in the preparatory mine working in the area falling into the front part of the bearing pressure zone. The stresses components analysis has been performed in each load-bearing element: frame support, roof-bolts and central prop stays of the strengthening support.

The final component, subjected to research in the fastening system of the preparatory mine working, is the stresses intensity σ.

At the beginning of the bearing pressure zone ahead of the longwall face, the field in the fastening system elements is characterized by the following features. In the frame support (as in the area outside the zone of the stope works influence) the prop stays are the most loaded. The cap board of the frame, thanks to the roof-bolting and bearing fastening and the central prop stay of the strengthening support, is in a relatively unloaded state:

– in the central part of the arch, the σ make up only 52 – 70% of the yield limit of steel St.5;

– near the spring the stresses increase to $\sigma = 200 - 260$ MPa and approach to the limiting state;

– in the local area of contact of the central prop stay of the strengthening support with the cap board, the concentrations $\sigma = 180 - 250$ MPa act, which further lead to active bending of the cap board with the loss of its initial form.

In the frame prop stays, especially along the length of the rectilinear part, $\sigma = 260 - 300$ MPa cause a plastic state (or close to it) of their material.

The nature of σ distribution along the SCP section indicates the areas of active bending of the prop stays:

– into the mine working cavity at a height of 1.8 – 2.2 m;

– in the direction of the lateral rocks at a height of 1.4 – 1.7 m;

– into the mine working cavity at a height of 0.7 – 1.3 m;

– in the direction of the lateral rocks in the area of the prop stay bearing at a height of 0.3 – 0.5 m.

Such an alternating bending causes the stability loss of the frame prop stay and requires its strengthening.

The central prop stay of the strengthening support is characterized by a fairly uniform distribution of σ along the length and in the cross section; the level of $\sigma = 20 - 40$ MPa indicates its stability in general, but the upper limit of σ have already reached the compressive strength of pine along the fibers.

The roof-bolts in the roof-bolting and bearing system do not yet experience a significant load ($\sigma = 50 - 90$ MPa) for the most part of their length and only in the areas close to the contour of mine working, $\sigma = 110 - 140$ MPa. The exception is an additional roof-bolt from the side of the longwall face, where the maxima $\sigma = 180 - 210$ MPa are already approaching the estimated yield limit of the roof-bolt reinforcement.

As the longwall face approaches (near the stope face), the following patterns of changes are observed in the stresses intensity in the fastening system of the preparatory mine working. The cap board of the frame support flattens out with appearing the limiting state ($\sigma = 260 - 290$ MPa) in the upper part of the SCP cross section, and in the area of contact with the central prop stay of the strengthening support, the plastic deformations develop even more intensively ($\sigma \geq 290$ MPa), which causes the bend of the cap board and indicates once again the negative effect of setting the rigid central prop stays of strengthening support.

The prop stays of the frame support almost along their entire length change to a plastic state and require strengthening in the lateral direction.

The stresses intensity in the central prop stay of the strengthening support is not changed significantly, however, in its upper part, where the maximum bending of the cap board is observed, the local areas with destructive stresses of $\sigma = 45 - 55$ MPa appear. The providing of the central prop stays of the strengthening support with constructive pliability would allow performing two positive functions: reduce stresses and flexure strain in the prop stays themselves and at the same time limit plastic deformations in the cap board and its bend in the area of contact with the prop stay of the strengthening support.

In the roof-bolts of roof-bolting and bearing support, especially peripheral ones, there is an increase of σ up to values of 200 – 260 MPa in the areas of their length adjacent to mine working. In the additional roof-bolt, the increased stresses (up to 290 MPa) act over a longer part of its length. The curve σ of stresses intensity in the roof-bolts (near the stope face) indicates almost complete loading of the peripheral roof-bolts, but the central vertical roof-bolt of the roof-bolting and bearing fastening is the least loaded.

3.4. THE STRESS INTENSITY OF BORDER ROCKS IN THE ZONES OF BEGINNING OF PROTECTIVE STRIP RESISTANCE AND STABILIZATION OF THE ROCK PRESSURE MANIFESTATION

After the first passage of the longwall face, the change in the stresses intensity field σ has been studied from the area of the beginning of the security system work to the area of stabilization of the rock pressure manifestations. After erection of the security system, it begins, as a more rigid element, to take up the increased load from the side of lowering rock layers of the roof and transfer it to mine working berm with the formation of appropriate concentrations σ in the rocks of the immediate roof and the bottom of the coal seam. In the immediate roof above the security system, the concentration of destructive stresses $\sigma = 30 - 80$ MPa act only in local areas above the more rigid rows of wooden prop stays. These places of σ_{max} do not present a significant danger, since beyond their limits the value $\sigma = 18 - 23$ MPa has not yet exceeded the compressive resistance of the rock.

With an increase in the rock pressure manifestations in these areas, the immediate roof begins to destruct with the development of increased vertical loads on the fastening system. The mechanism of this process is conditioned by the increased pliability of the weakened rock volume above the security system.

Therefore, the lowering layers of the main roof will predominantly load the fastening system. It is therefore important to limit the volumes of weakening the immediate roof above the part of the security system close to mine working, and transfer the main load into the depth of mined-out space.

Similar processes occur in the rocks of the immediate bottom under the security system. Here the concentrations of $\sigma = 30 - 50$ MPa still have an even more limited distribution, but a number of places arise in the area of the frame prop stays, which subsequently form an increased lateral load. That is, it is advisable to unload the border rocks of the berm again by introducing a variable rigidity into the design of the security system.

From the side of the virgin massif, the local areas of hazardous concentrations are formed only in the border rocks of the immediate bottom along the height of its ripping. And here there is the prospect of increasing the lateral load on the frame support prop stays.

These assumptions are confirmed by the calculation of the field σ in the adjacent massif in the area of stabilization of the rock pressure manifestations after the passage of longwall face. Indeed, there is an expansion of the bearing pressure zones with increasing the concentrations σ of the level 30 – 49 MPa. Thus, from the side of virgin massif, the noted concentration σ is distributed over the entire height of the bottom ripping and below the bearings of the frame prop stays and into the sides at a distance of 0.8 – 1.0 m. There definitely will be a destruction of even the water-free rock with the formation of an increased lateral load on the frame support prop stays. In the rocks of the immediate roof and bottom in the area of security system location, the above destructive concentrations repeatedly expand the area of their action. For example, in the roof, the width of the weakening zone reaches 2.0 – 2.2 m, and the height – 1.2 – 1.4 m. The weakened rock partially "overworks" the security system, penetrating into existing cavities and voids. The reaction to the rock layers of the main roof decreases, and when they are lowered, the rock pressure on the fastening system increases. In the rocks of the immediate bottom, along the entire height of the berm to a width of 1.8 – 2.0 m, it is destroyed with movement into the cavity of mine working and with the deformation of the frame prop stays.

3.5. THE STRESS INTENSITY ANALYSIS IN BASIC FASTENING AND PROTECTION SYSTEMS BEHIND THE LONGWALL FACE

After the passage of the longwall face and the erection of the protection system, the field of stresses intensity distribution in the fastening system is partially changed as compared with the bearing pressure area ahead of the longwall face, retaining the common features (Fig. 3.2).

The cap board of the frame is the least loaded with the formation of plastic contact with the prop stay of the strengthening support. The relative uniformity of

σ distribution in the frame prop stays is disturbed by occurring the flexure strains:

– from the side of virgin massif, the bending of the frame prop stays is less intensive and directed into the cavity of mine working in the central part, and in the bearing of the prop stay – towards the massif;

– from the side of mined-out space, a more active bending of the frame prop stay occurs with a regular change in the curvature sign: into the cavity of mine working at the level of extracted coal seam; towards the massif along the height of the berm and in the area of the prop stay bearing.

The stresses σ in the central prop stay of the strengthening support are sufficiently uniform and remain at the same level ($\sigma = 20 - 40$ MPa) as compared with the bearing pressure zone ahead of the longwall face.

Among the roof-bolts, an additional roof-bolt is the most loaded, set in the roof above the protection structure; in its reinforcement, the stress intensity exceeds the yield limit σ_{yield} of steel in the middle part of the length, and in the buried part they approach or equal to σ_{yield}. This is caused by the high intensity of stresses and displacements in the bearing pressure area above the protection structure.

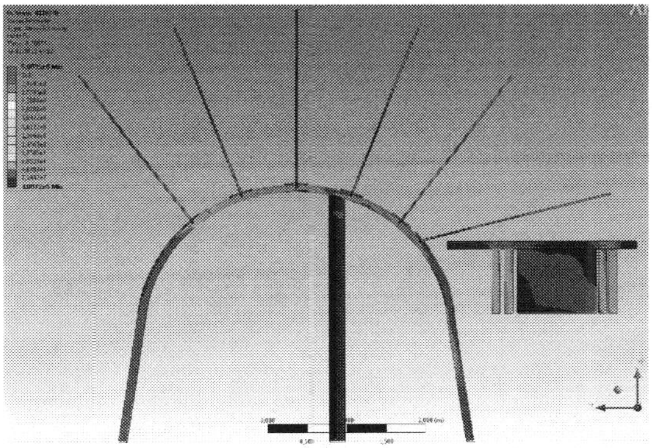

Fig. 3.2. The curve of stresses σ intensity in the basic fastening and protection systems in the initial period of its resistance

The rest of the roof-bolts are in an elastic state. The peripheral roof-bolts are the most loaded, where in the buried areas and in the areas close to mine workings contour, the value σ approaches the yield limit of steel of the reinforcement. Such a feature of σ distribution along the length of the roof-bolts corresponds to the existing ideas about the mechanism of resin-grouted roof bolts operation and indicates the process of their displacement into the cavity of mine working together with strengthened rocks of the immediate roof. These rocks are stratified more intensively in the border part of mine working and less intensively in the buried part of the length of the roof-bolts. It also follows that the active displacements of the roof rocks extend beyond the roof-bolts length of 2.4 m: the condition of the roof from the side of the longwall face (especially above the protection structure) requires not only a dense reinforcement grid, but also its deeper distribution.

One of the technological solutions to this problem is setting the rope bolts, which can perform the function of deeper strengthening of the roof rocks, limiting its stratification. If to set two rope bolts (with a step of $L = 1.6 - 2.0$ m) in the cross section symmetrically at 0.8 – 1.0 m from the vertical axis of mine working at an angle of 70 – 80° to the horizontal, so they strengthen the roof volume to the width of 7.5 – 8.0 m. This width of the strengthening zone is not enough to completely overlap the bearing pressure area above the protection structure, but the increased stability of the rock cantilevers does not require this, because of the dimensions' constraint of the limiting state zone.

In the security structure, a rolling-on chock and overlapping girders are only in a sufficiently stable state with a slight contortion of the wood. All wooden prop stays are exposed to stresses σ, by 1.5 – 3 times exceeding their compressive resistance. For this reason, they are actively deformed, and the process of pressing into the rocks of bottom and roof in a certain way protects the prop stays from total destruction.

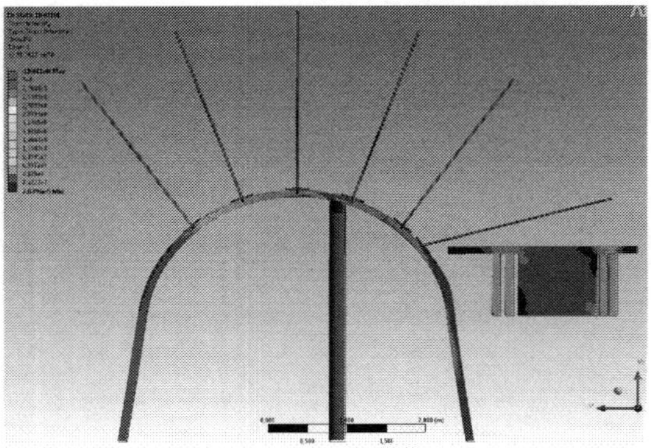

Fig. 3.3. The curve of stresses σ intensity in the basic fastening and protection systems in the zone of stabilization of the rock pressure manifestations

With further operation of mine working in the zone of stabilization of the rock pressure manifestations, a number of changes are observed in the stresses σ intensity field (Fig. 3.3) with a general tendency of a certain increase in the loading on the fastening and security systems. This relates to the following load-bearing elements. In the frame prop stay, from the side of mined-out space, at a height of up to 0.7 m from the bottom, the stresses σ and the corresponding dimensions of the area of single-valued plastic deformations ($\sigma \geq 300$ MPa) increase up to 10 – 15%. The load on the central prop stay of the strengthening support, which is almost in its entire volume in the limiting or close to the limiting state, increases most significantly. Also the σ increase significantly in two peripheral roof-bolts of the roof-bolting and bearing support and, especially, in the additional roof-bolt from the side of mined-out space. This once again emphasizes the expediency of intense strengthening the volume of the roof rocks above the security system, the shifts of which occur for quite a long period. This is also indicated by the increase in the load in all elements of the security structure:

– in a rolling-on chock the stresses σ are almost in entire volume close or equal to resistance of pine to compression, and in the corner parts they exceed it noticeably; however, because of the contortion of wood, a low ratio of the chock height to the rest its dimensions (less than unity), pressing into the rocks of bottom and roof, the rolling-on chock continues to function as a bearing part of the security system, but at the same time its height decreases and, as a consequence, the height of the preparatory mine working decreases;

– in all wooden prop stays, the already high stresses σ increase by 10 – 20% and are capable of destroying them, despite the influence of other factors.

Thus, a tendency has been established for the long-term development of the roof rocks lowering from the side of mined-out space, which results in an increase in the loading of the security system, the rocks of berm and frame prop stays, as well as the central prop stays of the strengthening support and roof-bolts from the side of mined-out space. The remaining load-bearing elements of the fastening system (central roof-bolts in the arch, the cap board of the frame and its prop stay from the side of virgin massif) practically do not change their stressed state after a long period after the passage of the longwall face.

CONCLUSIONS

In general, based on the research results of the effectiveness of the basic fastening and security systems, the following conclusions have been made:

1. The greatest danger to the stability of the fastening system is constituted by the frame prop stays, especially from the side of mined-out space – they are deformed plastically along the entire length; the cap board of the frame flattens out with a bend into the cavity of mine working in the central part and with arising the limiting state ($\sigma = 260 - 290$ MPa) in the upper SCP part, and in the area of contact with the central prop stay of the strengthening support, the plastic deformations develop even more intensively ($\sigma \geq 290$ MPa); among the roof-bolts, an additional roof-bolt, located in the roof of mine working above the protective strip, and partly the peripheral roof-bolts are the most loaded, but the central roof-bolts in the mine working arch are underloaded; the central prop stays of the strengthening support (wooden) are in the limiting state in their entire volume, and on contact with the cap board of the frame – in the superlimiting state.

2. The peculiarity of σ distribution along the length of the roof-bolts corresponds to the existing ideas about the mechanism of resin-grouted roof bolts operation and indicates the process of their displacement into the cavity of mine working together with strengthened rocks of the immediate roof, which are stratified more intensively in the border area of mine working and less intensively in the buried part of the roof-bolts length. It also follows that the active displacements of the roof rocks extend beyond the length of the roof-bolts of 2.4 m: the state of the roof from the side of the longwall face (especially above the security structure) requires not only a denser reinforcement grid, but also its deeper distribution. One of the technological solutions to this problem is setting the rope bolts, which can simultaneously perform the function of deeper strengthening of the roof rocks, limiting its stratification. Thus, if to place two rope bolts (with a step of 1.6 – 2.0 m, that is, through two frames along the length of mine working) in the cross section, symmetrically 0.8 – 1.0 m from the vertical axis of mine working at an angle of 70 – 80° to the horizontal, they strengthen the roof volume to a width of 7.5 – 8.0 m, which is enough to increase the stability of the rock cantilevers and to constraint the dimensions of the limiting state zone of border rocks.

4. RESEARCH AND ANALYSIS OF THE STRESS STATE OF THE LOAD-BEARING ELEMENTS IN THE PREPARATORY MINE WORKING "MASSIVE – FRAME – COMBINED ROOF-BOLTING SYSTEM"

4.1. STRESS STATE OF ROCKS ENCLOSING MINE WORKING IN THE AREA OF STABILIZATION OF ROCK PRESSURE MANIFESTATIONS

The carried out multivariate computational experiments are quite extensive, which does not allow putting all the obtained diagrams in the scope of this paper. Therefore, one of the variants of the combined roof-bolting systems (two resin-grouted roof bolts in the central part of the arch and two rope bolts in its peripheral area) has been selected, which is compared with the traditional scheme for maintaining the belt entry, for example, in the conditions of mining the seam C_6.

For this variant, the results have been stated of a comparative analysis of the three main stresses components (vertical σ_y, horizontal σ_x and stresses intensity σ) distribution in the massif and the fastening system; but, the utmost importance is placed on the peculiarities of the stresses intensity σ distribution curve, as a generalizing characteristic of t'he stressed state of any mining and engineering object.

Analysis of the stresses intensity curve in the massif surrounding mine working, with the use of a combined roof-bolting system instead of the basic scheme has shown a number of changes in the state of adjacent rocks.

In the roof of the belt entry, an unloading zone σ is formed, which is different from that of the basic fastening scheme in its form and dimensions. Firstly, the unloading practically does not affect the rocks of the immediate roof. From the side of mined-out space, a local area is formed with a value σ corresponding to the state of the virgin massif. Above the central part of the arch, there is an area in the rocks with concentration σ of level 1.6 – 2.0 from the initial state of the virgin massif. By absolute value, this concentration is not dangerous, since it is 3.0 – 3.8 times lower than the compressive resistance of the sandstone in the immediate roof. It is necessary to pay attention to this fact for the following reason. Usually, including the basic fastening scheme, an unloading zone is formed in the immediate roof above the mine working, where the lowered σ characterize the deflection of the rock layer into mine working with its corresponding horizontal shifts. If the rock layer is holistic, then its horizontal shifts do not reduce the resistance to the vertical rock pressure. As a rule, the immediate roof is broken by tension cracks into the rock blocks, and if the rock blocks come out of contact with each other, then at their deflection the thrust structure is not formed, and the rock blocks collapse on the frame support [11, 15, 72]. If there is a horizontal thrust, the rock blocks resist vertical rock pressure, namely, a small concentration of σ

indicates an increased horizontal thrust of rock blocks. The described state of an immediate roof was formed owing to its connection with the main roof by means of rope bolts, – a very thick armored and rock plate has been formed with a high load-bearing capacity.

Secondly, the nature of σ distribution in the first layer of the main roof (represented by a thick siltstone) indicates the possibility of a partially independent deflection of the main and immediate roof with some horizontal shifts relative to each other owing to the limited rope pliability by the effect of its elongation under tension. The possibility of horizontal displacements of the layers of the immediate and main roof relative to each other contributes to occurrence in the main roof of the local unloading area with dimensions in the plane of the section 0.7×1.0 m. The unloading level is 0.4 – 0.8 of the initial state of the virgin massif. This indicates the preservation of a part of the horizontal thrust within this local area, above which the siltstone state returns to its initial state. At half the thickness of the rock layer of the main roof, the concentration σ of the level 1.2 – 2.0 acts. This concentration characterizes the action of thrust forces in the resistance to the deflection of the lower layer of the main roof, and the absolute values of σ are 1.75 – 2.91 times lower than the compressive resistance of siltstone.

Thus, the following conclusion can be made: it was not possible to achieve the absolute rigidity of the armored and rock plate from the main and immediate roof. But this is not necessary because of their joint and active resistance to the rock pressure within the range of stresses σ, which are very far from destructive values. At the same time, in the basic fastening scheme, the unloading zone passes through the entire thickness of the immediate roof with a very likely its collapse.

A negative circumstance is the occurrence of weakened rocks area to a depth of 0.15 – 0.20 m near the contour from the side of virgin massif. However, this cannot significantly affect the resistance of the immediate roof to the rock pressure. The concentration of σ once again emphasizes the active involvement of the immediate roof rocks into resistive action against the displacement processes of the coal-overlaying formation.

In the adjacent lateral rocks from the side of virgin massif, one more peculiarity is observed in the stresses intensity distribution – the lowered σ in comparison with the basic variant of fastening. Thus, the concentration of σ is 1.6 – 2.0 across the width into massif of up to 2.1 m, which is many times less than the compressive resistance of both the immediate roof rocks and the coal seam; in weak rocks of the immediate bottom, such a concentration causes their weakening and induces the intensification of the swelling process.

If to compare the lateral bearing pressure when using a combined roof-bolting system, it is 2 – 3 times lower than in the basic fastening variant. Obviously, this is because the formed armored and rock plate in the roof has an increased area of bearing on the side rocks; then, their stress intensity decreases. Higher concentrations of σ are observed only in border rocks with a width of up to 0.3 – 0.7 m, but neither for the immediate roof rocks, nor for the coal seam, they constitute a danger in terms of weakening.

From the side of mined-out space, above and under the protection structure, high stress concentrations are formed, conditioned by a relatively small area of bearing in the form of a protection system, which induces the weakening of certain volumes of rocks in the immediate roof and bottom.

The increased hardness of the immediate roof rocks restricts the area of weakening. On a contact with the protection structure, the weakening is predicted in full its width, and at the border with the main roof the width of the rocks weakening does not exceed 0.3 m. In the immediate bottom rocks, the depth of weakening reaches 6.5 m with a width of up to 3.1 m, which, of course, generates an active displacement of the bottom rocks of the seam into the cavity of mine working.

From the above information, it follows that in order to increase the rigidity of the bearing from the side of mined-out space, it is necessary to restrict the area of weakening above and under the protection structure. But, this is a different, separate task, which is studied in the work [80]. It should be noted here that across the width of the rope bolt influence (from the side of mined-out space) in the immediate roof rocks adjacent to mine working, the stresses concentration is not destructive, as well as in the lower layer of the main roof. The integrity of the armored and rock plate, formed by the combined roof-bolting system, is predicted. Therefore, this plate will actively resist to vertical rock pressure.

In the soft bottom rocks of mine working, across its width, an unloading zone is formed, which extends to a depth of 5.5 m. In the corner parts of mine working, destructive stresses act in vast areas of bottom rocks. These areas will contribute to the movement of rocks into the unloading zone and further into mine working. Therefore, it is possible to predict a sufficiently intensive swelling of the bottom rocks of the drift. Here the effect of the combined roof-bolting system is observed to a lesser extent. The main factor of swelling is a very low hardness of the bottom rocks. Nevertheless, the dimensions of weakening areas in the bottom is noticeably lower than in the basic variant of fastening, and the main reason for this is not the intensive bearing pressure in the sides of mine working.

Summarizing the peculiarities of stresses intensity distribution in an adjacent rock massif, it should be noted the positive effect of the combined roof-bolting system on the roof rocks state, where an armored and rock plate with a high load-bearing capacity is formed. A satisfactory state of mine working is predicted under its protection.

Positive peculiarities, which emphasize the process of formation with the help of a combined roof-bolting system of an armored and rock plate in the roof, are quite clearly manifested in the curve of horizontal stresses σ_x. The area of their concentrations is distributed across the entire width of the arch and the thickness of the immediate roof. The value $\sigma_x \leq 18-25$ MPa is high in comparison with the initial state of the virgin massif (concentration coefficient is $k_x = 5.2-7.3$). It is not caused by the weakened rocks of the immediate roof. The increased horizontal compressive stresses act throughout the thickness of the layer, creating such a thrust in the rock blocks of the immediate roof, which does not allow its significant deflections into the cavity of mine working. In the lateral areas (in relation to mine working) of the immediate roof, there is also a relatively non-uniform distribution field of the compressive σ_x through the thickness – here, again, there is a thrust of the rock blocks. As a result, a thrust block system in the immediate roof is not only stable, but also actively resists to the load from the side of the main roof.

In the rock layers of the main roof, there is a bend through the thickness of each lithotype. From the side of mined-out space, a change in the curvature sign of the layers determines the direction of the complete displacements surface (downward movement of the main roof blocks to the collapsed rocks of the immediate roof), in accordance with the existing ideas about the displacement processes in the coal-overlaying formation in the stope extraction [29, 86]. The beginning of this surface is outside the width of the protection system and removed towards the mined-out space at a distance of 2.5 – 3.0 m. In this area, the bending of the rock layers of the main roof with the occurrence of small tensile σ_x up to 1.0 – 2.5 MPa in the upper part of each layer is observed. The tension is limited and amounts to 5 – 10% of the layers' thickness. Consequently, the probability of the tension cracks development in the layers of the main roof is also limited. The compressive stresses, which ensure the stable state of the layers by means of the thrust, act throughout their thickness. In most areas of the main roof, the stress varies in a range close to the initial state of the virgin massif, and is lower than the resistance of lithotypes to compression.

Thus, according to the factor of the component σ_x distribution in the rocks of the roof, their stable state is predicted; this especially concerns an immediate roof, where the combined roof-bolting system creates high thrust forces between the rock blocks, then connects them with the rocks of the main roof and thus provides sufficient resistance to vertical rock pressure.

The armored and rock plate, formed in the roof, has an expanded area of bearing on the rocks in the sides of mine working. This is evidenced by the almost uniform stresses field σ_x with the exception of local areas of border rocks to width of up to 0.8 m. In the basic variant, the anomalies of horizontal stresses extend to 4.0 – 4.5 m in the side of mine working.

From the side of mined-out space, in the rocks of the immediate bottom (in depth of the lower ripping of the drift), the perturbations of the horizontal stresses are not as developed as in the basic fastening variant and in most parts of this area they have small deviations from the initial state of the virgin massif. Here, it is possible to assume a lowered lateral bearing pressure owing to the fact that the rock blocks of the main roof, which move down into mined-out space, have a certain thrust with the rock layers above mine working and, by means of it, they retain a part of the load-bearing capacity in order to resist the vertical rock pressure.

In the bottom rocks area, directly under the mine working, the stresses anomalies are also not so high, but there is an area of tension with a depth to 0.6 m, where the stratification and movement of rocks into mine working occurs.

Based on the results of analysis of the horizontal stresses field distribution, it can be confirmed that the combined roof-bolting system contributes to the creation of sufficient thrust forces to maintain the stability of the immediate roof, on the one hand, and, on the other, by combining the immediate roof with the main roof, it creates an armored and rock plate with high load-bearing capacity.

In conclusion of the SSS analysis of the adjacent rock massif, the peculiarities of vertical stresses σ_y distribution have been studied.

In the roof of the drift, the zone of unloading from vertical stresses σ_y differs significantly from the basic fastening scheme in the direction of reducing the dimensions of the zone and the degree of unloading:

– the unloading of $k_y = 0.47 - 0.83$ level extends into the roof of up to only 1.8 m against 4.3 m in the basic fastening scheme;

– the deeper degree of unloading ($k_y = 0.10 - 0.47$) occurs in the border rocks to a height of 0.6 m and a width of up to 1.9 m. In the basic fastening scheme, these dimensions are 2.0 and 3.5 m, respectively. This indicates the distribution of very deep unloading throughout the thickness of the immediate roof and for the most part of the width of mine working arch;

– the complete unloading ($\sigma_y = 0$) with the occurrence of tensile vertical stresses is practically not observed, unlike in the basic fastening scheme.

The above data indicate the absence of arch of ultimate equilibrium, which is formed during the stratification of the roof rocks due to the action of tensile stresses. This is conditioned by the formation (with the help of rope bolts) of a thick armored and rock plate, which prevent significant deflections above mine working, and the local stratifications in the central part of the arch are prevented by two resin-grouted roof bolts.

Another peculiarity is the less evident bearing pressure zone from the virgin massif. Here, the concentration of compressive ($k_y = 1.2 - 1.9\,\gamma H$) stresses is observed, and only in the local areas of border rocks to a width of 0.2 – 0.5 m the increased concentrations act of the level ($k_y = 3.0 - 4.2\,\gamma H$). Such a behavior of rocks in the side of mine working is explained by the increased bearing area of the armored and rock plate in the roof, which distributes the vertical load more uniformly. From the side of mined-out space, there is a high concentration ($k_y = 3.0 - 4.2\,\gamma H$) above and below the protection structure, but this area extends in height into the roof by 1.7 times and in depth into the bottom by 2.1 times less than in the basic fastening scheme.

The main reason here is in the partial preservation of the thrust forces between the rock blocks above mine working and from the side of mined-out space. These forces make it possible to create a resistance to the vertical component of the rock pressure and distribute it over a larger area, including the area towards the mined-out space. A large area of its distribution reduces the concentration.

In the bottom of mine working, the unloading zone σ_y is also smaller, both in dimensions and in value, if compared with the basic fastening scheme. This is conditioned by a decrease in the parameters of bearing pressure zones in the sides of mine working.

Thus, the analysis of the state of the coal-bearing massif around the preparatory mine working convincingly proves that the application of combined roof-bolting systems makes it possible to use more effectively the load-bearing capacity of the roof rocks by means of preservation of thrust forces between the rock blocks and an armored and rock plate formation. Acting upon analogy with the "chain reaction", the intensity of the SSS anomalies in the sides and bottom of the preparatory mine working decreases. The conclusion follows from this about the positive effect of combined roof-bolting systems on the stability of the surrounding rock massif, which restricts the rock pressure manifestations in mine working.

4.2. THE STRESS STATE OF FASTENING SYSTEM ELEMENTS IN THE PREPARATORY MINE WORKING AND COMPARISON WITH THE BASIC VARIANT OF ITS FASTENING

Analysis of the behavior of massif in the vicinity of the preparatory mine working made it possible to assume that the combined roof-bolting system provides for the creation of such thrust forces that allow to consider the immediate roof and the lower layer of the main roof as an armored and rock plate, which takes up the most of the vertical rock pressure and, thereby, protects the frame support from excessive loads.

The next stage of the research consisted in a comprehensive assessment of the state of all the fastening elements of compared variants (basic and proposed) of mine working, guided by the curves of vertical σ_y, horizontal σ_x stresses distribution and there intensity σ. First of all, the stresses intensity field has been studied, since this parameter includes a certain combination of all stress components and is an integral indicator of the state of one or another fastening element. The stresses intensity curve is shown in Fig. 4.1.

The cap board of the frame support is in an unloaded state, which is very often determined when performing the computational experiments for the conditions of the Western Donbas mines. However, the degree of underloading of the cap board when using the combined roof-bolting system differs significantly from that for the basic fastening variant, which being studied in more detail. The peripheral area of the cap board near the yielding joists is characterized by minimal intensity ($\sigma \leq 20 - 25$ MPa), which is quite conditioned by normal mode of their operation. In the basic fastening variant, in the end part of the cap board from the side of mined-out space, the stresses intensity value is already 47 – 93 MPa, that is, it increases several times. From the side of virgin massif, the end part of the cap board is exposed to action of $\sigma = 185 - 270$ MPa; the upper limit of this range already corresponds to the estimated yield limit of St.5 steel, from which the special SCP profile is usually made. Such a high intensity of stresses indicates the occurrence of a rigid operation mode of the junction node between the cap board and the prop stay due to depletion of the vertical pliability of the frame. Further, when moving to the central part of the arch (in the basic fastening variant), the stresses intensity decreases to 70 – 185 MPa from the side of virgin massif and increases from the side of mined-out space. In the place where the cap board is bearing on the central (wooden) prop stay of the strengthening support, there an area arises of its plastic state, which extends up to 0.7 m. Thus, along the length of the cap board in the basic fastening variant, together with the unloading areas, the areas of its limiting state are observed, where it is possible the cap board plastic bending with a change in its initial form. A different situation is observed in the cap board when setting a combined roof-bolting system. Here there is a smooth increase in the stresses intensity to 83 – 104 MPa in the central part of the arch. It should be noted that these values are significantly lower than in the basic fastening variant and is only 31 – 39% of the estimated yield limit of SCP steel, that is, the cap board state is far from the limiting state over its entire length. In our opinion, the main reason for the revealed differences in the cap board state is the action of a reduced vertical load caused by the armored and rock plate formation (combined roof-bolting system) with the proper thrust forces in the immediate roof and in the lower layer of the main roof.

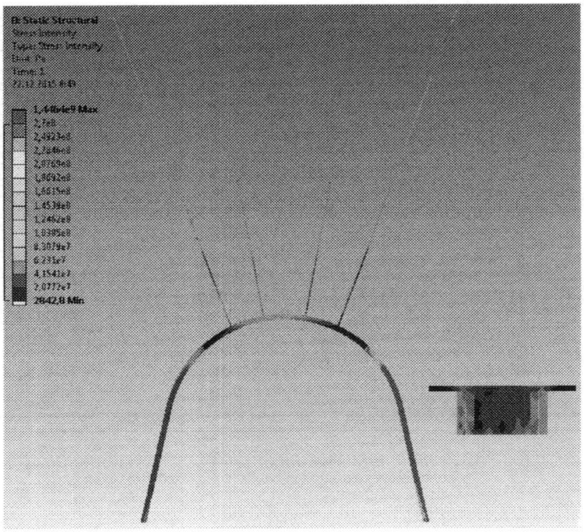

Fig. 4.1. The curve of stresses σ intensity in the fastening elements of the preparatory mine working when using a combined roof-bolting system

In contrast to significant differences in the state of the frame cap board, the distribution of σ in its prop stays is quite similar for both compared fastening variants. The main feature of this similarity is the high loading (at the level of the limiting state) of the frame prop stays, but there are also some differences in the stresses intensity distribution.

Firstly, in the frame prop stay from the side of mined-out space, the stresses intensity is close to the estimated yield limit of SCP steel (92 – 98%), but does not exceed it; that is, a state close to the limiting state is observed. For the most part of the prop stay length, the stresses intensity distribution is uniform in its cross section, which indicates the absence of any significant bending moment. The exception is the lower part of the prop stay at a height of up to 0.7 m from the bottom, where it bends into the cavity of mine working. In the basic fastening system there are several such areas of the prop stay bending, and they have different direction and are located along the entire height of the prop stay from its curvilinear part to the bearing. Furthermore, in the basic variant, the value of the stresses intensity is higher by 10 – 40%, so that the prop stay is in the limiting state throughout its height and requires strengthening to resist the lateral rock pressure.

Secondly, in the frame prop stay in the proposed variant, from the side of virgin massif, the stresses intensity is either equal to, or exceeds the estimated yield limit of SCP steel. The stresses intensity is distributed uniformly, and the bending moment acts only in the area of the prop stay bearing to a height of 0.4 m from the bottom. In the basic fastening variant, a bending moment of variable sign (although reduced value) arises along the entire length of the prop stay, and the value σ increases to 30%.

Thus, it should be noted the increased load of the frame prop stays in the basic fastening variant, which confirms the claim of reducing the vertical load on the frame support when using a combined roof-bolting system in the roof of mine working.

Consider the roof-bolts state in the mine working roof in the basic fastening scheme – the five resin-grouted roof bolts, set along the arch contour within the length of the frame cap board, have different degrees of load. In the central part of the arch, three roof-bolts are exposed to stresses ranging from 8 – 26% to 34 – 51% of the estimated yield limit of steel of the roof-bolt reinforcement. Two peripheral roof-bolts are loaded to a greater extent:

– in the border part of the roof-bolt length up to 0.4 m (from the side of the virgin massif) the σ act of the level 68 – 85% of its load-bearing capacity. This indicates active stratification of the rock massif in the area of the roof-bolt setting, and the rest of its length is loaded to a lesser extent;

– the peripheral roof-bolt from the side of mined-out space, on the contrary, experiences an increased load (77 – 100% of the yield limit of steel) in the buried area of up to 0.5 m in length. This is conditioned by the resistance of the roof-bolt to horizontal shifts of the rock blocks in the immediate roof, which move down on the protection structure, and the rest part of the roof-bolt length also remains low loaded.

Thus, in the basic fastening scheme, there is an increased resistance to rock pressure only in the peripheral roof-bolts set in the area of the frame yielding joists.

In the combined roof-bolting system, in the mine working arch, the following peculiarities of the stresses intensity distribution are observed. Two resin-grouted roof bolts, set in the central part of the arch, are predominantly loaded in the buried area of their length, which is about 52% for the roof-bolt from the side of mined-out space, and the roof-bolt from the side of virgin massif is loaded along its entire length. The stresses level in these areas corresponds to 31 – 77% of the estimated yield limit of the roof-bolt steel. From the side of virgin massif, there is a limited area, where the stresses develop up to 84 – 92% of its load-bearing capacity. If to compare the level of resistance of the central resin-grouted roof bolts, then it is much higher in the combined roof-bolting system, which indicates their more efficient work in creating a thrust structure from rock blocks in the immediate roof.

The setting of rope bolts also actively resists to the rock pressure development, and this refers to parts of their length located in the lower layer of the main roof. Thus, for a rope bolt from the side of virgin massif, in the area up to 75% of its length, the stresses from 60 to 90% of the load-bearing capacity act. For a rope bolt from the side of mined-out space, the length of the area with active resistance is slightly lower (up to 70%), but at its buried part, the area of almost full load of the roof-bolt expands to 100% of the load-bearing capacity.

Here it is possible to compare the intensive work of the combined roof-bolting system and the reduction of the SSS anomalies in the rock massif, especially in the roof of mine working. These processes, of course, are interrelated and prove the validity of the application of combined roof-bolting systems in the roof of mine working.

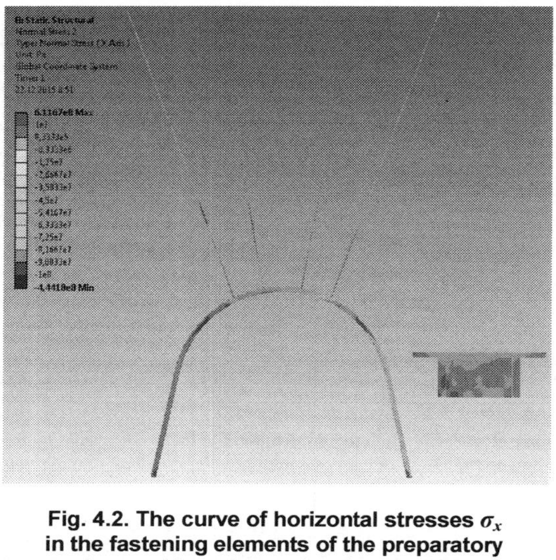

Fig. 4.2. The curve of horizontal stresses σ_x in the fastening elements of the preparatory mine working when using the combined roof-bolting system

There is one more indirect argument in favor of combined roof-bolting systems – the degree of loading of the protection construction of mine working. Individual wooden prop stays on the berm of the drift and the prop stays of the breaker-prop rows are loaded by 1.5 – 1.7 times less than in the basic variant of fastening. This confirms the previously stated assumption about a partial preservation of the thrust between the rock blocks above mine working and blocks that move down on the collapsed rocks into mined-out space. Such a partial thrust makes the rock blocks of the main roof to take up a certain fraction of the vertical rock pressure, which reduces the load on the protection structure.

To develop an evidence base of objectivity for the expedient use of combined roof-bolting systems, the peculiarities have been studied of horizontal stresses σ_x (Fig. 4.2) distribution in the fastening elements of mine working.

In the cap board of the frame, the distribution of σ_x is the most informative in terms of identifying the bending moments. In this regard, it should be noted, that different horizontal stresses act in each part of the cap board, but in the cross section of the SCP the σ_x are distributed rather uniformly. Therefore, it is possible to define the absence of any significant bending moment along the length of the cap board. This is one of the signs of a reduced vertical load acting on the frame. By their value, the decreased σ_x up to 8 – 15 MPa are distributed in the peripheral areas of the cap board, which indicates the development of small horizontal forces near the yielding joists and indirectly indicates the limited horizontal shifts of the immediate roof due to resistance of the combined roof-bolting system. In the central part of the arch, the compressive σ_x increase from 25 – 35 MPa to 70 – 80 MPa in the arch keystone. The stresses values are many times lower than the limiting values of SCP steel for support, but nevertheless, they characterize a certain load of the cap board, despite the creation of a load-bearing armored and rock plate in the roof.

If to compare the represented data with the basic fastening variant, then there are no significant differences in the value of compressive σ_x, however, there are two areas in semi-arch from the side of mined-out space, where the action was observed of an average value of the bending moments with different signs. Nevertheless, these perturbations do not play a decisive role in the occurrence of the limiting state of the cap board of the frame.

In the prop stays of the frame, the reduced horizontal stresses act for both compared fastening variants of mine working with differences that do not make a significant contribution to the stability of the frame support; nevertheless, we note the main ones.

Firstly, in the curvilinear part of the prop stay from the side of virgin massif in the basic fastening variant, the compressive $\sigma_x = 30 - 70$ MPa act, and in the recommended variant – the alternating σ_x of tension from 10 – 15 MPa near the yielding joist and up to 15 – 35 MPa of compression from below of the curvilinear part of the prop stay. These results indicate a reduced lateral pressure (when using a combined roof-bolting system) due to a decrease in the concentrations of bearing pressure in the side of mine working. In the rectilinear part of the prop stay, the value of compressive stresses turned out to be somewhat lower in the basic variant than in the recommended variant. Their absolute value has almost no effect on the stability of the prop stay.

Secondly, in the prop stay of the frame from the side of mined-out space, the distribution curve σ_x is changed insignificantly (in comparison with the opposite prop stay) for the application of the combined roof bolting system. For the basic fastening variant in the curvilinear part of the prop stay, a curvature sign is changed actively due to the impact of the alternating bending moment. This is caused by the high concentrations of bearing pressure in the area of the protection structure, which are partially smoothed when using a combined roof-bolting system in the roof.

The regular reference to the efficiency of the combined roof-bolting system is confirmed by the analysis of σ_x distribution in resin-grouted roof bolts and rope bolts. For both types of roof-bolts, despite their predominantly vertical arrangement, there are small tensile σ_x within the thickness of the immediate roof, which indicates the restriction of its horizontal shifts by means of a combined roof-bolting system. In the basic fastening variant, three resin-grouted roof bolts in the central part of the arch do not participate in the resistance to horizontal shifts of the immediate roof, and the tension of the reinforcement by σ_x forces is manifested only in the border area of the peripheral resin-grouted roof bolts.

In general, the results of analysis of the horizontal stresses distribution in the fastening elements of mine working confirm the more favorable state of the frame support when setting the combined roof-bolting system.

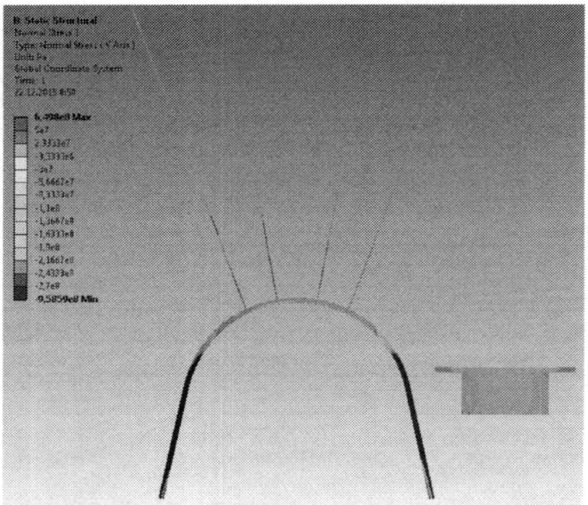

Fig. 4.3. The curve of vertical stresses σ_y in the fastening elements of the preparatory mine working when using the combined roof-bolting system

In conclusion of the SSS analysis of the fastening structures elements of mine working, the vertical stresses σ_y distribution has been studied, the diagram of which is shown in Fig. 4.3.

In both of the considered fastening structures, the cap board of the frame is in a relatively unloaded state, but there are differences in both the degree of unloading and the nature of σ_y distribution along the length of the cap board and in its cross section.

The minimal loading of the cap board with rock pressure from the side of rocks is observed when setting a combined roof-bolting system in it. Throughout the length of a cap board, the σ_y varies from 15 – 20 MPa of tension and up to 25 – 30 MPa of compression. This is manifested in a limited area of up to 0.6 m in the central part of the arch, somewhat displaced in the direction of a virgin massif. There is a small bending moment, directed towards the roof rocks. On the remaining length of the cap board, the distribution of σ_y in its cross-section is sufficiently uniform without the occurrence of any significant bending moment. In the area of yielding joists location, there are local centers (up to 100 – 150 mm in size) of tensile $\sigma_y \leq 40 - 50$ MPa, which is caused by elastic deformations of the cap board junction with the prop stays of the frame.

In the basic fastening variant, for the most part of cap board length, the value σ_y is also small – from 25 MPa of tensile to 75 MPa of compression, and its deflection into the cavity of mine working is observed almost throughout the central part of the arch. The anomalies of σ_y act in the area of the cap board bearing on the central prop stay of the strengthening support. Here, the concentration of compressing σ_y increases to 150 – 200 MPa, and in certain local centers it reaches the estimated yield limit of SCP steel, which induces the plastic bending of the cap board. In the peripheral areas, there is also an increase in the compressive σ_y to the level of 100 – 150 MPa. This fact is very significant. The compression of the cap board in the joist part of yielding indicates a flattening when the central part of the arch is deformed into the cavity of mine working. Although this process is mainly at the elastic stage of deformation, it characterizes the increased vertical load on the cap board of the frame from the side of the roof rocks. Therefore, it can be noted that, by the factor of the cap board state, there is a decrease in the vertical load on the frame when using the combined roof-bolting system.

The prop stays of the frame are very loaded with vertical stresses for both fastening variants, and there are some differences. In the variant of application of the combined roof-bolting system, the substantial part of the prop stays length is loaded enough uniformly in their cross section. That is, the bending moment is practically absent, except for the areas of prop stay bearings up to 0.4 m in height from the side of mined-out space and up to 0.3 m from the side of virgin massif. Here, there is a bending of the bearing part of the prop stays into the mine working cavity.

In the basic fastening variant, the distribution of σ_y in the prop stay from the side of virgin massif is very similar to the above described. Throughout the height of the prop stay, from the side of mined-out space, there are four areas of plastic bending towards the mine working. These areas can significantly change the geometry of the prop stay in the direction of reducing its stability and a significant loss of load-bearing capacity of the frame support as a whole. Thus, the analysis of the state of the frame prop stays by the factor σ_y, also confirms the beneficial effect of the combined roof-bolting system.

Further, the distribution of σ_y is studied in the roof-bolts placed along the contour of the mine working arch. In all five resin-grouted roof bolts of the basic fastening scheme, a more or less identical stresses pattern is observed. In the border part of the roof-bolts, there are tensile σ_y act up to 50 – 75 MPa along the armature length of 0.25 – 0.70 m. Moreover, the lower value refers to the central roof-bolt, and the higher – to the peripheral roof-bolt from the side of mined-out space. These data indicate the resistance to stratification of the border part of the rocks in the immediate roof, since the buried part of the roof-bolts is free of tensile forces.

In the combined roof-bolting system of the arch fastening, the tensile σ_y act in the areas adjacent to mine working, both in resin-grouted roof bolts and rope bolts. The length of these areas is much larger and varies in the range of 1.1 – 1.8 m with some increase in the roof-bolts located from the side of virgin massif. Here, the system resists stratification on almost entire or the most part of thickness of the immediate roof. Consequently, its continuity and the ability to resist vertical rock pressure are preserved to a greater extent.

Summarizing the performed SSS analysis, we state a comprehensively substantiated fact of reducing the stress intensity of the frame support when setting a combined roof-bolting system in the roof. This fact is quite understandable, taking into account the patterns of changes in the stress state of an adjacent massif in the direction of increasing the stability of the roof rocks due to the formation of an armored and rock plate with high load-bearing capacity.

CONCLUSIONS

As a result of the performed SSS studies of the coal-bearing stratum surrounding the preparatory mine working, as well as the fastening and protection systems, the following conclusions have been made.

1. In roof rocks, by means of a combination of rope bolts and resin-grouted roof bolts, an armored and rock plate is formed, the high load-bearing capacity of which is achieved by retaining the horizontal thrust forces even in case of partitioning of rock layers into rock blocks. The concentrations of all stress components are reduced to a level below the strength characteristics of lithotypes, and the occurrence of tensile stresses σ_y and σ_x has a local nature. Therefore, a thick armored and rock plate in the roof protects the fastening system of mine working from excessive vertical rock pressure.

2. A decrease has been established in the stress intensity of the cap board and the prop stays of the frame support when using the combined roof-bolting system in the roof. This decrease has been substantiated by the fact that the armored and rock plate formed in the roof takes up a fraction of the vertical rock pressure owing to to the active resistance of resin-grouted roof bolts and rope bolts to the displacement processes of the coal-bearing massif.

3. A decrease in the stresses concentrations and dimensions of weakening areas in the sides of mine working has a positive effect on the level of bottom rock stress intensity, which makes it possible to predict less intensive development of the swelling process.

5. ESTABLISHING AND ANALYZING THE PATTERNS OF GEOMECHANICAL FACTORS INFLUENCE ON THE RESISTANCE OF THE COMBINED ROOF-BOLTING SYSTEM TO ROCK PRESSURE

5.1. SUBSTANTIATION OF THE CRITERION FOR ASSESSING THE LEVEL OF THE ROOF-BOLTS RESISTANCE TO THE ROCK PRESSURE MANIFESTATIONS

The resistance of the combined roof-bolting system is determined by the level of action of the stresses σ intensity and the extent of the most loaded areas of resin-grouted roof bolts and rope bolts set in the roof of mine working. The greater σ and longer the area of its distribution, the more actively the system resists the displacement of the roof rocks into mine working. The patterns of connection between the resistance parameters of a combined roof-bolting system and the geomechanical factors, which characterize the conditions for maintaining the preparatory mine workings, are established on the basis of a series of multivariate computational experiments that have the ultimate goal of determining the rational parameters of the combined roof-bolting system setting, depending on the mining and geological conditions for maintaining mine working. In this connection, the first stage of research is to determine the degree of influence of a number of geomechanical factors on the resistance level of resin-grouted roof bolts and rope bolts. With their low resistance, the question is legitimate about the expediency of using the combined roof-bolting system in these mining and geological conditions. It is quite likely that it is enough the use of traditional fastening schemes.

At the second stage of research, it is necessary to determine the number of roof-bolts, sufficient to form an armored and rock plate in the roof, which protects reliably the preparatory mine working from the manifestations of rock pressure. Therefore, it seems extremely important to outline the terms of effective application of combined roof-bolting systems based on the established patterns of the degree of the roof-bolts loading and the influence of geomechanical factors that characterize the conditions for maintaining mine working.

On the basis of a general algorithm for the rational parameters search of setting the combined roof-bolting systems as a part of the scheme for maintaining mine working in general, a number of geomechanical factors have been identified and substantiated that have the greatest influence on the degree of loading the roof-bolts. These include: the depth of mine working location H, the average calculated compressive resistance R of adjacent coal-bearing rocks, the ratio of the calculated compressive resistance of the immediate roof rocks of the coal seam to its thickness $\frac{R_1^R}{m_1^R}$. For a better perception, the accepted ratio is expressed through b_m^r, which is applied in the obtained dependences.

To assess the degree of the roof-bolts loading, it is necessary to substantiate the corresponding criterion, which is a difficult task due to significant fluctuations in stresses intensity σ along the length of the roof-bolts. The most objective characteristic is the length of an area where the roof-bolt is loaded at the level of its load-bearing capacity, that is, it works with maximum efficiency. Of course, it is necessary to recognize the sufficient efficiency of the roof-bolt performance when it resists to rock pressure at a level, for example, 70 – 90% of its load-bearing capacity. But, then there is a multi-criteriality and a certain ambiguity in assessment of its strengthening action. In addition, the areas of the roof-bolt length with the specified degree of load are always adjacent to the areas of the limiting (plastic) state (100% loading) of its reinforcement (bearing member). Therefore, the condition for occurrence of plastic state areas of the bearing member automatically includes the areas of prelimiting state with a high degree of resistance.

Also, it should be taken into account the different lengths of the roof-bolts in the combined roof-bolting system. The use of such a relative parameter as the ratio of the length of the plastic state area of the bearing member to the entire length of the roof-bolt will be the most objective. As a criterion for assessing the effectiveness of the roof-bolt performance, as part of the combined roof-bolting system, the relative length of the plastic state area of the bearing member (the reinforcement in resin-grouted roof bolts and the rope in rope bolts) is proposed, which is expressed as a percentage.

Among other things, the question arises about the grouping of roof-bolts, especially when there are quite a large number of them in the mine working roof. This is done on the basis of results of the previous studies on the stresses components distribution along the length of the roof-bolts. The grouping (where it was possible) was performed by the condition of the degree of their SSS similarity and the similarity of tendencies in the geomechanical factors influence. As a result, the following criteria parameters have been accepted for study:

– Δ – the relative length of the plastic state of the bearing member of the resin-grouted roof bolts in the central part of the mine working arch;

– $\Delta_{l,m}^{rope}$ – the relative length of the plastic state of the bearing member of the rope bolts in the roof from the side of mined-out space and virgin massif, respectively.

5.2. THE INFLUENCE OF THE DEPTH OF MINE WORKING LOCATION

The tendencies of influence of the preparatory mine working location depth H on the degree of loading the roof-bolts in the central part of the arch are shown in Fig. 5.1 for one of the series of SSS calculations of the studied geomechanical system. The general pattern for all roof-bolts in the combined roof-bolting system is an increase in the relative length of the plastic state areas with an increase in parameter H. All patterns are close to linear functions $\Delta(H)$ and $\Delta_{l,m}^{rope}(H)$.

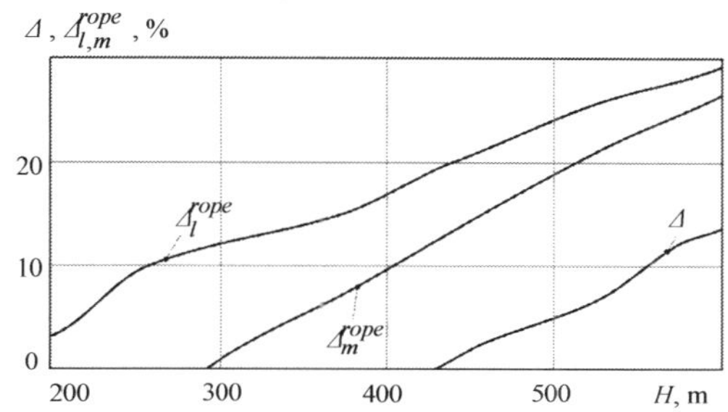

Fig. 5.1. Dependences of change in the relative length Δ and the plastic state $\Delta_{l,m}^{rope}$ areas of the roof-bolt bearing member on the depth H of mine working location

There is a different degree of loading of resin-grouted roof bolts and rope bolts. Thus, in this series of calculations, mentioned as one of the examples, the areas of plastic state in resin-grouted roof bolts appear, beginning with the depth $H = 428$ m, and develop to $\Delta = 13.7\%$ at the boundary depth of $H = 600$ m performing the computational experiments. For a rope bolt, located from the side of mined-out space, the areas of plastic state with value $\Delta_m^{rope} = 3.3\%$ are already present at a depth $H = 200$ m and increase to $\Delta_m^{rope} = 29.5\%$ at $H = 600$ m. From the side of virgin massif these areas occupy an intermediate position. The plastic state of the rope occurs at a depth of $H = 293$ m and the length of the area increases to $\Delta_m^{rope} = 26.7\%$ with the boundary value of calculations $H = 600$ m.

The above data allow to make a number of conclusions in addition to the already noted tendency of increasing the plastic state areas of the bearing member with increasing the depth of mine working location.

Firstly, the rope bolt from the side of mined-out space is more loaded than from the side of virgin massif, which is explained by a more intensive displacement of the coal-overlaying formation above the mined-out space.

Secondly, there is a stable tendency for a reduced load on resin-grouted roof bolts. This can be explained by their shorter length (2.4 m) compared with the rope bolt (6 m). It is known that the roof-bolt is loaded with tensile forces due to the difference in the displacement of the massif on the mine working contour and in the buried area. The longer the roof-bolt, the farther the buried area from the contour of mine working, and the greater is the indicated difference in displacements. But, there is one more constantly repeating factor. The closer the resin-grouted roof bolt to the vertical axis of mine working, the less it is loaded. Thus, at different coordinates of the resin-grouted roof bolts location along the length of the frame cap board, the line $\Delta(H)$ will take a different position. The tendency of a reduced value Δ with respect to Δ_m^{rope} remains stable.

Thirdly, summing up the above, it should be noted a high level of resistance of rope bolts at any depth of mine working location (within 200 m $\le H \le$600 m the range of performing the computational experiment), while resin-grouted roof bolts actively resist to rock pressure, starting from the depth $H \approx 400$ m for the described series of calculations.

5.3. INFLUENCE OF THE AVERAGE CALCULATED ROCKS RESISTANCE TO COMPRESSION OF THE ADJACENT MASSIF

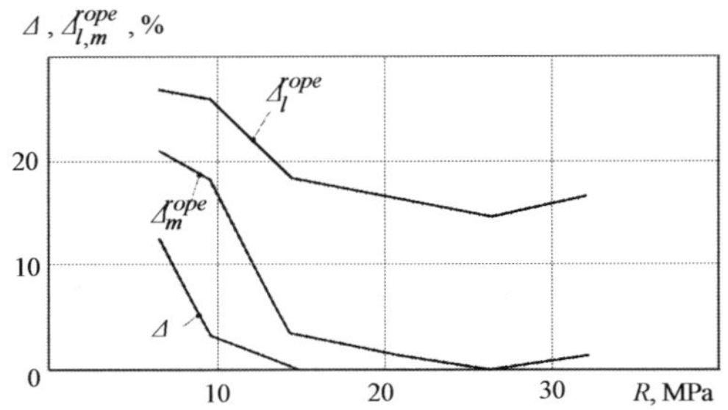

Fig. 5.2. Dependences of change in the relative length Δ and the plastic state $\Delta_{l,m}^{rope}$ areas of the bearing member of the roof-bolt on the average calculated resistance to compression R of the adjacent massif

The second geomechanical factor significantly influencing the degree of load of the roof-bolts as a part of the combined roof-bolting system, is the value of the average calculated resistance to compression R of the rock layers of the adjacent coal-bearing stratum; one of the fragments of the influence patterns is shown in Fig. 5.2.

The general tendency for all roof-bolts is an increase in the relative length of the plastic state areas Δ, $\Delta_{l,m}^{rope}$ when the parameter R decreases, but the dependences $\Delta(R)$ and $\Delta_{l,m}^{rope}(R)$ differ significantly from each other. Thus, for resin-grouted roof bolts, the plastic state area occurs only at $R \leq 14.5$ MPa and with a further decrease in R, there is a rather intensive growth of Δ by an obviously nonlinear dependence: at $R = 6.6$ MPa the value Δ is already 12.7%.

A different pattern of the function $\Delta_{l,m}^{rope}(R)$ change is observed for a rope bolt set from the side of mined-out space. In this example, in the range of $14.3 \leq R \leq 32.1$ MPa, the value Δ_l^{rope} is sufficiently stable and ranges in the interval $\Delta_{l,m}^{rope} = 14.9 - 18.1\%$. At $R < 14.3$ MPa, there is some increase in the relative length of the plastic state area with a maximum of $\Delta_l^{rope} = 26.7\%$ at $R = 6.6$ MPa.

For a rope bolt set from the side of a virgin massif, the pattern of connection Δ_m^{rope} and R is qualitatively similar, but quantitatively the graph is below the dependence $\Delta_{l,m}^{rope}(R)$. So, in the range of $14.3 \leq R \leq 32.1$ MPa, the value Δ_m^{rope} is close to zero, with a maximum of 3.3% at $R = 14.3$ MPa. If R is below the specified value, then there is an increase in the parameter Δ_m^{rope}, which reaches a value of 21.2% at $R = 6.6$ MPa.

When revealing the described patterns, it was confirmed that the degree of loading the resin-grouted roof bolts decreases when they are set closer to the vertical axis of mine working and, on the contrary, Δ increases when the coordinates of setting the resin-grouted roof bolts are shifted to the peripheral areas of the cap board.

Noteworthy is another conclusion on the analysis results of dependences $\Delta(R)$ and $\Delta_{l,m}^{rope}(R)$. There is a certain value of the parameter R, below which the load of all the roof-bolts in the combined roof-bolting system develops quite intensively. In this example it is $R = 14.3$ MPa. Such a phenomenon can be explained by the partial loss of stability of the thrust system from the rock blocks of the immediate roof. The created thrust cannot keep its weight from collapse due to the thrust forces. Additional efforts are required from the roof-bolts to maintain the stability of the thrust block system in the immediate roof.

5.4. INFLUENCE OF THE RATIO OF THE AVERAGE CALCULATED RESISTANCE TO COMPRESSION OF THE IMMEDIATE ROOF ROCKS TO ITS THICKNESS

Analysis of the computational experiments results has shown that the degree of loading the roof-bolts in the combined roof-bolting system is significantly influenced by the ratio of b_m^r calculated resistance to compression R_1^R of the immediate roof rocks to its thickness m_1^R. The fragments of dependences Δb_m^r are shown in Fig. 5.3.

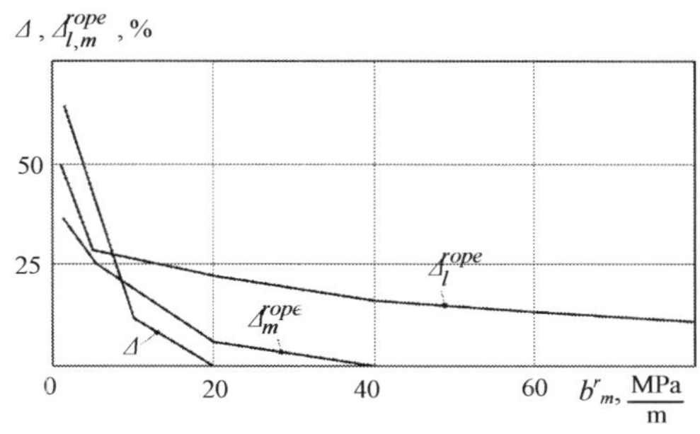

Fig. 5.3. Dependences of relative lengths Δ and $\Delta_{l,m}^{rope}$ of the plastic state areas of the bearing member of the roof-bolt on the parameter b_m^r change

Here, the already traditional ratio between the parameters Δ and $\Delta_{l,m}^{rope}$ has been partially disturbed. Thus, up to a value of $b_m^r \geq 7.5 - 8.3$ MPa/m, the relative length of the plastic state area of the resin-grouted roof bolt is noticeably less than that for the rope bolts. The plastic state areas in resin-grouted roof bolts appear only at $b_m^r \leq 20$ MPa/m. When the parameter b_m^r decreases below the marked boundary of 7.5 – 8.3 MPa/m, the relative length of the plastic state area of the reinforcement of the resin-grouted roof bolts Δ

increases sharply, exceeding the corresponding values $\Delta_{l,m}^{rope}$ and reaching a value of 64.4% at $b_m^r = 1.4$ MPa/m; with that, the dependence Δb_m^r itself is highly nonlinear. The explanation for such rapid growth of Δ, in our opinion, is quite obvious. Low values of parameters b_m^r characterize the conditions for the occurrence of a very weak immediate roof at a sufficiently high its thickness, which induces an intensive stratification of the rocks adjacent to mine working to a considerable distance into the roof. Therefore, the resistance of resin-grouted roof bolts to the active lowering of the roof rocks affects most of the roof-bolt length with the occurrence of extended areas of the plastic state of their reinforcement.

For rope bolts set from the side of mined-out space, the dependence Δ_l^{rope} is somewhat different. Here, the plastic state area of the rope occurs at any value of b_m^r from 10.6% at $b_m^r = 80$ MPa/m and till 22.0% at $b_m^r = 20$ MPa/m. In this range of b_m^r values, the relatively slow growth of Δ_l^{rope} at decreasing the parameter b_m^r; with a further decrease in the ratio b_m^r, the growth gradient Δ_l^{rope} increases significantly nonlinearly, reaching $\Delta_l^{rope} = 50\%$ at $b_m^r = 1.4$ MPa/m.

In rope bolts set from the side of virgin massif, the dependence of Δ_m^{rope} on b_m^r takes, as usual, an intermediate position, both in value of Δ_m^{rope}, and in the degree of nonlinearity of these parameters ratio. The plastic state areas of the rope appear at $b_m^r \leq 40$ MPa/m and their length increases rapidly to 36.1% at $b_m^r = 1.4$ MPa/m. The marked intensive growth of the plastic state areas of the ropes from both sides of mine working is explained by the same instability of the immediate roof rocks at low values of b_m^r.

The active stratification of a soft and sufficiently hard immediate roof increases the load on the rope bolts fastened in the main roof. This component of the roof experiences smaller movements in the direction of mine working and, due to the ropes tensioning, it keeps the rocks of the immediate roof from collapsing. Therefore, a part of decrease in the stability of immediate roof is shifted to an increase in the load on the rope bolts, and the length of the plastic state areas of the ropes increases.

The revealed patterns give an assessment of the degree of geomechanical factors influence on the loads of all roof-bolts included into the combined roof-bolting system, which is one of the bases when developing a method for calculating its rational parameters, depending on the conditions for maintaining the preparatory mine workings.

CONCLUSIONS

1. The dependence has been determined of the coordinates of setting and the gradient angle of the resin-grouted roof bolts in the arch of the preparatory mine working from the intensity of the rock pressure b^r with account of its manifestation in the immediate roof rocks, depending on the extracted seam b_m^r thickness. The effectiveness of the resistive action of the armored and rock plate against vertical rock pressure increases by reducing the rock stress in the sides of mine working, which are bearings for the thrust system in the roof.

2. The criterion for assessing the level of resistance of roof-bolts as part of a combined roof-bolting system has been substantiated, which is used to determine the most influencing geomechanical factors in terms of the degree of the system loading: the depth of mine working location, the average calculated compressive resistance of adjacent rock massif, the ratio of the calculated compressive resistance of the immediate roof rocks to its thickness.

3. The dependences have been established of the degree of the roof-bolts loading (differentiated) in the combined roof-bolting system from the main influencing geomechanical factors. The gradation has been proved of the resistance level of the roof-bolts to the rock pressure manifestations: the maximum resistance is created by rope bolts set from the side of mined-out space; the minimum resistance – by resin-grouted roof bolts.

4. The stresses components σ, σ_x and σ_y in all elements of fastening schemes, are practically unaffected by the reaction of the central pliable prop stays of the strengthening support, with the exception of local areas in the frame cap board near the place of their joint contact. This indicates the expediency of excluding the central prop stay of the strengthening support with a rigid performance.

5. The rational coordinates of the roof-bolts location in the central part of the mine working arch and the gradient angles have been established, depending on the standard size of its section. The obtained patterns are the basis for the search for rational parameters of a combined roof-bolting system with the ultimate goal of developing a methodology for selecting its parameters depending on the mining and geological conditions for maintaining mine working.

6. DEVELOPMENT AND SUBSTANTIATION OF THE PARAMETERS CALCULATION OF THE ROOF ROCKS STRENGTHENING BY A COMBINED ROOF-BOLTING SYSTEM IN THE PREPARATORY MINE WORKINGS

In accordance with the general strategy of selecting the parameters of the fastening system elements in the preparatory mine working, depending on the geological and mining conditions for its maintenance, the main task of this chapter is to determine the rational parameters for the resin-grouted roof bolts and rope bolts location that strengthen the roof rocks of the in-seam working. The rational parameters of resin-grouted roof bolts and rope bolts are those that realize two conditions: on the one hand, the roof-bolts together with the strengthened roof rocks should maximally unload the frame support to effectively limit the loss of the mine working section; on the other hand, the roof-bolts in the roof should work with the maximum resistive reaction to the rock pressure, that is, to be set in a minimum sufficient quantity.

6.1. SUBSTANTIATION AND CALCULATION OF THE PARAMETERS OF THE RESIN-GROUTED ROOF BOLTS LOCATION IN THE ROCKS OF MINE WORKING ARCH

A separate consideration of resin-grouted roof bolts and rope bolts in the roof of mine working is dictated by the following circumstance.

Firstly, the length of the rope bolts two or more times exceeds the length of the resin-grouted roof bolts, which predetermines the different areas of the border rocks strengthening.

Secondly, their functions differ significantly from each other. When strengthening the rocks of immediate roof of the mine working arch, the resin-grouted roof bolts restrict its stratification and together with the side roof-bolts create connections between the individual rock blocks. Ultimately, a rock plate is formed, which takes up a part of the rock pressure, thus reducing the load on the frame support.

The rope bolts connect the immediate and main roof, which increases the stability of both one and the other rock volumes above mine working, and they protect the support from the rock pressure.

The set of parameters for setting the resin-grouted roof bolts, located in the roof of mine working along the contour of the frame cap board, consists of determining the gradient angle β_i of the roof-bolt to the horizontal axis of mine working and its load-bearing capacity (expressed through the diameter d_i of the “reinforcement”). The length of the resin-grouted roof bolts l_i (2.4 m) is specified, which is conditioned by their function in the formation of the load-bearing armored and rock plate in the roof of mine working. The dimensions of the arch of ultimate equilibrium in the roof (in the zone of stabilization of the rock pressure manifestations) are such that in most cases this length is not enough to fasten their joists in stable rocks, thus, this function is delegated to rope bolts. The fourth parameter is the coordinate of setting each roof-bolt x_i, which is measured horizontally from the vertical axis of mine working.

When identifying the relation of the coordinates x_i of the resin-grouted roof bolts setting with geomechanical factors according to the conditions formulated above, the patterns of change in the degree of the roof-bolts loading were used. The strength potential of resin-grouted roof bolts in the roof is realized only with the intensive manifestation of the vertical component of the rock pressure. This is important for the conditions of Western Donbas, where the depth of mine working depositing is $H \geq 400-450$ m and the soft rocks of the adjacent coal-bearing massif ($R \leq 10-15$ MPa), as well as low compressive resistance of rocks of the immediate roof with its increased thickness ($b_m^r \leq 15-20$ MPa/m). Here, the resin-grouted roof bolts actively resist to the processes of stratification and lowering the rocks of mine working roof, which implies an increased density of their arrangement. In practice, this mining and geological situation is characterized by placing in the mine working roof of about 7 – 9 roof-bolts. In more favorable geological conditions, the number of roof-bolts in the roof is significantly reduced (up to 3 – 5 roof-bolts), since they remain unloaded, as it is shown by the results of computational experiments [20].

The question of the minimum required number of resin-grouted roof bolts in the roof is combined with the search for the coordinate x_i of each roof-bolt location. At the same time, the search of calculation variants is simplified significantly based on the following considerations.

Firstly, it has been established that the degree of the roof-bolts loading is directly proportional to the depth H of mine working location and inversely proportional to the value of the average calculated compressive resistance R of the adjacent coal-bearing strata. This term refers to the radius of mine working influence on the surrounding rocks, which is 3 – 5 times of the mine working width. Thus, the known tendencies are confirmed of the rock pressure development, the intensity of which is traditionally associated with the ratio $\frac{H}{R}$, which for simplicity is indicated as b^r. In addition, due to the gradual increase in the parameter H (according to the technology for calculating the elastic-plastic problem), in a single computational experiment, a set of values can be obtained by the degree of the roof-bolts loading.

Secondly, using the research data [20], the variants with an increased number of resin-grouted roof bolts in the roof are calculated only for a limited range of complex mining and geological conditions mentioned above.

Thirdly, with a significant underloading of any roof-bolts, they are excluded from consideration while maintaining their results. At the same time, the load taken up by the underloaded roof-bolts is taken into account when increasing the intensity of stresses due to the reserve of resistance conditioned by the segment and the stage of strengthening of any steel of the roof-bolt reinforcement.

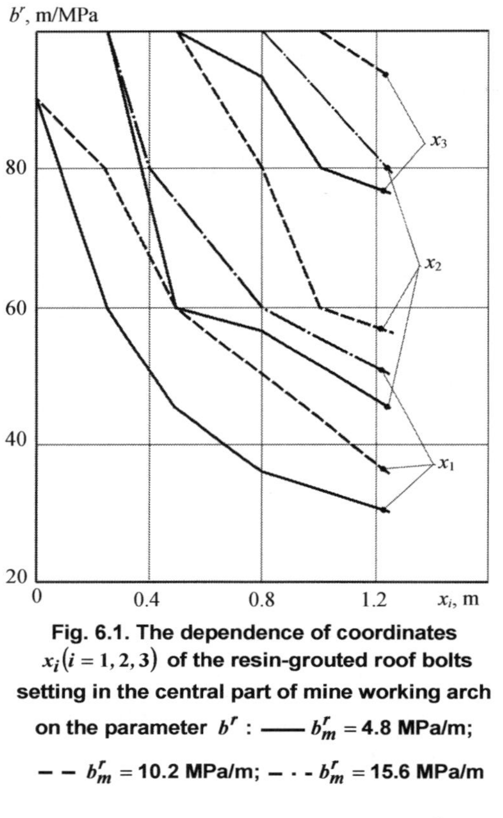

Fig. 6.1. The dependence of coordinates $x_i(i = 1, 2, 3)$ of the resin-grouted roof bolts setting in the central part of mine working arch on the parameter b^r : —— $b_m^r = 4.8$ MPa/m; – – $b_m^r = 10.2$ MPa/m; – · · $b_m^r = 15.6$ MPa/m

Using this approach, the graphs of dependences have been obtained of the coordinates x_i of the resin-grouted roof bolts setting on the parameters b^r and b_m^r, which are shown in Fig. 6.1. It has been preliminary established that it is not rational to set more than six roof-bolts in the central part of the roof because of their low loading. This is confirmed both by a series of multivariate SSS calculations of the fastening system [20], and by the experience of maintaining the preparatory mine workings. Then, at a symmetric (with respect to the vertical axis of mine working) scheme for the roof-bolts setting, it is necessary to analyze the change in three coordinates $x_i\left(i=1,\ 2,\ 3\right)$ depending on the mining and geological conditions.

There is an inverse proportional relation between the coordinates x_i and the value of the ratio b^r, which is consistently repeated for all calculation variants, regardless of the number of the roof-bolts in mine working arch. With a large ratio of b^r (the increased depth of mine working location with a reduced hardness of rock of the surrounding massif), it is advisable to set the resin-grouted roof bolts closer to the arch keystone, which is dictated by the very intensive stratification of the roof mainly in the middle of the mine working span, that is, in the central part of its arch. With an increased stability of the enclosing rocks (a decrease in the value b^r), the coordinates of the roof-bolts setting move closer to the spring, and the number of roof-bolts in the roof decreases, which does not contradict the existing concepts and recommendations. The roof-bolts, for which the coordinate x_i reaches or exceeds the coordinate of location of the frame yielding joist, and then changes to a different quality – the upper side roof-bolts, and their rational parameters have already been determined in [35]. In this case, the number of roof-bolts in the roof is reduced.

In a different way, the coordinates of the resin-grouted roof bolts setting are influenced by the ratio b_m^r. With an increase in the hardness of the roof rock, the load on the roof-bolt decreases, and their number is advisable to reduce, which corresponds to an increase in the distance between the roof-bolts (the displacement of coordinates of their setting from the arch keystone to the spring).

The revealed patterns of relation between the rational coordinates of the resin-grouted roof bolts setting in mine working arch and the geomechanical parameters of its maintenance made it possible to obtain a set of regression equations for calculating the values x_1, x_2, x_3

$$x_1 = -5.8 \cdot 10^{-2} b_m^r \ln\left(0.015b^r - 0.52\right), \text{ m}; \tag{6.1}$$

$$x_2 = 10^{-2} b_m^r \left[4.8 - 10.9 \ln\left(0.015b^r - 0.52\right)\right], \text{ m}; \tag{6.2}$$

$$x_3 = 10^{-2} b_m^r \left[12.6 - 17.2 \ln\left(0.015b^r - 0.52\right)\right], \text{ m}. \tag{6.3}$$

In the above expressions, the index “1” denotes the coordinate x_1 of setting the resin-grouted roof bolt at a minimum distance from the vertical axis of mine working, and the index “3” – at the maximum distance. If the value x_3 exceeds the horizontal coordinate of location of the yielding joist of the frame support, then the outer roof-bolt is excluded from the roof-bolt setting scheme in mine working arch. Similarly, the expediency is considered of setting the resin-grouted roof bolts with coordinates x_2 and x_1 in mine working arch. Then, in a certain domain of geomechanical parameters ratio, there is no need to set the roof-bolts along the contour of the arch in the area between the yielding joists of the frame. The validity of this statement is that a stable roof does not load the roof-bolts to the level of their active work on resisting to rock pressure. The length of the roof-bolts also influences, when there are different displacements of the roof rocks in the tail joint and in the area of the yielding joist. With an increased length of the roof-bolts, the difference in the indicated displacements increases, which contributes to their full load. The peculiarities should be noted of work of the frame support for maintaining the roof of mine working. The cap board is less loaded than its support prop stays, especially in favorable mining and geological conditions of coal mines [20].

Thus, there is an area of mining and geological conditions, when there is a low efficiency of strengthening the roof rocks with resin-grouted roof bolts within the central part of the mine working arch. To determine the boundaries of this area, the following ratios of geomechanical parameters for the preparatory mine workings with TSYS support have been obtained:

– for TSYS-11.0 (11.7)

$$b^r \leq 34.7 + 66.7\exp\left(-17.2b_m^r\right), \text{ m/MPa}; \quad (6.4)$$

– for TSYS-15.0

$$b^r \leq 34.7 + 66.7\exp\left(-19.8b_m^r\right), \text{ m/MPa}; \quad (6.5)$$

– for TSYS-17.7

$$b^r \leq 34.7 + 66.7\exp\left(-22.4b_m^r\right), \text{ m/MPa}. \quad (6.6)$$

If the value b^r is less than calculated by formulas (6.4) – (6.6), it is not recommended to set the resin-grouted roof bolts in the central part of an arch. Otherwise, the coordinates of the roof-bolts location and their number are determined by the formulas (6.1) – (6.3).

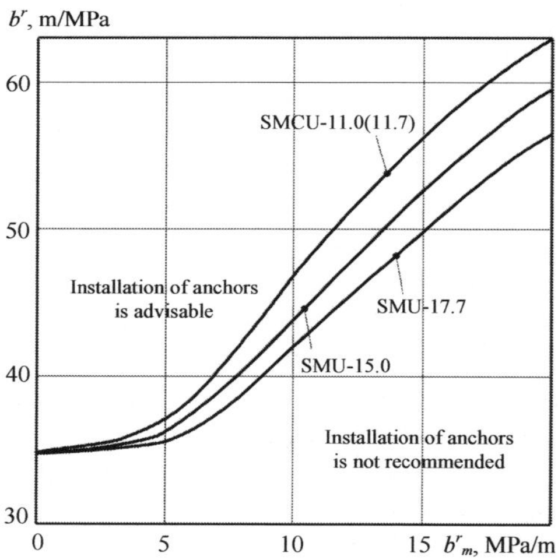

Fig. 6.2. he area of expedient location of the resin-grouted roof bolts in the central part of the preparatory mine working arch

For the operational assessment of the expediency of the roof-bolts setting in the central part of mine working arch, the graphs were constructed that determine the boundaries of these areas according to the ratios of geomechanical parameters b^r and b_m^r, which are shown in Fig. 6.2.

The following parameters for the resin-grouted roof bolts setting in the mine working arch are their gradient angle β_i and diameter d_i of the bearing rod. During the study, the fact of underloading of the considered group of roof-bolts was constantly noted. It is inappropriate here to use the roof-bolts of a conventional construction, the load-bearing capacity of which is estimated to be of the order of 200 kN [4, 87, 88]. The constructions with a reduced diameter of "reinforcement" in the range of $d_i = 15 - 18$ mm are recommended, the tear resistance of which is up to 2 – 2.5 times lower than that of conventional structures. This is quite enough for the formation of load-bearing armored and rock plate in the roof of mine working. Therefore, for this group of roof-bolts, it is inappropriate to perform a study on identification of the patterns of relation between the "reinforcement" diameter and the geomechanical parameters of maintaining mine working. The diameter value d_i is chosen to be constant for the whole group of roof-bolts within the above-mentioned range.

The situation is different with the choice of rational gradient angles β_i of resin-grouted roof bolts to the horizontal axis of mine working. This parameter is associated with the coordinate x_i and is variable for the roof-bolts, set in the central part of the mine working arch. The maximum effect of strengthening the massif is achieved when the longitudinal axis of the roof-bolt coincides with the vector of roof rocks displacements at a given coordinate x_i [42]. In accordance with the general tendency of changes in the vector of rocks displacement from the vertical in the arch keystone to oblique in its spring, the patterns have been obtained of relation between the gradient angle β_i of the roof-bolt and the coordinate x_i of its setting, which are shown in the graphs of Fig. 6.3. There is a stable alternating decrease in β_i, when the point of roof-bolt setting is displaced (the increase in the coordinate x_i) towards the yielding joist of the frame. This stability is confirmed by the insignificant influence of such geomechanical parameters as the ratio of b^r and b^r_m.

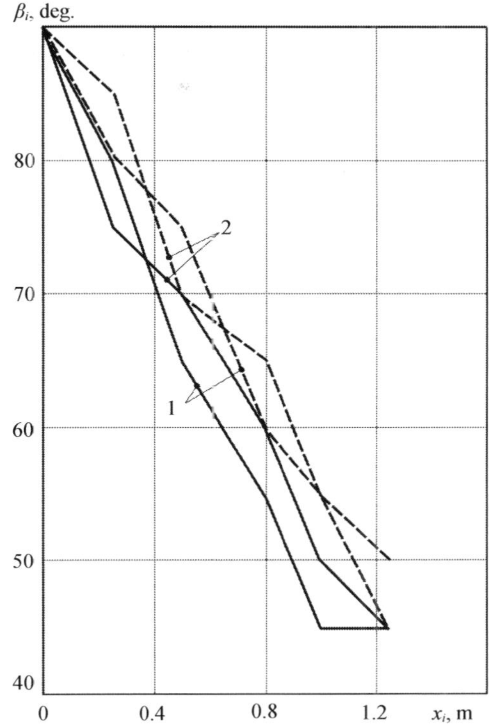

Fig. 6.3. The relation of the gradient angle β_i of the resin-grouted roof bolt to the coordinate x_i of its setting in the arch of the preparatory mine working for different values of b^r: —— $b^r = 40$ m/MPa; – – $b^r = 80$ m/MPa; 1 – $b^r_m = 4.8$ MPa/m; 2 – $b^r_m = 15.6$ MPa/m

Thus, if the parameter b^r is increased by 2 times, the range of variation of the gradient angle of the roof-bolt is 0 – 10% with a maximum deviation of 11.8%. When the parameter b^r_m is reduced by 3.25 times, the maximum deviation β_i does not exceed 10%. The above data indirectly emphasize the objective criterion for choosing the gradient angle of the roof-bolt along the vector of the surrounding massif displacement in the direction of mine working cavity. The use of the correlation-dispersion analysis of the computational experiment results made it possible to obtain the following regression equation

$$\beta_i = 90\left(1 - 0.41x_i^{0,8}\right), \text{ deg.} \tag{6.7}$$

The usefulness of the expression (6.7) in combination with the formulas (6.1) – (6.3) is in the possibility of a reasonable choice of the parameters for setting the resin-grouted roof bolts along the arch contour of the preparatory mine working.

6.2. SUBSTANTIATION AND CALCULATION OF STRENGTHENING THE ROCKS IN THE ARCH OF MINE WORKING WITH ROPE BOLTS

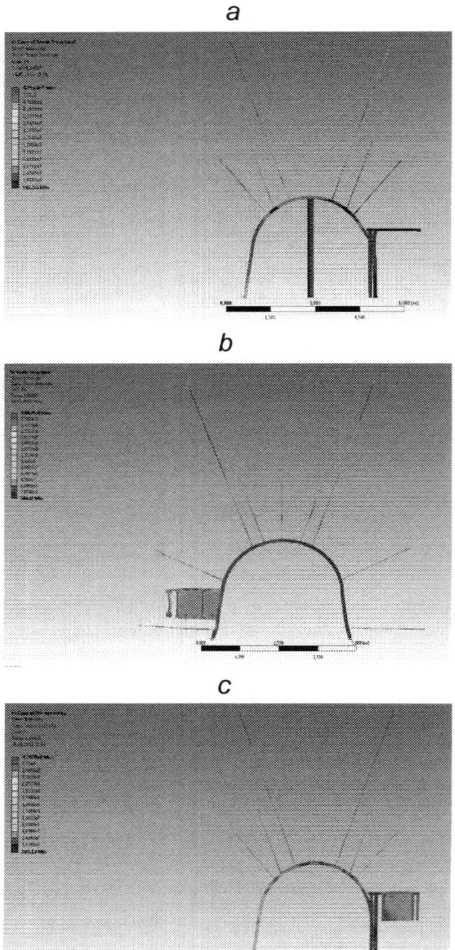

Fig. 6.4. The stresses intensity σ curves in fastening and security systems of the preparatory mine workings in Western Donbas: *a* – Samarska Mine; *b* – Heroiv Kosmosu Mine; *c* – Stepova Mine

The rope bolts perform the main function in the process of maintaining the preparatory mine working in the area of stope works influence. The deep strengthening of the roof rocks, the main effect of which is to restrict the stratification of a thick structure of rocks of the immediate and main roof, not only prevents their lowering into mine working, but also increases the resistance to the rock pressure.

The above function of rope bolts fundamentally distinguishes them from the resin-grouted roof bolts. To confirm this thesis, the stresses intensity curves σ of fastening and security systems of the preparatory mine workings in three mines of the Western Donbas are shown in Fig. 6.4. In the mine working arch, a higher loading is observed of rope bolts as compared to the resin-grouted roof bolts, regardless of the geological conditions for maintaining mine working, the scheme of the roof-bolts setting and the security method used.

Based on the above, a method for calculating the rational parameters has been developed for rope bolts, including: rope bolt length l^{rope}, rope diameter d^{rope}, coordinate x^{rope} and gradient angle β^{rope} of its setting. It is assumed that the rope bolts should be set symmetrically in the cross section relative to the vertical axis of mine working. This decision is substantiated by the fact that the main argument is the prospect of reusing the preparatory mine working and the leveling the asymmetry of the geomechanical processes during the approach of the second longwall face.

The rational length l^{rope} of the rope bolt is chosen by the condition of fastening its joist part outside the arch of ultimate equilibrium, that is, in relatively stable rocks of the main roof. This condition is written in the form

$$l^{rope} \geq l_{joist} + b_l^r + l_{tail}, \quad (6.8)$$

where l_{joist} – the length of the roof-bolt joist; l_{tail} – the length of the tail part of the roof-bolt that extends into the cavity of mine working; usually it is taken as $l_{tl} = 0.1$ m; b_l^r – the width of the area of the lateral rocks weakening, which is determined by the formula [58]

$$b_l^r = \frac{4.1 l_{p.s.}^{0.73}}{R^{0.1}\left(R_1^r\right)^{0.21}}\left[1 - \exp\left(-5.9 \cdot 10^{-3} H\right)\right]. \quad (6.9)$$

There $l_{p.s.}$ – is the width of the protective strip.

In formula (6.8), only the length of the joist remains of the undetermined parameters. By means of calculations and tests, it has been established that the load-bearing capacity of a resin-grouted roof bolt of about 200 kN is provided by a joist with a length of $l_{joist} = 0.6 - 0.7$ m. At least not less load-bearing capacity of the rope bolts is achieved with a joist length of not more than 1.0 – 1.1 m.

In order to determine the minimum sufficient length of the rope bolt, it is necessary to know the boundaries of the contour of the arch of ultimate equilibrium, which has been the subject of many studies, beginning with prof. M.M. Protodiakonov [89], prof. P.M. Tsymbarevych [90] and ending with modern methods of geomechanical processes modeling [55, 73, 91]. These classical concepts are so well reasoned that they are included into a number of normative documents [71, 72, 92, 93], including the industry standard [71]. However, in this document, along with the calculation of the height of an arch, the width of the area of weakened rocks in its sides is not taken into account, which has been proven in the work and determined by the regression equation (6.9). Taking into account the areas of weakening in the sides of mine working, the dimensions of the arch of ultimate equilibrium increase significantly, which is substantiated by prof. P.M. Tsymbarevych. To clarify the dimensions of the arch, it is suggested to use the following combination:

– the height h of the arch of ultimate equilibrium is determined by document [71];

– the width of the area of weakening in the sides of mine working is determined by (6.9);

– the shape of the contour of the arch is determined by the parabola equation according to prof. M.M. Protodiakonov

$$y = h\left[1 - \frac{4x^2}{\left(B + 2b_l^r\right)^2}\right], \tag{6.10}$$

where y and x – the vertical and horizontal coordinates with a reference point located in the arch keystone of mine working; B – the width of mine working during drifting; b_l^r – is determined by the formula (6.9); this parameter is accepted for calculation as the maximum width of the weakening area in the side of mine working from side of the longwall face, which goes to a certain margin of reliability of calculations, and when the mine working is reused, the dimensions b_l^r and b_m^r will become equal in their values.

The condition for fastening the joist part of the rope bolt beyond the boundary of the arch of the ultimate equilibrium is transformed with respect to formula (6.8) without changing its physical significance. The calculation of the minimum length l^{rope} of the rope bolts (set symmetrically relative to the vertical axis of mine working) is made by the formula

$$l^{rope} = l_{joist} + l_{tail} + \frac{h_w - h^{rope}}{sin\beta^{rope}} + \frac{Z_2 - Z_1}{\cos\beta^{rope}}, \tag{6.11}$$

where h_w – the height of mine working during drifting; h^{rope} – the height of setting the rope bolt from the bottom of mine working; β^{rope} – the gradient angle of the rope bolt to the vertical axis of mine working; Z_1 and Z_2 – the auxiliary geometric parameters determined by formulas

$$Z_1 = x^{rope} + \left(h^w - h^{rope}\right) ctg\beta^{rope}, \tag{6.12}$$

$$Z_2 = \frac{\left(B + 2b_l^r\right)^2}{8h}\left[\sqrt{tg^2\beta^{rope} + \frac{16h}{\left(B + 2b_l^r\right)^2}\left(h + Z_1 tg\beta^{rope}\right)} - tg\beta^{rope}\right]. \tag{6.13}$$

In the formula (6.12), the symbol x^{rope} indicates the distance from the vertical axis of mine working to the place of the rope bolt setting. The problem of choosing the rational values of the parameter x^{rope} is solved by a search of variants of SSS calculation of the studied geomechanical system, taking into account the previously established patterns. By analogy with the choice of parameters x_i for resin-grouted roof bolts, the criterion of the rationality of values is the combination of conditions for the maximum resistance of rope bolts and the minimum loading of the frame support. The patterns of x^{rope} change are such that the influence of geomechanical parameters is assessed ambiguously due to the action of opposite tendencies.

Firstly, some noticeable influence has been revealed of only two geomechanical parameters that determine the dimensions of the ultimate equilibrium, – the depth H of mine working location and the average calculated compressive resistance R of an adjacent massif. This fact seems quite logical:

– the depth of mine working location determines the total stress of massif, on which the dimensions of the arch of ultimate equilibrium and the degree of the rope bolts loading depend;

– parameter R characterizes the resistance of an adjacent massif to the boundaries distribution of the arch of ultimate equilibrium.

The parameters R_1^r and m_1^r of the immediate roof are included as components in a more general indicator R, since the increased length of rope bolts also affects other rock layers of the roof to a technically developed height of up to 9 m. Therefore, the ratio b^r most fully characterizes both the dimensions of the arch of ultimate equilibrium, and the degree of resistance of rope bolts to lowering the roof of mine working.

Secondly, the influence of the ratio b^r on the coordinate x^{rope} is determined by the action of a number of opposite tendencies. The greater the height of the arch, the greater the load in its central part and for the active resistance of the rope bolts they should be concentrated closer to the arch keystone (decrease in the coordinate x^{rope}). With an increase in the height of the arch, its width also increases, that is, the volumes of unstable rocks increase in the sides of mine working – thus, the load in this part of the arch contour increases, as well as an expediency of location here of the rope bolts (the increase in the coordinate x^{rope}).

The result of the action of these opposite tendencies leads to some mutual compensation and stabilization of the rational coordinate of rope bolts setting in a rather narrow range of $x^{rope} = 0.8 - 1.1$ m. The smaller of the values x^{rope} corresponds to the smaller standard size of mine working section (for example, for the TSYS-9.5 support), and the larger of the values x^{rope} – to an increased mine working dimension (e.g., for TSYS-15.0 support). In general, it is recommended to place the tail joint of the rope bolt at a distance of 1.0 m from the vertical axis of mine working for the section of mine workings for TSYS-11.7 (11.0) support.

A similar situation has been revealed when determining the rational gradient angle β^{rope} of the rope bolt to the vertical axis of mine working. There are two main factors. The first is the narrow range of variation of the rational coordinate x^{rope} of the rope bolt setting predetermines a limited range in the values of its gradient angle. The second factor is conditioned by the increased length of the rope bolt about 6 – 9 m. The vector of its movements is changed. It was noted earlier that the roof-bolt most effectively resists the displacement of massif when its longitudinal axis coincides with the displacement vector of the enclosing rocks. The peculiarity of the roof rock displacements along its height is that the change in their vector (across the width of arch) from vertical to oblique is most active in the border rocks. With the removal from the contour into the depth of the roof, the intensity of the change in the displacement vector direction is weakened. At the buried end of the rope bolt (taking into account the coordinate of its setting $x^{rope} = 0.8 - 1.1$ m) the displacement vector is directed closer to the vertical position than on the mine working contour. It is not technically possible to bend the roof-bolt (by drilling a curved bore hole) to such an extent that its longitudinal axis always coincides with the displacement vector. It is advisable to specify the average direction to the rope bolt as the arithmetic average of the vectors of the massif displacement at its buried end and on the mine working contour. This procedure was carried out for the above range of the coordinate x^{rope} changes and the interval of variation of the rational gradient angle $\beta^{rope} = 65 - 80°$ of the rope bolt has been obtained for different mining and geological conditions of maintaining the preparatory mine workings. This interval is significantly different from the gradient angles β_i of resin-grouted roof bolts for the same coordinate x_i. A lower value β^{rope} corresponds to $x^{rope} = 1.1$ m and to more favorable mining and geological conditions; a higher value β^{rope} corresponds to $x^{rope} = 0.8$ m and to more complex mining and geological conditions.

The last of the required parameters of rope bolts is the rational value of their resistance to the roof rocks displacement, which is expressed through the minimum sufficient diameter d^{rope} of the “reinforcement” of the rope bolt. The criterion for choosing the rational values of d^{rope} remains the same (as for resin-grouted roof bolts) and its essence is to realize the maximum possible loading of the rope of a bolt within the limits of permissible tensile stress.

An analysis of the results of multivariate computational experiments has shown that two geomechanical parameters most significant influence on the degree of the roof-bolts loading – the depth H of mine working location and the value of the average calculated compressive resistance R of the adjacent coal-bearing strata. The ratio b_m^r has a certain influence in the range of $b_m^r \leq 0 -$ 15 MPa/m. Here, the value R is changed significantly, since the immediate roof of the increased thickness is the nearest to mine working. According to the results presented in [71], this parameter has the maximum influence on the value of the average calculated compressive resistance of massif. It contains the effect of low values of the ratio b_m^r and there is no need to separate the influence of this parameter on the degree of the rope bolts loading. The specified patterns are confirmed by the results of the performed series of computational experiments, as well as the fact that the influence of each of the remaining geomechanical parameters (H and R) must be determined by separate dependences, since the ratio b^r does not give a close correlation.

Based on the results of the performed studies, the dependences have been obtained of the relation of the minimum sufficient diameter of the rope of a bolt with geomechanical parameters H and R, which are shown in Fig.6.5. With an increase in the depth H, the diameter d^{rope} of the rope increases close to a linear law and inversely proportional to the parameter R. The main range of variation of the minimum sufficient diameter of the rope bolt is limited by the value $d^{rope} = 15$ – 16 mm, which is well correlated with mine observations for assessing the stability of mine workings, in the roof of which the rope bolts are set with a diameter $d^{rope} = 15.2$ mm. However, at a very low value of $R \leq 5-7$ MPa, the load on the rope bolts sharply increases (the boundaries of the arch of ultimate equilibrium are expanding intensively), which requires an increase in the diameter of their rope to 20 – 28 mm in the studied scheme of setting.

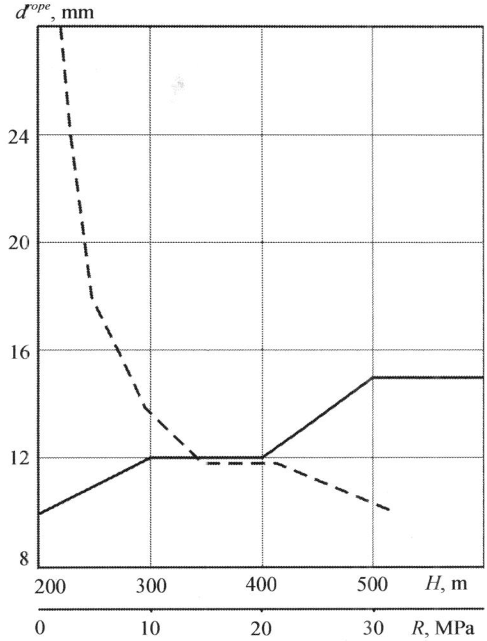

Fig. 6.5. The patterns of change in the minimum sufficient diameter of the rope d^{rope} of a bolt from the depth H of mine working location (——) and the average calculated compressive resistance R (– –) of the adjacent coal-bearing massif

The processing of the results of multivariate calculations made it possible to obtain a regression equation for predicting the minimum sufficient rope diameter

$$d^{rope} = \frac{139}{\sqrt{\sigma^t}}\left[\left(1.7 + 9.1 \cdot 10^{-3} H\right)\left(\frac{14.3}{R^{1.8}} + 0.66\right)\right]^{0.5}, \text{ mm.} \tag{6.14}$$

The equation (6.14) has been obtained when considering the variant of setting two rope bolts with a steps of $L^{rope} = 3.2$ m along mine working, that is, in the interframe space for every three frames to the fourth, which, in turn, being set with a step of $L = 0.8$ m. The rope bolts are available in standard diameters d_{st}^{rope} (the most common $d_{st}^{rope} = 15.2$ mm) of a limited number of standard sizes. Therefore, if the calculated value d^{rope} (6.14) allows to choose the nearest larger standard size d_{st}^{rope}, then it is that which is accepted for use in the fastening system of mine working. If the calculated value d^{rope} exceeds the standard value d_{st}^{rope} of available ropes, the number n^{rope} of rope bolts for 1 m of mine working increases according to the formula

$$n^{rope} = 0.625 \frac{d^{rope}}{d_{st}^{rope}}, \text{ pcs/m}, \tag{6.15}$$

where d_{st}^{rope} – the diameter of rope bolt used in a certain mine working, mm.

The results of multivariate computational experiments and their correlation-dispersion analysis made it possible to determine all the parameters necessary for the effectively maintenance of the roof of the preparatory mine working with rope bolts.

6.3. ALGORITHM FOR CALCULATING THE PARAMETERS OF THE COMBINED ROOF-BOLTING SYSTEM FOR STRENGTHENING THE ROCKS OF THE ARCH IN THE PREPARATORY MINE WORKINGS

The complex of performed research has created all the prerequisites necessary for the development of methodology for calculating and selecting the parameters of the fastening system of the preparatory mine workings in conditions of intensive coal seams mining. The methodology provides the following sequence of calculations:

1. Grouping the source data used in the calculations performing. This item is aimed to determine the calculated value of each of the parameters involved in the calculations.

The group of geomechanical parameters includes:

– the calculated depth H of mine working location;

– the average calculated compressive resistance R of adjacent rocks of roof and bottom;

– the thickness m_1^r and calculated compressive resistance R_1^r of rocks of the immediate roof;

– the thickness m_1^b and calculated compressive resistance R_1^b of rocks of the immediate bottom.

The listed initial parameters are determined from the data of geological service of the mine, by studies [58, 59] and by normative documents [67, 68]. The group of mining and technical parameters:

– the standard size of section of the preparatory mine working and its support;

– the width $l_{p.s.}$ of protective system;

– the width B and height h_w of mine working during drifting.

The specified initial parameters are determined by the normative and technical documentation when preparing the passports for drifting and fastening of mine workings.

From the mechanical characteristics of the fastening materials, the yield limit of the bearing rod of the roof-bolt is used, which is determined from the reference data [16, 19], depending on the steel grade of the manufacturer of the used roof-bolts.

2. The determination of the dimensions of the unstable rocks area around mine working is characterized by parameters h and b_l^r. The height of arch h of ultimate equilibrium is calculated by document [71]. The width b_l^r of unstable rocks area in the roof of mine working is calculated by the formula (6.9).

3. The choice of the type of frame support and calculation of the step of its setting is made according to document [67] for the area of mine working outside the zone of the stope works influence.

4. The calculation of the parameters of the combined roof-bolting system in the arch of the preparatory mine working is made in the following positions:

– the resin-grouted roof bolts in the central part of mine working arch;

– the rope bolts in the immediate and main roof.

5. The calculation of location parameters of the resin-grouted roof bolts in the central part of mine working arch.

The assessment of the expediency of the resin-grouted roof bolts setting for the most widely used standard sizes of frame support in the preparatory mine workings is carried out by the values of ratio b^r and b_m^r, and in accordance with the graphs in Fig. 6.2, or by the formulas (6.4) – (6.6): if the actual value b^r is less than the calculated, the setting of the resin-grouted roof bolts in the central part of the arch of mine working is not recommended.

The coordinates $X_{1,2,3}$ of resin-grouted roof bolts setting along the contour of the arch (horizontal distance from the vertical axis of mine working to the tail joint of the roof-bolt; i – number of the roof-bolt with distance from the arch keystone) are determined by formulas (6.1) – (6.3). If the calculated coordinate X_i exceeds the horizontal coordinate of the yielding joist location of this standard size of frame support, then this pair of resin-grouted roof bolts is excluded from their setting scheme in the arch of mine working.

The gradient angle $\beta_{1,2,3}$ of resin-grouted roof bolts to the horizontal axis of mine working is calculated by the expression (6.7) separately for each pair ($i = 1$, 2, 3) of symmetrically located roof-bolts. For the diameter of "reinforcement" of the resin-grouted roof bolts in the central part of the arch of mine working, it is recommended to take lower values $d_i = 15 - 18$ mm.

6. The choice of rope bolts parameters includes the following items.

The rope bolts in the cross section of mine working are recommended to be set at a distance of $X^{rope} = 0.8 - 1.1$ m from its vertical axis. The smaller values of the interval correspond to the reduced standard size of the cross section, and the upper limit of the interval corresponds to the increased standard size of the section of mine working.

The gradient angle of rope bolts to the horizontal is recommended to take $\beta^{rope} = 65 - 80^{\circ}$. A smaller value corresponds to $X^{rope} = 1.1$ m, a greater – to $X^{rope} = 0.8$ m.

In the longitudinal section of mine working, the rope bolts are set with a basic step of $L^{rope} = 3.2$ m, that is, in every fourth interframe space. With the complication of mining and geological conditions, the number of rope bolts per 1 long meter of mine working length is calculated by the formula (6.15) and the nearest larger whole number is taken. To perform the calculations by formula (6.15), the required diameter of rope bolts is calculated by formula (6.14).

The length of the rope bolt is determined by the formula (6.11) with the calculation of the intermediate parameters by the formulas (6.12) and (6.13), which include the dimensions of the arch of ultimate equilibrium in paragraph 2 of this algorithm for selecting the rational parameters of the fastening system for the preparatory mine working.

7. The represented calculation algorithm provides a choice of rational parameters of load-bearing elements of the combined roof-bolting system depending on the mining and geological, as well as mining and technical conditions for maintaining the arch of the preparatory mine workings in mines that develop thin-bedded coal seams in soft rocks.

CONCLUSIONS

1. By means of multivariate calculations the efficiency has been proved of using the rope bolts in the fastening system of the preparatory mine workings, which have a multifunctionality of work in the "deep" strengthening of roof rocks, increasing the stability of the frame support and intensifying the terminal operations in the junction area of the longwall face and mine working.

2. It has been established that the ratio b_m^r influences the coordinates of the resin-grouted roof bolts setting. With an increase in the thickness of the roof rock,

the load on the roof-bolts decreases, and their number should be reduced, which corresponds to an increase in the distance between the roof-bolts (the displacement of coordinates of their setting from the keystone to the spring of the arch).

3. The recommendations have been developed and calculation expressions have been obtained for choosing all the necessary parameters for the combined roof-bolting system setting in the preparatory mine workings.

4. The area of expedient location has been determined of the resin-grouted roof bolts in the central part of mine working arch, as well as the relation of the gradient angle β_i of the resin-grouted roof bolt with the coordinate x_i of its setting in the arch of mine working for different values of the parameter b^r.

5. The evidence base has been substantiated for cardinal limitation of the rock pressure manifestations by formation in the roof of a thick armored and rock plate using a combined roof-bolting system. This plate is designed to protect the frame support from excessive loads and to create the conditions for reuse of the preparatory mine workings.

7. THE INFLUENCE OF THE CENTRAL PROP STAYS OF THE STRENGTHENING SUPPORT ON THE FASTENING SYSTEM STRESS-STRAIN STATE OF MINE WORKING

7.1. ANALYSIS OF THE OPERATION MODES OF THE CENTRAL PROP STAYS OF THE STRENGTHENING SUPPORT IN THE PREPARATORY MINE WORKINGS

In the mines of Donbas, in particular in its western region, the central prop stays of the strengthening support (hereinafter – prop stays) are widely used to maintain mine workings in the zone of the stope works influence. They differ in construction: wooden, hydraulic and friction prop stays from the special SCP profile. The main function of the central prop stays (single and double) is the strengthening of the frame support to resist to the vertical rock pressure, which increases dramatically from the bearing pressure zone ahead of the longwall face and up to the area of stabilization of geomechanical processes in mine working behind the stope face. Moreover, an important function of the central prop stays is to maintain the cap board of the frame in the area of dismantling its stump near the 'window' of the longwall face.

When analyzing the stress-strain state of the prop stays, the criterion has been substantiated for assessing their loading by the resistance reaction when lowering the rocks of the coal-overlaying formation. The most objective parameter here is the deformation-strength characteristic of the prop stay, which describes the relation of its resistance to longitudinal pliability.

The friction prop stays from the special SCP profile and hydraulic prop stays work in a mode close to the mode of constant resistance: to a certain value of resistance they are in a rigid mode with an insignificant pliability, mainly conditioned by the elastic deformations of the material; after, the yielding mode occurs and then the resistance of the prop stay varies within a certain range around the average value, which practically does not depend on the lowering of the roof rocks. This operation mode is provided by the yielding node in the friction prop stays from the special SCP profile or by the periodic actuation of the safety valve in the hydraulic prop stays.

As a rule, the yielding mode of the prop stays structures occurs even before the approach of the longwall face, in the zone of its front bearing pressure, which is caused by the processes of lowering the roof. Therefore, in the zone of stabilization of the rock pressure manifestations, the friction prop stays and hydraulic prop stays operate in the mode of constant resistance, at which the load on them is enough constant. This means that the stress state of the prop stays will also be relatively stable regardless of the geomechanical processes occurring around the preparatory mine working and conditioned by the depth of its location, by the structure and properties of the adjacent coal-bearing massif.

Thus, it is impossible to observe the influence of geomechanical factors on the state of the prop stays, working in the mode of constant resistance. The formulated conclusion is consistently confirmed by the calculation results of the stress-strain state of geomechanical models with different variants of fastening and security systems, where the central prop stays of the strengthening support with the studied structures are used. Thus, the range of stresses intensity variation is up to 20 – 25%.

This variation depends on the action of the oblique load (bending with compression), as well as on the occurence of local areas of stresses concentration due to the influence of contact stresses in the places, where the prop stays bear on the footing rocks of mine working and their junction with the cap board of the frame. In the process of research, there were no significant changes in the stress-strain state of the prop stays when they were set in one or two rows both in the central part of the cross-sectional area of mine working and in some their displacement towards the longwall face.

a

b

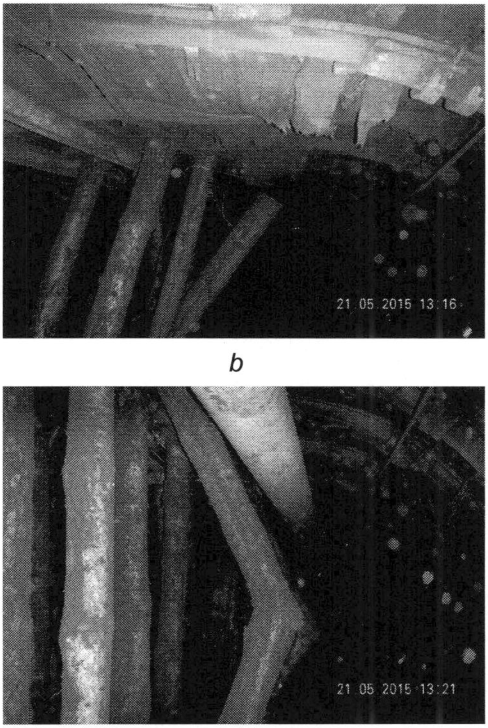

Fig. 7.1. The general view of the fastening system in the preparatory mine working behind the longwall face: *a* – flattening and plastic deformations of the frame cap boards; *b* – destruction of the central (wooden) prop stays of the strengthening support

A slightly different situation arises in the case of setting the wooden central prop stays of the strengthening support. They also have some pliability due to:

– elastic-plastic longitudinal deformations of wood;

– contortion of wooden substrates under the prop stays in the contact with the mine working footing and the frame cap board;

– pressing the prop stays into the mine working footing;

– bending of the prop stays with loss of stable shape.

The last component of pliability is observed in cases of setting the wooden prop stays with a diameter of not more than 15 – 16 cm in the conditions of intense vertical rock pressure. The deformation of the wooden prop stays occurs in accordance with the classical ideas of loss of the rod stability under the action of a compressive load. The specific load Q induces the occurrence of both the first and the second forms of stability loss according to Euler. In both cases, the areas of the prop stays section with the maximum bending represent the greatest danger due to loss of stability with subsequent destruction. The process is promoted additionally by the oblique load, which, as a rule, acts at an angle to the vertical $\alpha = 10 - 20°$. The bending of the central wooden prop stays of the strengthening support with their destruction is often observed in the practice of maintaining the preparatory mine workings (Fig. 7.1). To limit this phenomenon, the central prop stays are often set at a slight angle to the vertical.

The considered components of the wooden prop stays pliability cannot be adjusted (in comparison with the friction prop stays from the special profile or hydraulic prop stays), therefore their operation mode remains uncontrollable, which complicates the development of any recommendations for choosing the parameters of the prop stays. Nevertheless, for the prop stays with a diameter of 20 – 22 cm, some patterns of relation are observed with such geomechanical parameters as the depth H of the mine working location and the average calculated compressive resistance R of rock layers of the adjacent coal-bearing stratum.

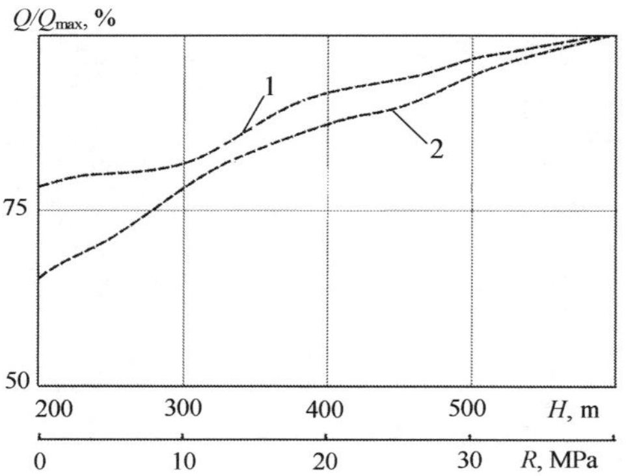

Fig. 7.2. The patterns of influence on the degree of loading Q/Q_{max} of the central wooden props of the strengthening support: *1* – depth H of mine working location; *2* – average calculated compressive resistance R of adjacent coal-bearing stratum

As can be seen from Fig. 7.2, these patterns are not so evident and are limited by the possibility of destroying the prop stay, on the one hand, and its uncontrollable pliability, on the other. At the same time, the general tendency to increase the load Q on wooden prop stays with respect to their maximum load-bearing capacity is manifested with an increase in the depth of mine working location and a decrease in the strength properties of an adjacent coal-bearing massif.

The performed analysis of the stress-strain state of the central prop stays allows to substantiate the calculated positions for determining their rational parameters: the distances x horizontally from the vertical axis of mine working; the gradient angle β of the prop stays to the vertical axis of mine working and diameter d of wooden prop stays of the strengthening support; for all types of prop stays structure, a minimum sufficient reaction of a prop stay is defined.

The need to set the central prop stays of the strengthening support is increased by intense rock pressure manifestations in complex mining and geological conditions, for example, in the presence of a weak, water-flooded thin-bedded massif of the main roof with a thickness of 8 – 10 m and more. Under the conditions of a very unstable main roof, a vast arch of ultimate equilibrium is formed, the rocks of which create a great load on the fastening system of mine working. The situation is complicated by the removal of the boundaries of the arch of ultimate state from the mine working contour, which makes it impossible to fasten the joist of the roof-bolt (of the required length) in more stable rocks. This weakens the effect of strengthening the roof, especially in the area of dismantling the frame stumps near the 'window' of the longwall face.

The processing of the results of a series of computational experiments by the methods of correlation-dispersion analysis made it possible to obtain a regression equation that determines the critical depth H_{cr} of mine working location, more than which the prop stays setting is recommended:

$$H_{cr} = 74.2R^{0.8}. \tag{7.1}$$

In studies, the asymmetry has been noted of the fastening system loading in the zone of the stope works influence, which is manifested in the formation of an increased oblique load from side of approaching or receding the longwall face. This phenomenon is confirmed by the experience of conducting the terminal operations in strengthening the fastening system in the place of junction of the long face with the preparatory mine working, when, as a rule, the prop stays are shifted towards the longwall face by some distance relative to the vertical axis of mine working. The search for the rational coordinate x of the prop stays setting was carried out by the condition of maximum unloading of the frame support, according to which the stresses intensity $\sigma.$ was the determining parameter. Moreover, it is not the cap board that needs unloading more, but the stumps of the frame support, especially from the side of the longwall face. The necessity to shift the strengthening support of the prop stays in the direction of the longwall face with a sufficiently stable value of $x = 0.4 - 0.6$ m has been determined by calculations. A lower value corresponds to a reduced standard size of the section area of mine working and to a reduced overhang of the rock cantilevers from the side of mined-out space, and an increased value – to an increased standard size of the cross section area of mine working and to an increased length of the hanging rock cantilevers.

The gradient angle β of the central prop stay of the strengthening support to the horizontal axis of mine working is important, since the oblique load from the side of the roof rocks (eccentric compression) activates the loss of its stability, so the load-bearing capacity of the prop stay decreases sharply. To avoid eccentric compression, the longitudinal axis of the prop stay should coincide with the direction of the displacement vector of the roof rocks in the place of their junction. The calculations of the stress-strain state of the fastening system have shown, that with a recommended coordinate of $x = 0.4 - 0.6$ m, the value β ranges within $75 - 85^{\circ}$. The value $\beta = 80^{\circ}$ is recommended when setting the central prop stays of the strengthening support, regardless of their structure as well as mining and

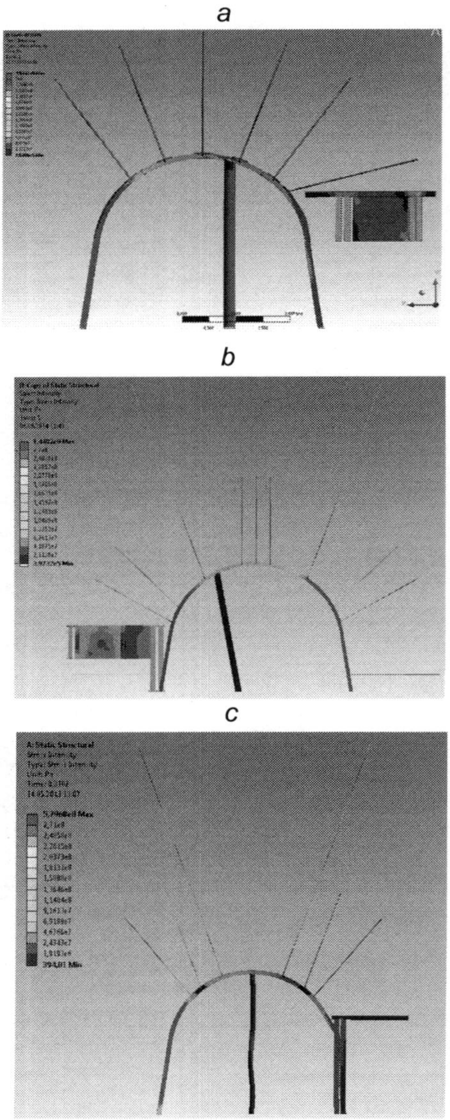

Fig. 7.3. The curves of stresses intensity σ in the fastening system of the preparatory mine working with an increased (*a*), moderate (*b*) and reduced (*c*) rigidity of the wooden prop stay of the strengthening support of the frame

geological conditions of mine working operation.

The last of the required rational parameters of the prop stays is their minimally sufficient load reaction, the rationality criterion of which is the condition of maximum unloading of the frame, on the one hand, and exclusion of bending of the frame cap board in the place of its contact with the prop stay, on the other. This circumstance (the possibility of occurrence of a plastic hinge in the cap board) is important for maintaining the stable shape of the mine working arch, especially in the area of dismantling the frame prop stays in the area of 'window' of the longwall face.

The revealed peculiarity of stress-strain state (SSS) of the frame cap board on the contact with the prop stay of the strengthening support is shown in the curves of Fig. 7.3, where the schemes are presented for fastening and security systems for different mines in Western Donbas. It has been established by multivariate computational experiments that the occurrence and development of plastic deformation areas in the place of contact between the cap board and the prop stay is possible in different mining and geological conditions, and is mainly determined by the deformation-strength characteristic of the prop stay, that is, by the dependence of the load reaction on the pliability of the support during the resistive action to the roof lowering.

With an increased rigidity of the wooden prop stay ($d = 24$ cm), its increased reaction causes, if pliability is low, an active bending of the frame cap board at the point of contact, as evidenced by the occurrence of a sufficiently extensive area of stresses intensity σ, exceeding the yield limit of steel of St.5

grade, – the plastic flow of steel occurs (in the entire cross section area of the SCP profile) with a sharp decrease in the cap board resistance (see Fig. 7.3, *a*).

If the diameter of the wooden prop stay is reduced to 20 cm, its resistance reaction decreases, and pliability increases, also due to significant deformations (see Fig. 7.3, *b*). The area of the plastic state of the frame cap board is noticeably reduced and no longer extends over the entire cross section area of the SCP, – the reduction in the load-bearing capacity of the cap board is low. If the central prop stay of the strengthening support has a low diameter ($d = 16$ cm), then it works on the verge of stability loss (see Fig. 7.3, *c*), and the two half-waves of transverse bending are observed in its lower part – the second form of stability loss according to Euler. Moreover, the pliability of the prop stay increases sharply with a decrease in its resistance reaction. As a result, in the place of contact of the prop stay with the cap board, the stresses intensity σ is only 50 – 60% of the yield limit of steel St.5 and such operation mode of the prop stay does not generate additional perturbations of the stress-strain state in the cap board of the frame, and the intermediate bearing increases sharply the load-bearing capacity of the frame support as a whole.

The results of the conducted studies confirm the expediency of applying the rigid and pliable operation mode of the prop stays of the strengthening support with the maximum approximation to the mode of their constant resistance, which is characterized by the following conditions:

– minimum pliability (within the elastic deformations of the material) with the growth of the resistance reaction;

– changing to the constant resistance mode after reaching the value Q_1 with a slight reaction fluctuation with an increase in pliability until the equilibrium state of the entire geo-mechanical system to maintain the mine working.

In addition, the problem has been solved of establishing a rational reaction Q_1 depending on the mining and geological conditions of maintaining the preparatory mine working. According to the results of multivariate computational experiments (Fig. 7.4), two main parameters make a significant influence on the formation of the dimensions of the arch of ultimate equilibrium of mine working: the depth H of mine working location and compressive resistance R of adjacent roof rocks.

The patterns of relation are such that with an increase in the depth of mine working location, a practically linear law of the required reaction Q_1 growth of the prop stay, and with the parameter R an inversely proportional dependence is revealed that tends to zero (the setting of central prop stays of the strengthening support is impractical) with an average calculated compressive resistance of the adjacent roof rocks $R \geq 12 - 15$ MPa. With highly unstable roof rocks ($R \leq 5 -$ 7 MPa), the required load-bearing capacity of the prop stay increases to 100 – 200 kN. In practice this is achieved by selecting the diameter of the wooden prop stay, the standard size of the friction prop stay or hydraulic prop stay.

By static processing of the calculation results of the stress-strain state of the fastening system, an equation has been obtained for determining the required

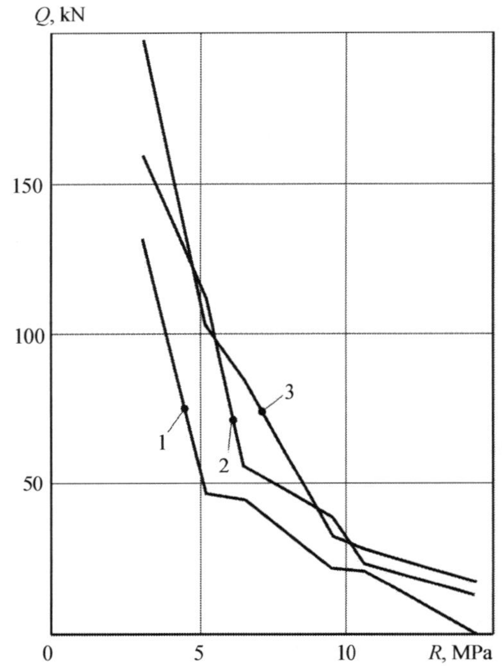

Fig. 7.4. The relation of Q load on the central prop stay of the strengthening support with an average calculated compressive resistance R of adjacent roof rocks at the depth of mine working location: 1 – $H = 200$ m; 2 – $H = 400$ m; 3 – $H = 600$ m

resistance reaction of the prop stay:

$$Q = 620 / R^{1.7}\left(1 + 2.1 \cdot 10^{-3} H\right). \quad (7.2)$$

When calculating, it is also necessary to take into account the recommended constant-resistance mode during the prop stays operation.

In order to provide a reliable mode of constant resistance, it is recommended to use the friction prop stays made of SCP or hydraulic prop stays with a load-bearing capacity corresponding to the loading from the side of rock massif, and it is rational for them to control the resistance Q_1.

The structural features of the friction prop stays from the special SCP profile and the hydraulic prop stays, which provide the operation in the constant resistance mode, give them versatility in the management by the frame support loading in the optimal mode. The non-controllable rigid operation mode of the wooden prop stays of the strengthening support often leads to the loss of their stable form and to destruction, that is, their original functions for strengthening the fastening system are lost.

7.2. THE RESEARCH OF THE FASTENING SYSTEM STRESS-STRAIN STATE OF MINE WORKING WITH THE YIELDING PROP STAY IN THE STRENGTHENING SUPPORT

In the course of performing the previous stages of the SSS analysis of various variants of the fastening systems of the drifts being overworked, a negative effect was observed (on the frame cap board) of the work of a rigid wooden prop stay of the strengthening support. To check the recommendations for improving the stability of the cap board of the frame support, the computational experiments have been carried out with the setting of a central hydraulic prop stay while maintaining all other equal conditions of the fastening system "frame-resin-grouted roof bolts and rope bolts in the roof".

The analysis of the fastening system state has been carried out in three main components: vertical σ_y, horizontal σ_x and stresses intensity σ.

7.2.1. ANALYSIS OF VERTICAL STRESSES COMPONENTS

The curve of vertical stresses σ_y distribution in the elements of fastening system with a central hydraulic prop stay is shown in Fig. 7.5 and characterized by the following features.

The cap board is exposed to the strongest influence of the operation mode of the central prop stay of the strengthening support. This influence has a different intensity along the length of the cap board. In its peripheral areas (near the yielding joists), the curve σ_y is changed insignificantly in comparison with the variant of setting the wooden prop stays of the strengthening support. Thus, the main range of σ_y variation is from 1 – 2 MPa of tension to 20 – 25 MPa of compression, which does not exceed 7 – 9% of the value of estimated yield limit of the SCP steel. The bending moment is minimal: in the area below the coordinates of the rope bolt tail joint location, the bending is directed towards the roof rocks; in the area above the tail joint – there is a deflection into the drift cavity. In general, a sufficient uniformity of σ_y distribution in the cross section of SCP and their low absolute values indicate a high degree of the cap board stability (by the factor of vertical stresses action) in peripheral areas with a length of up to 0.6 m from the side of the stope face and up to 0.7 m from the side of virgin massif.

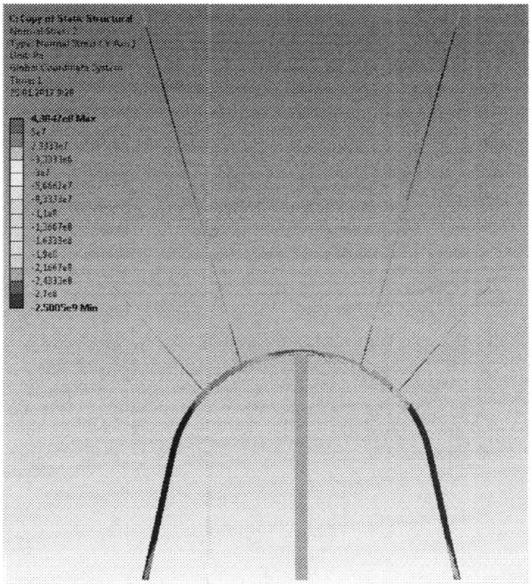

Fig. 7.5. The curve of vertical stresses σ_y for the variant of the fastening system with central hydraulic prop stay

In the central part of the cap board, the situation is changed drastically: certain disturbances σ_y appear with some asymmetry in the direction of the stope face of the longwall face; this area occupies a length of 0.7 – 0.8 m. When approaching a central hydraulic prop stay, the tensile stresses σ_y of up to 40 – 70 MPa increase, which is only 15 – 26% of the estimated yield limit of steel and does not constitute a danger to the stability of the cap board. Moreover, there is a differently vectored bending: from the side of virgin massif – into the cavity of mine working; from the side of the stope face – in the direction of the roof. Directly in the area of contact with a hydraulic prop stay (across the width of up to 300 mm), the direction of bending is observed towards the roof, however, the concentration of compressive σ_y is low – only up to 50 – 80 MPa (19 – 30% of the estimated yield limit of steel).

Thus, it is established that when setting a hydraulic prop stay (yielding mode of operation), there is no 'traditional' bending in the cap board of the frame, as it was observed in the variants with a rigid wooden prop stay of the strengthening support. The level of acting concentrations σ_y is more than 3 times less than the estimated yield limit of the SCP steel, which not only indicates the stable state of the cap board of the frame, but also confirms the expediency of using the yielding prop stays of the strengthening support, for example, hydraulic prop stays, during the drifting of the longwall face. Consequently, the account of the modes of the fastening system elements interaction (the peculiarities have been revealed by the results of the computational experiment) allows to change radically the state of the cap board by the factor of the vertical stresses action.

In the prop stays of the frame, there are no significant changes in the curve σ_y when comparing the variants with the wooden prop stay and hydraulic prop stay of the strengthening support in the upper part of the prop stays (in the area of the yielding joists) with a length of 300 – 350 mm, where the σ_y increases from the low tensile (up to 20 – 30 MPa) to significant stresses (80 – 190 MPa) of compression. And with a transition to the rectilinear part of the prop stays σ_y, it reaches 80 – 90% of the estimated yield limit of the SCP steel. In the rectilinear part of the prop stays, the vertical stresses are stable and close, but do not reach the estimated yield limit of the SCP steel. Such a state, close to the limiting state of the prop stays of the frame, indicates their high degree of loading. However, an almost uniform distribution of σ_y in the cross section of the SCP minimizes the bending moment until its complete disappearance; this argument somewhat improves the prediction of the prop stays stability. In their lower part, at a height of up to 0.5 m, a significant bending moment occurs, and this makes possible to predict a high probability of bending in the area of the prop stays bearings into the cavity of mine working.

The hydraulic prop stay, due to its work in the mode of almost constant resistance, is characterized by extreme homogeneity of the curve σ_y. The uniformity of the vertical compressive stresses up to 30 MPa indicates a complete absence of bending moment and any concentrations of σ_y: the hydraulic prop stay works in favorable loading conditions, and its resistance reaction corresponds to the technical characteristics.

The combined roof-bolting system actively resists the lowering of the roof rocks, and the distribution parameters σ_y are slightly different from those in the variant of setting the rigid wooden prop stays of the strengthening support.

It is quite natural that a yielding hydraulic prop stay, providing a certain pliability of the cap board of the frame, promotes the redistribution of a part of its resistance to a combined roof-bolting system. There is an increase of 15 – 20% in the tensile stresses σ_y of resin-grouted roof bolts and rope bolts, but this is manifested only in the adjacent rock layer of the roof at a distance of 0.6 – 0.9 m, and further along the length of the roof-bolts the differences in the curves σ_y decrease. This fact can be explained as follows: it is well known that the most active roof stratifications occur near the mine working contour and it is here that if the cap board of the frame 'deviates' from the increased rock pressure, this difference should be compensated by the resistance of other fastening elements, which in this case are resin-grouted roof bolts and rope bolts.

In general, based on the results of the analysis of the curve of vertical stresses distribution, two factors have been established related to the replacement of a rigid wooden prop stay with a yielding hydraulic prop stay of the strengthening support:

– the state of the cap board of the frame is radically improved, and its bend in the area of contact with the central prop stay is not predicted, since the maxima σ_y are more than 3 times lower than the estimated yield limit of the SCP steel;

– the distribution of σ_y in the remaining fastening elements is changed either insignificantly, or in the limited areas of the roof-bolts length.

7.2.2. THE STUDY OF THE CURVE OF THE HORIZONTAL STRESSES COMPONENTS

To illustrate a comparative analysis of the horizontal stresses distribution with two variants of the central prop stay of the strengthening support, the curve σ_x is shown in Fig. 7.6.

The most significant differences are observed in the cap board of the frame, which is quite predictable, since it is in contact with the central prop stay, the resistance mode of which has been changed fundamentally; in addition, the horizontal stresses most informatively describe the process of bending of the cap board under the influence of rock pressure. The local areas of the cap board (with length of 200 – 250 mm) near the yielding joists of the frame are exposed to the action of a bending moment directed into the mine working cavity. However, the level σ_x (from a tension of about 10 – 30 MPa to compression in the range up to 50 – 70 MPa) does not allow to judge any danger from the point of view of the plastic flexure strains occurrence, since σ_x does not exceed 19 – 26% of the estimated yield limit of the SCP steel. Further on, when moving to the central part of the cap board, only compressing stresses act: in the lower part of the SCP section, about 40 – 75 MPa, in the upper part – 70 – 90 MPa. The height difference σ_x of the SCP section is low, which indicates a reduced bending moment, and the maxima σ_x themselves are 3 times less than the estimated yield limit of steel.

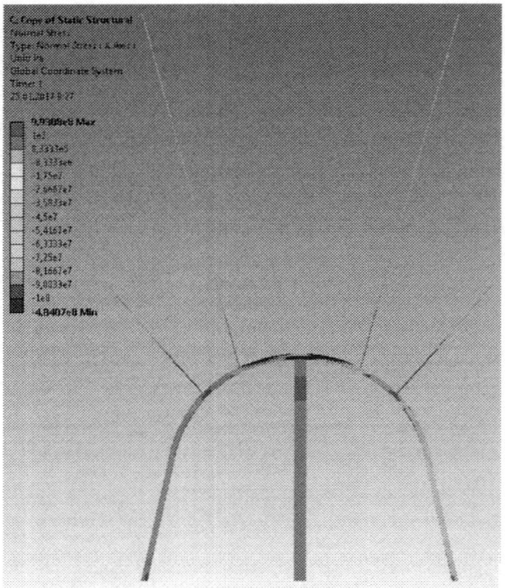

Fig. 7.6. The curve of horizontal stresses σ_x for a variant of the fastening system with the central hydraulic prop stay

In the central part of the arch, in the area of contact with the hydraulic prop stay, the direction of bending of the cap board is changed: the tendencies of its bending around the hydraulic prop stay are observed. Thus, in the upper part of the SCP section, there are the compressing $\sigma_x = 55 - 70$ MPa, in the lower part of the section $\sigma_x = 90 -$ 110 MPa. These data point out two facts: firstly, the relatively low difference σ_x on the outer and inner surfaces of the SCP indicates the action of a very limited bending moment; secondly, the maxima σ_x do not exceed 41% of the estimated yield limit of steel. As a result, the state of the cap board (by the factor of σ_x action) should be assessed as very stable without any signs of the occurrence of its plastic bending relative to the central prop stay. Consequently, a hydraulic prop stay, due to its work in the constant resistance mode, eliminates the dangerous concentrations σ_x and the very possibility of the occurrence of plastic deformations in the cap board of the frame support.

There were no significant changes in the curve σ_x of the prop stays of the frame support connected with the replacement of a rigid wooden prop stay with a yielding hydraulic prop stay. There is a relatively uniform (in the cross section of SCP) distribution of σ_x, with the exception of an area of prop stays to a height of 0.6 m. In the curvilinear part of the prop stays, the compressing σ_x are changed in the range of 30 – 45 MPa and the separate local decreases of σ_x up to 0 – 10 MPa are observed only in the area of yielding joists of the frame. In the rectilinear part of the prop stays, the compressing σ_x decrease to 20 – 25 MPa and only in the area of prop stays bearings, there are some disturbances σ_x with the occurrence of tensile σ_x up to 10 – 30 MPa. In general, the level of acting horizontal stresses is very low and by the distribution factor σ_x, it should be pointed out the stable state of the frame support.

A very uniform distribution of σ_x is observed in the cross section of hydraulic prop stay, which once again emphasizes the peculiarity of "self-balancing" of the load across the section area due to the pliability of the prop stay and its "deviation" not only from the increased vertical load, but also from its oblique action. In the most part of the prop stay height (89%), the value σ_x is minimal in the range from 0.8 MPa of tension to 8 MPa of compression and only in the rest area the tensile stresses up to 5 – 6 MPa appear, which can be explained by the known effect of the transverse deformations development at the longitudinal compression.

In the combined roof-bolting system, the effect of replacing the wooden prop stay with a hydraulic one is insignificant as a whole and is observed only in the border area of the roof-bolts length:

– in rope bolts, the minimum level of σ_x (from 0.8 MPa of tensile to 8 MPa of compression) is observed in the border area with length 0.8 – 1.0 m; further on, as it approaches to the buried area of the rope bolt, the compressing σ_x increase to 17 – 25 MPa, which is obviously due to an increase in horizontal stresses in the rock walls of the bore hole and the surrounding massif as a whole;

– in the resin-grouted roof bolts (due to their decreased gradient angle), the tensile σ_x are more evident, which indicates the active resistance of the roof-bolts to lateral movements of the border massif; so, at the length of 0.9 – 1.1 m of the border area of the resin-grouted roof bolts, the tensile σ_x reach 30 – 50 MPa and prevent stratification of the rock.

In general, based on the analysis of the horizontal stresses distribution in the main elements of the fastening system, it is possible to make such conclusions regarding the central hydraulic prop stay setting:

– the maximum influence is made on the cap board of the frame in terms of eliminating the concentrations σ_x and high bending moments; the probability of occurring the plastic deformations is completely excluded;

– the remaining fastening elements are not affected by the operation mode of the hydraulic prop stay in terms of a significant change in the horizontal stresses.

7.2.3. ANALYSIS OF STRESSES INTENSITY

The result of changes in the curves of stresses components when setting the hydraulic prop stay is summarized by studying the stresses intensity field σ in the main elements of the recommended fastening system of mine working (Fig. 7.7).

As approaching the central part of the mine working arch, the stresses intensity increases to 140 – 180 MPa (52 – 67% of the estimated yield limit of the SCP steel), but these values are far from dangerous. In addition, a high degree of uniformity of σ distribution over the SCP section area indicate the absence of any significant bending moment in these areas of the cap board.

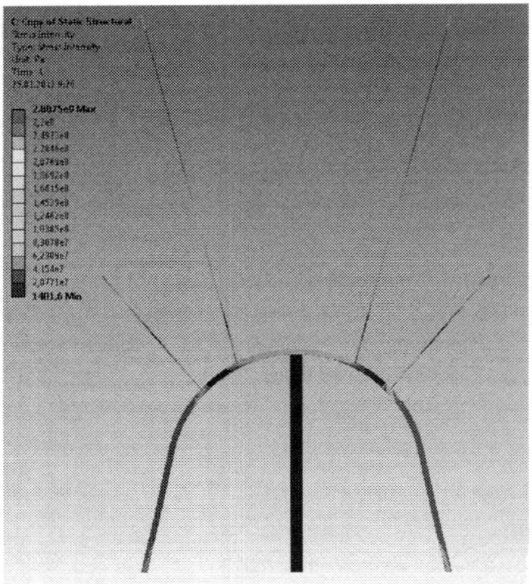

Fig. 7.7. The curve of stresses intensity σ for a variant of the fastening system with the central hydraulic prop stay

A very moderate bending moment appears in two areas of the cap board: directly in the area of contact with the hydraulic prop stay and in an area of 300 – 350 mm in length, adjacent to the hydraulic prop stay from the side of the stope face. Here the difference σ in height of the SCP section is only 50 – 60 MPa, which is not able to form a significant bending moment, but a change in its sign is clearly observed: from the side of the stope face, the bending is directed into the cavity of mine working, and directly on contact with the hydraulic prop stay – towards the roof rocks. In these areas, the maximums are $\sigma = 190$ – 210 MPa, which is 70 – 78% in relation to the estimated yield limit of SIP steel. Summarizing the represented data, it is possible to state about a sufficient degree of stability of the cap board with a considerable safety factor, which is not observed when setting the rigid wooden prop stays of the strengthening support.

The prop stays of the frame support are practically not affected by the operation modes of the hydraulic prop stay in terms of a change in the curve of stresses intensity σ distribution. In the upper area of the prop stays, near the yielding joists of the frame, there is a reduced value of σ in the range of 40 – 120 MPa, but as approaching the rectilinear area of the prop stays, the stresses intensity reaches the dangerous values, accounting for 85 – 96% of the estimated yield limit of the SCP steel. However, the distribution of σ in the SCP cross section is very uniform (up to the area of the prop stays bearings), which indicates the absence of any significant bending moment and allows to predict the stable state of the prop stays, although close to the limiting one. In the area of bearings to a height of up to 0.4 m, the disturbances σ appear in the SCP cross section, which fix the occurrence of a significant bending moment, deforming the prop stay into the cavity of mine working. In general, the noted parameters of σ distribution practically do not differ from the variant of setting the rigid wooden prop stay of the strengthening support. This fact is explained by the fact that the hydraulic prop stays (regardless of the yielding operation mode) provide an active resistance to vertical rock pressure, helping to the prop stays of the frame support at the same level as the central wooden prop stay.

The distribution of σ in a hydraulic prop stay is characterized by extremely high uniformity both in cross section and in its height. The yielding operation mode contributes to a more complete "adaptation" of the prop stay to the nature of the rock pressure manifestations, which almost eliminates the occurrence of any stresses concentrations. This peculiarity enables the hydraulic prop stay to develop a "normal" resistance force in accordance with its technical characteristics.

In the combined roof-bolting system, the distribution of the stresses intensity is very similar to that of the central wooden prop stay. It should be noted the high efficiency of rope bolts in resisting the displacements of the roof rocks of the drift. In the buried area (with length 52 – 57% of the total length of the roof-bolt), the stresses intensity is 77 – 98% of the calculated yield strength of the rope, and in the remaining its length it also actively prevents the stratification of the roof rocks.

The resin-grouted roof bolts in the boundaries of mine working arch work to restrict the stratifications of the lateral rocks and create relatively holistic end areas of the armored and rock plate above the entire mine working and beyond its width in the sides.

Thus, the SSS analysis results of the fastening system with the use of a central hydraulic prop stay as a strengthening support prove convincingly the expediency of its applying in order to increase the stability of the cap board, without damaging the rest fastening elements.

CONCLUSIONS

As a research result of changes in the SSS of fastening system when setting a hydraulic prop stay (yielding operation mode), the following conclusions have been made:

– the plastic bending of the cap board of the frame, which was previously observed (when setting rigid wooden prop stays of the strengthening support), does not occur; the level of acting concentrations of vertical stresses σ_y is more than 3 times less than the estimated yield limit of the SCP steel, which proves the expediency of using the yielding prop stays of the strengthening support. The changes in σ_y distribution in other elements of the fastening structure are not dangerous for its stable resistance to vertical rock pressure;

– a hydraulic prop stay, due to its work in the constant resistance mode, eliminates the dangerous concentrations of horizontal stresses σ_x and the very possibility of occurrence of plastic deformations in the cap board of the frame support. There was no significant influence of a change in the operation mode of the prop stay of the strengthening support on the curve of σ_x distribution in the frame prop stays and in combined roof-bolting system;

– the resulting assessment of the fastening structure in terms of the stresses intensity confirmed the conclusion about the sufficient stability of the cap board of the frame (it resists to rock pressure with a considerable safety factor), which is not observed when setting the rigid wooden prop stays. In general, the obtained results prove convincingly the expediency of using the central yielding prop stays of the strengthening support with their setting in the stope face ahead of the zone of frontal rock pressure manifestation, as well as their dismantling behind the longwall face in the zone of rock pressure anomalies stabilization.

SYNTHESIS OF THE RESEARCH RESULTS TO CHAPTER I

Based on the established patterns of change in the stress state of the rocks around the preparatory mine working, depending on the geomechanical factors and the location of the resin-grouted roof bolts and rope bolts in the system of combined roof-bolting, the technological parameters of mine workings reuse have been substantiated.

1. It has been established that in the adjacent roof rocks, by means of a combination of rope bolts and resin-grouted roof bolts, an armored and rock plate is formed, the high load-bearing capacity of which is achieved by maintaining the horizontal thrust forces even when the roof layers are being broken into rock blocks. Due to the formation of a thrust system in the roof, the concentrations of all stresses components are reduced to a level many times lower than the strength characteristics of lithotypes composing it, and the occurrence of tensile vertical and horizontal stresses is of exclusively local nature. Therefore, a thick armored and rock plate in the roof protects the fastening system of the preparatory mine working from excessive vertical rock pressure.

2. The geomechanical model has been substantiated of the computational experiment of the fastening systems in the preparatory mine workings with resin-grouted roof bolts and rope bolts, which made it possible to establish the formation of armored and rock plate which resists to the rock pressure with concentration criterion of stresses intensity no more than 2.0. This provides the holistic state of most of the lateral rocks of the roof and coal seam, reducing the stresses concentration and expanding the area of the armored and rock plate bearing in the roof of mine working.

3. The patterns have been established of the influence of the depth of mine working location, the average calculated compressive resistance of the adjacent rock massif, the ratio of the calculated compressive resistance of the immediate roof rocks to its thickness on the degree of loading the system of resin-grouted roof bolts and rope bolts. The criterion has been substantiated for assessing the level of resistance of roof-bolts as part of a combined system, which is used to establish the most influencing geomechanical factors.

4. The dependence has been determined of the coordinates and the gradient angle of setting the resin-grouted roof bolts in the arch of mine working on the intensity of rock pressure manifestation, when it is considered in the rocks of the immediate roof depending on the thickness of the extracted seam. The decrease in the stresses concentrations and sizes of the rocks weakening areas in the sides of mine working has a positive effect on the level of the bottom rocks stress – the area of its probable weakening is being reduced. The range of

changes has been specified for the parameters of the rope bolts setting in the combined system, depending on the standard size of the section of the preparatory mine working.

5. The criterion for assessing the level of resistance of roof-bolts as part of a combined roof-bolting system has been substantiated, which is used to determine the most influencing geomechanical factors in terms of the degree of loading the system: the depth of mine working location, the average calculated compressive resistance of adjacent rock massif, the ratio of the calculated compressive resistance of the immediate roof rocks to its thickness. Thus, the increased loading of the elements of the combined roof-bolting system protects the frame support from the rock pressure, thereby reducing the cross section loss of mine working.

6. The patterns have been established of the roof-bolts loading (differentially) in the combined fastening system depending the main influencing geomechanical factors. The gradation value of the level of the roof-bolts resistance to the rock pressure has been established: the maximum resistance s created by rope bolts set from the side of mined-out space; minimum – the resin-grouted roof bolts. The pattern has been revealed of increasing the resistance reaction of the resin-grouted roof bolts as the coordinates of their setting are changed in the direction of the peripheral areas of the cap board of the frame support. The received results became the basis for substantiating the rational parameters of the combined roof-bolting system.

7. The dependences of a change in the relative length Δ and $\Delta_{l,m}^{rope}$ of the plastic state areas of the bearing member of the roof-bolt on the average calculated compressive resistance R of an adjacent massif have been established. The criterion has been specified for assessing the level of resistance of resin-grouted roof bolts and rope bolts in the combined system from the depth of mine working location and parameters b^r and b_m^r. The calculated expressions have been obtained that determine all the necessary parameters for strengthening the roof of the preparatory mine workings with a combined roof-bolting system: the coordinates of the roof-bolts setting, their gradient angles, the diameter of the bearing rod and the length of the roof-bolt. The boundary of the areas of expedient use of resin-grouted roof bolts has also been substantiated.

8. The underground investigations of the rock pressure manifestations in the preparatory mine workings, fastened by the recommended combined roof-bolting system, have proved its advantages compared with the traditional roof-bolting fastening, which are as follows:

– the convergence of roof and bottom rocks reduced by 35%, the convergence of the prop stays of frames decreased by 43%, the displacements along the diagonal dimension from the side of working flank of mine working reduced by 34%, from the side of non-mining flank – by 27%;

– the asymmetry of the diagonal dimensions is reduced by 2 times, and the vertical – by 3.62 times;

– the value of bottom rocks heaving is reduced by 1.5 – 2 times.

8. GEOMECHANICAL AND METHODICAL ASPECTS OF RESEARCH INTO STABILITY OF THE REUSABLE MINE WORKINGS WHEN OVERWORKING AND UNDERWORKING A PARTING IN SEAMS C_8^l AND C_9

8.1. GEOMECHANICAL PECULIARITIES OF C_8^l SEAM MINING

One of the main peculiarities of C_8^l seam mining, is its overworking when extracting the overlaying seams C_9 and C_{10}^u. The distances between these seams vertically vary within a fairly narrow range of 22 - 28 m with the occurrence of mainly soft rocks between the partings with a strength coefficient $f = 1 - 2$ at a periodic occurrence of a harder sandstone ($f = 3 - 6$) with different thickness. The sandstone, coal seams and interlayers are water-flooded and are the main source of mine water supply to the stope faces and development faces. This is additionally facilitated by the moderate fracturing of sandstones (3 – 5 cracks per meter) and the very strong fracturing of coal seams (18 – 25 cracks per meter). The argillites and siltstones are predominantly horizontally layered, a part of which belongs to unstable rocks, reducing sharply the strength characteristics when being moistened and are prone to swelling. The contacts between adjacent lithological varieties are soft and easily lose adhesion when being outcropped or under intense deformations of the massif.

Due to a combination of foregoing, it is possible to predict the unstable state of the overworked parting at its undermining with intensive stratification and the formation of extensive zones of the weakened massif. However, it is premature to make a conclusion about the significant influence of overworking on the stability of extraction mine workings of the C_8^l seam. Here, the main reason is the sufficient parting thickness, which "absorbs" the rock pressure disturbances while the stope works operation in the overlying seam. This is substantiated by similar studies of identifying the degree of influence of the longwall face drifting through the C_{10}^u seam on the stability of the preparatory mine workings of the C_9 seam: at about the same mining and geological situation, there was not revealed any significant influence of the overworking in studies by method of computational experiment. The same results have been obtained in sufficiently equivalent mining and geological conditions of overworking in mine workings in other mines of the Western Donbas. Therefore, it is possible to assume reasonably that the disturbance of the rock pressure is localized (during the stope extraction) within the thickness of a parting and exclude the influence of the overworking when mining the overlying seams.

Another main peculiarity is the variability of the geological structure of the coal-bearing strata along the length of the panel, for example, the 861 longwall face; wherein, the variation of the structure mainly concerns the roof rocks of the seam C_8^l. Therefore, an important task is to assess the structure of the coal-overlaying formation along the length of a panel in order to identify the most dangerous areas of the 861 belt entry in terms of predicting the most intense manifestations of rock pressure.

According to the mining and geological survey along the length of the panel of the 861 longwall face, in the technical documentation several zones have been distinguished with increased rock pressure (IRP), in which the most unfavorable situation in terms of the 861 belt entry stability will obviously develop. Based on the analysis of the mining and geological sheet, the following conclusions have been formulated.

Almost along the entire length of the panel of the 861 longwall face, there is a sufficiently mature bedding of C_8^l and C_8^u seams with a relatively low thickness fluctuation of a parting (argillite) in the range of 3.6 – 3.85 m. The compressive resistance of argillite in the sample is $\sigma_c = 9 - 20$ MPa, but, enclosed between two coal seams, it is likely to be moist with a decrease in strength of 2 – 2.5 times according to [53]. With account of fracturing of the argillite and its tendency to creeping [62, 63], the value of calculated compressive resistance decreases by 4 – 5 times, making up $R_c = 2.0 - 5.0$ MPa. Such low strength characteristics, the presence of complexity of the structure and poor contact with coal seams C_8^l and C_8^u allow to predict an unstable state of argillite (constituting the immediate and lower parts of the main roof), which will collapse immediately after its outcropping. To this thickness of unstable rocks, the thickness of a highly fractured coal seam C_8^u (0.66 – 0.8 m) should be added; then the total height of the probable roof arch will be 4.3 – 4.65 m. From the point of view of the vertical load on the fastening system of the 861 prefabricated drift, the weight of the specified rock volume creates a pressure of about 100 – 110 kPa, which is equivalent to a load of 450 – 500 kN per 1 long meter of mine working and corresponds to the boundary of the maximum load-bearing capacity of the TSYS supports. Thus, even without taking into account the stope works influence, the frame support needs to be strengthened with other fastening elements.

In the main roof of the seam C_8^l, a more significant change is observed in the structure along the length of the panel of the 861 longwall face with a very significant fluctuation in thickness of lithotypes and their replacement. In the lower part of the panel under the thick sandstone (4.3 – 15.6 m), siltstone occurs (with thickness of 5.2 – 9.1 m) replaced by argillite with thickness of up to 5.5 m. Here, the overlying sandstone serves as a source of moistening of argillite and siltstone, but, most likely, only for some part of their thickness, as the properties of waterproof rocks make it possible to leave a part of the thickness of argillites and siltstones in the naturally wet state without loss of strength properties. On the other hand, a fairly hard sandstone ($\sigma_c = 30 - 60$ MPa), due to its thickness, is able to restrict the displacement of the coal-overlaying formation, especially if it concerns the volume of the massif above the 861 belt entry (on the extraction panel boundary). Under the protection of sandstone, the relatively stable state of the underlying argillite and siltstone can be predicted. Thus, in the lower part of the panel, there are no developments of the arch of ultimate equilibrium (above the 861 belt entry) above the seam C_8^u. The exception is the area of replacement of argillite and siltstone in the area of PK 185 – PK 195, where it is quite likely the loosening of the argillite with thickness of up to 1 m thick and the load of the fastening system due to it up to 20 – 23%.

In the upper part of the panel in the main roof of the seam C_8^l, argillite with sufficiently high thickness (up to 8.6 m) and siltstone (up to 7.2 m) occur under the protection of sandstone with an medium thickness of 2.0 m. This part of the main roof rocks is predicted as stable at the lateral border of the panel in the area of location of the 861 belt entry.

The least stable state of the main roof is predicted in the area of the PK 128+1 – the PK 132+8, which was identified by the geological service of the mine in the mining-geological sheet (Fig. 8.1) and an explanatory note to it. This area is located in the region of periodic replacement of siltstone with water-flooded sandstone. Here above the seam C_8^u there is a layer of argillite with medium thickness (about 2.5 m) and sandstone (up to 2.5 – 3.0 m), but in the peripheral area of the sandstone it is of a lens-shaped form. The adhesion between the layers is weak, and the layer of argillite is also moistened with two watery lithotypes (bottom layer C_8^u, above – with sandstone). Its low strength characteristics with account of the effect of factors weakening the rock, allows to predict an unstable state with a probable load of the fastening system of the 861 drift up to 50 – 60%. To this the probability of an unstable state of sandstone at the edge of its lens should be added, which increases the possible load on the fastening system by half with the spread of the arch of ultimate equilibrium to 9 – 10 m. Such loads will obviously require the so-called 'deep' strengthening of the massif using the rope bolts and the creation of a fastening system with high load-bearing capacity.

This area of the 861 prefabricated drift, as the most dangerous from the viewpoint of intensity of the rock pressure manifestations, has been adopted for modeling by performing a computational experiment.

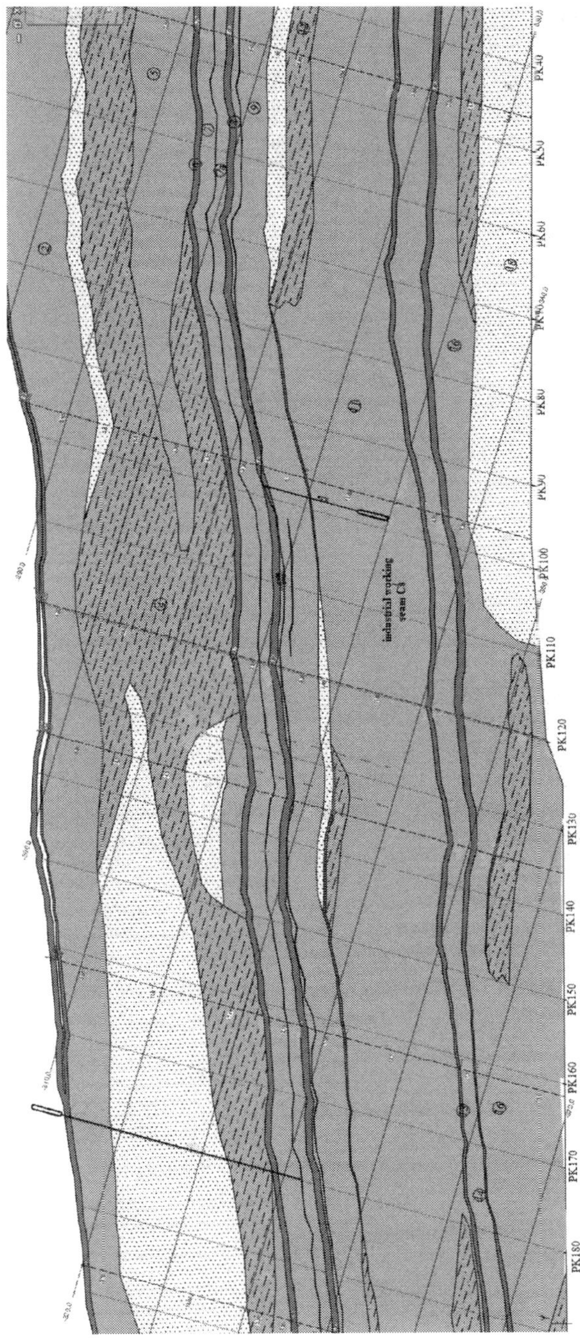

Fig. 8.1. Fragment of the mining and geological sheet of the area with the most unstable state of the roof rocks in the area of location of the 861 belt entry

8.2. DEVELOPMENT AND SUBSTANTIATION OF THE GEOMECHANICAL MODEL OF THE 861 BELT ENTRY

In accordance with the developed method for performing a computational experiment to study the displacement of the coal-bearing massif in the vicinity of the drifts in the zone of the stope works influence [55 – 58], a geomechanical model has been constructed (Fig. 8.2), which includes all necessary and sufficient positions for reflecting the state of the object:

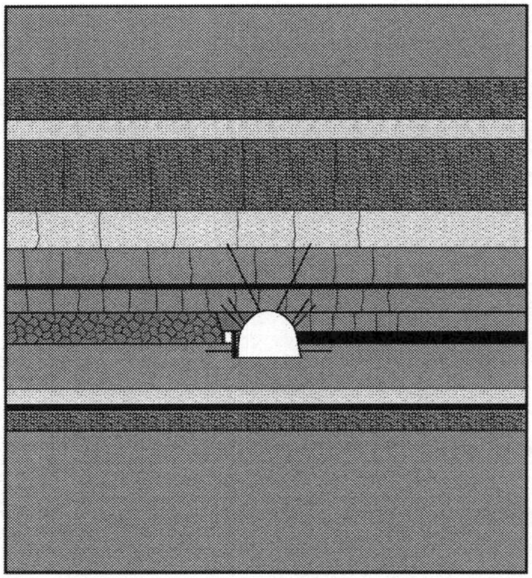

Fig. 8.2. The general geomechanical model of research into stability of the 861 belt entry

– the coal-bearing massif to the height in the roof, to the depth into the bottom, width along the strike of the seam, sufficient to reflect completely the parameters of the rock pressure anomalies in the zone of the stope works influence;

– the zone of scattered collapse from the side of mined-out space with parameters substantiated in [38, 55, 7, 94];

– the zone of hinged-block displacement above the mined-out space with thickness according to the recommendations [38, 94] and peculiarities of blocks interaction according to the studies [55 – 58];

– the zone of smooth deflection of the layers without discontinuity [38, 94], and modelling of contact disturbances along the laminations surfaces of adjacent lithological varieties, which is substantiated in [55, 67];

– the location of the 861 belt entry relative to the seam C_8^l according to the technical documentation for the panel mining of the 861 longwall face;

– adequate reflection of the structural and technological peculiarities of the fastening and security systems of the 861 drift according to the methodological developments [55 – 58, 67, 95].

The calculation of the stress-strain state (SSS) has been performed in an elastic-plastic formulation with the representation of the "stress – relative deformation" diagram of each lithological variety and fastening materials by a bilinear function. This allowed, when considering the plastic deformations of the geomechanical system elements, to avoid significant failures in the calculation technology and to increase the reliability of its implementation.

The mechanical characteristics of lithological varieties are taken from the mining-geological survey for the 861 belt entry; some of the missing deformation properties have been replenished according to the studies [62, 63, 74]. The influence of factors of fracturing, watering and rheology, weakening the rock, have been taken into account in accordance with the normative documents [53] and the results of studies [62, 63]. The mechanical characteristics of the fastening materials were calculated in accordance with the data of reference books [19, 96].

The scheme of fastening and security of the 861 belt entry is shown in Fig. 8.3. It has both common and distinctive features in comparison with the maintenance passport (the so-called "basic" scheme), which is conditioned by the following considerations.

Firstly, a particular way of mine working security has been assessed, providing for the setting of only one row of wooden prop stays with small diameter (10 – 12 cm) on the berm of the drift; it is approximated to the mine working contour at a distance of 0.7 – 1.0 m. There are no other security structures (for example, chocks, cast strips, etc.). Obviously, the low load-bearing capacity of this security method implies its short-term operation in a rigid mode in the near area behind the longwall face with the main task to facilitate (induce) the collapse of the immediate roof and the lower layers of the main roof. The preliminary assessment shows that a single breaker-prop row is able to withstand the collapse of the roof to the coal seam C_8^u inclusive. The further development of the main roof lowering (as the longwall face recedes) leads to the contortion (destruction) of a single row, but already at some distance behind the stope face. At the same time, the thickness of the collapsed rocks (including the seam C_8^u) is quite enough [94] to create an active bearing to the overlying layers of the roof, that is, a similarity of a rubble band with extensive dimensions is formed, which resists further to vertical rock pressure.

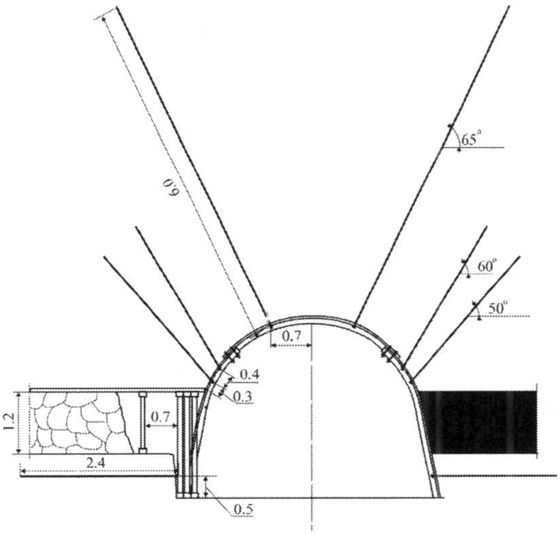

Fig. 8.3. The model of the recommended scheme for maintaining the 861 belt entry

Secondly, in order to bear the rocks of the main roof to the height of the hinged-block displacement zone in the border zone of massif from the side of mined-out space, three rows of lateral wooden prop stays of strengthening support should be set. Their limited-yielding mode of operation is intended to collapse the rock cantilevers in the layers above the seam C_8^u and, thereby, to reduce the concentration of both vertical and lateral rock pressure in the area of location of mine working.

These two positions in the scheme of maintaining the 861 belt entry are considered quite reasonable and correspond to the predicted (so far at the expert assessment level) nature of displacement of the coal-overlaying formation. Therefore, these technical solutions are used in the recommended scheme of fastening and security of the drift. Now let us focus on the specific aspects of the recommendations in relation to the basic scheme for maintaining the drift for its reuse.

Thirdly, the studied most dangerous area of the drift is characterized by an unstable main roof with a very likely development of stratifications and intensive deformation to a height of up to 12 – 15 m into the roof. In the case of formation of an extensive arch of limiting state, it is possible to predict high vertical and oblique rock pressure, since the width of the arch will definitely spread beyond the mine working limits. Therefore, it is evident the intention to strengthen maximally the frame support (along the contour of its arch) with the central wooden prop stays of the strengthening support (with a diameter 18 – 20 cm) in the amount of 4 pieces per a frame. With a high-quality setting of the central prop stays of the strengthening support, the load-bearing capacity of such a fastening system will be increased (presumably) by no less than 2 times. However, there is a number of disadvantages caused by the technological difficulties of high-quality setting of the prop stays, their dismantling during the subsequent bottom ripping, providing the necessary distances and clearances, reducing the effective section for air supply and, finally, high consumption of timber.

Fourthly, despite the significant strengthening of the frame support, there is no confidence in preserving the required drift section for its repeated use. The reason for this is the probability of exceeding by the vertical load (from the weight of unstable rocks) of total load-bearing capacity of the basic fastening system. An alternative to direct resistance to vertical loads is the strengthening of the roof rocks by combined roof-bolting systems that, on the one hand, "remove" part of the rock volume from the process of load formation, and, on the other hand, to create an armored and rock structure with high load-bearing capacity, resisting to the remaining volume of rock pressure. In this regard, the recommended combination of resin-grouted roof bolts and rope bolts implies the creation of an armored and rock plate in the roof with a thickness of up to 6 m, which (even when partitioning the roof into blocks) is capable of keeping up a load many times exceeding the total load-bearing capacity of the basic fastening system.

Fifth, the location parameters of the roof-bolting system (see Fig. 8.3) are designed to perform a number of tasks:

– to create a load-bearing thrust system, resting on a rock massif outside the width of mine working and protecting it from excessive rock pressure;

– for "expanding" the bearings of the load-bearing structure beyond the width of mine working, certain gradient angles of the resin-grouted roof bolts and rope bolts are provided so as to strengthen the rock volumes above the lateral prop stays of the strengthening support and above the breaker-prop row; then the wooden prop stays will not be "overworked" and the most likely collapse of rock cantilevers will occur beyond the boundary (vertically) of a breaker-prop row, while a relatively monolithic rock plate (slab) will be preserved in the roof of mine working;

– the rope bolts, connected with the peripheral part of the frame cap boards by means of flexible longitudinal crown runners, increase sharply a load-bearing capacity of the cap boards and allow them to be stable in the area of the longwall face "window" when dismounting the frame prop stays; this makes possible not to use the central prop stays of the strengthening support;

– the lateral horizontal roof-bolts in the immediate bottom of the seam C_8^l (in depth of bottom ripping of the drift) strengthen the rock volumes of the soft bottom to create a more rigid prop stays for load-bearing plate in the roof; moreover, the lateral roof-bolts in the immediate bottom of the coal seam contribute to the reduction of swelling intensity as a factor of section loss of the drift.

In view of the above considerations, a scheme has been developed for fastening and security of the 861 belt entry, included into the model of the general geomechanical system, which has become an integral part in the model of general geomechanical system, by which a computational experiment was carried out.

9. RESEARCH AND STRESS-STRAIN STATE ANALYSIS OF THE ROCK MASSIF ENCLOSING THE 861 BELT ENTRY AT THE BASIC AND RECOMMENDED SCHEMES OF MAINTAINING

It is well known [38, 55 – 57, 62, 63, 79] that the most intensive rock pressure manifestations develop after passing the stope face and this very area of the drift is very relevant in terms of its reuse. The algorithm for studying the behavior of the geomechanical system includes the primary assessment of anomalous SSS zones in the rock massif surrounding mine working and a subsequent study of the elements state of the basic fastening and security schemes in order to create the basis for the recommendations development.

9.1. ANOMALIES OF THE ROCK PRESSURE MANIFESTATIONS IN THE ROCK MASSIF IN THE VICINITY OF THE 861 BELT ENTRY AT THE BASIC SCHEME OF MAINTAINING

This paragraph presents a consistent study of the vertical σ_y and horizontal σ_x stresses components, as well as the stresses intensity σ.

9.1.1. ANALYSIS OF VERTICAL STRESSES COMPONENTS DISTRIBUTION

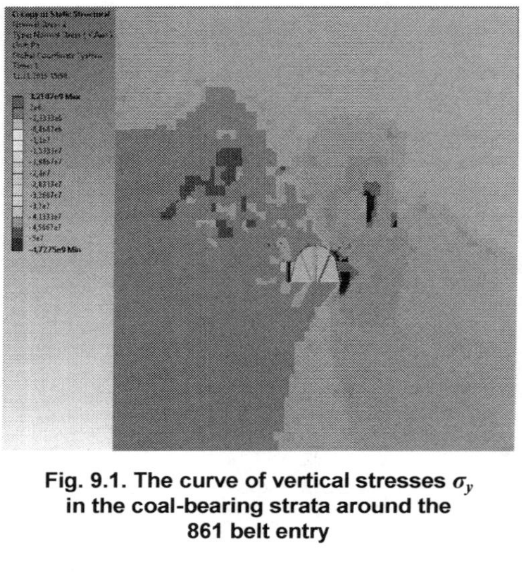

Fig. 9.1. The curve of vertical stresses σ_y in the coal-bearing strata around the 861 belt entry

The curve of vertical stresses is shown in Fig. 9.1 and is characterized by a number of parameters that qualitatively correspond to modern results of modeling the behavior of massif around extraction mine working behind the stope face [55 – 58]. In quantitative terms, the following peculiarities are observed.

From the side of the mined-out space, there is an extensive area of unloaded rocks of the roof and bottom of the seam C_8^l with a range of $\sigma_y = 2.3 - 6.7$ MPa. In relation to the initial non-hydrostatic state of virgin massif $\sigma_y = \gamma H$, the degree of unloading is 0.21 – 0.61. This area extends into the roof up to 16.6 m, affecting the top layer of sandstone, which is the boundary between the hinged-block displacement zone and the zone of smooth deflection of layers without discontinuity. Into the bottom of the layer, the zone of unloading extends to the entire depth of the model, i.e., more than 18.4 m. Inside the specified area, within the thickness of the hinged-block displacement zone, the places occur of full unloading ($\sigma_y \approx 0$) with the changeover to tension up to 2 MPa. Such a phenomenon does not contradict the existing concepts [38, 55, 94] on the displacement of the coal-bearing strata in the vicinity of the stope face.

The weight of stratifying and collapsing roof rocks creates a certain concentration σ_y in the areas of security and support elements location that resist the process of displacement: a single-row breaker props, the rows of lateral and central prop stays of the strengthening support. However, the concentration value is relatively low ($\sigma_y / \gamma H = 1.40 - 1.79$), which is caused by a small resistance reaction of only one row of wooden prop stays of the breaker props and due to the pliability the prop stays of the strengthening support; this behavior of a very 'lightweight' security structure was predictable in the technical documentation for maintaining the 861 drift.

From the side of virgin massif, a zone of lateral bearing pressure is formed. Its dimensions with a relatively low concentration coefficient ($\sigma_y / \gamma H = 1.40 - 1.79$) extend to the roof up to 14.2 m, and to the bottom – to the entire depth of the model. The concentrations of 1.79 – 2.18 level reach a height of 10.0 m in the roof and a depth of 5.3 m in the bottom. Such concentrations are already destructive (according to the factor of σ_y action) for argillite of the immediate roof and bottom, as well as argillite of the main roof, occurring below the sandstone. The width to the massif of the area of possible weakening reaches 6.6 m and it is quite likely the formation of considerable oblique and lateral loads. In this regard, it becomes relevant the combined strengthening (rope and resin-grouted roof bolts) of the roof rocks and the rocks of mine working sides, and also the strengthening of the frame prop stay in the lateral direction.

The concentrations of $\sigma_y / \gamma H = 3.4 - 6.0$, which definitely destroy the border rocks, are located in the side of mine working to a width of up to 1.8 m and to a height of up to 4.5 m. Of particular danger is the limited area with $\sigma_y / \gamma H \geq 4.5 -$ 5.0, the dimensions of which are 1.9 m in height and 0.8 m in width. Its negative impact is in the location – under the coal seam at the depth of the bottom ripping and further under the bearings of the frame prop stays. Here, the destroyed rocks create (from lateral pressure) a high bending moment, acting on the lower part of the prop stays, on the one hand, and, on the other hand, they create conditions for the intensive pressing of bearings of the prop stays into the bottom of the drift with a corresponding loss of its section. The revealed factor also indicates the expediency of strengthening the rocks of the immediate bottom (in depth of drift ripping) by horizontally placed roof-bolts and their connection by means of pliable binders with the lower part of the prop stays to increase their resistance to lateral loads.

Of interest is an area of concentration $\sigma_y / \gamma H = 3.8 - 6.0$, which is somewhat remote from mine working, located in the sandstone at the lateral boundary of the zone of inelastic deformations (according to the data of the mining and geological survey), approximately coinciding with the lateral boundary of the hinged-block displacement zone and the stope works. Sandstone, as a more rigid and hard rock, is a stress concentrator because of its increased resistance, and this is in complete agreement with the existing concepts in rock mechanics. But, this area with a height of up to 4.5 m and a width of up to 1.4 m, with its destruction induces a loss of continuity of the higher-lying rock layers; they can be deformed quite freely and the thrust system of rock blocks (in the hinged-block displacement zone) sharply loses its resistance to the rock pressure. Then, the entire vast volume of rocks in the main roof (up to 10 – 12 m high) will create with its weight the load on the fastening system (several times higher than the load-bearing capacity of the frame support), rather than resists to it at least partially. Therefore, it is critically important to preserve the conditions for joint deformation of rock zlayers and blocks by means of their binding (for example, with rope bolts) into a single load-bearing system. According to the analysis of the curve σ_y, the strengthening of the roof rocks (up to the boundaries of the identified area σ_y) can fully be provided by rope bolts of 6 m length, located at an angle of 60 – 70° to the horizontal. The established factor also testifies the use of the so-called "deep" strengthening of the rocks of the immediate and main roofs of the seam C_8^l.

In terms of assessing the stability of bottom rocks (according to the factor of σ_y action), it is possible to predict an increased swelling intensity from the side of virgin massif, where lateral bearing pressure plays a primary role in this process.

All the noted features of σ_y distribution indicate the need for significant strengthening the frame support of the drift, which has been done in the basic variant of its maintenance; however, in our opinion, it is more expedient here not to withstand the rock pressure with the help of numerous central wooden prop stays of the strengthening support, but to strengthen the adjacent massif with combined roof-bolting systems, which excludes vast rock volumes from the process of forming the load on the fastening system of the 861 belt entry.

9.1.2. DISTRIBUTION OF HORIZONTAL STRESSES COMPONENTS

Further on, the distribution of horizontal stresses σ_x is analysed, the diagram of which is shown in Fig. 9.2. The horizontal component most clearly reflects the deflection of the rock layers and its direction, which is quite clearly observed in the main roof of the drift. The sandstone layers are intensively loaded as more rigid and hard lithotypes. Thus, in a sandstone located at a height of 5.0 – 5.5 m from the contour of mine working arch, an extensive area of concentrations of compressive stresses σ_x is formed (by 8.7 – 13.2 times higher than the initial state of virgin massif), which extends along the thickness of the layer up to 1.8 m, and along the plane of bedding – up to 8.4 m. The deflection of the sandstone layer is directed towards mine working, and the high concentration of σ_x indicates its active resistance to rock pressure. In case of the sandstone stability loss, the overlying siltstone (with a high probability of water saturation) can also be intensively weakened (there will not be sufficient backing from below). And then a variant of the development of height of the arch of ultimate equilibrium to the next (in height) sandstone layer is possible, and this distance is 14.0 – 14.5 m to the mine working contour. In such a negative case, not a single, even strengthened, frame support can resist to the rock pressure.

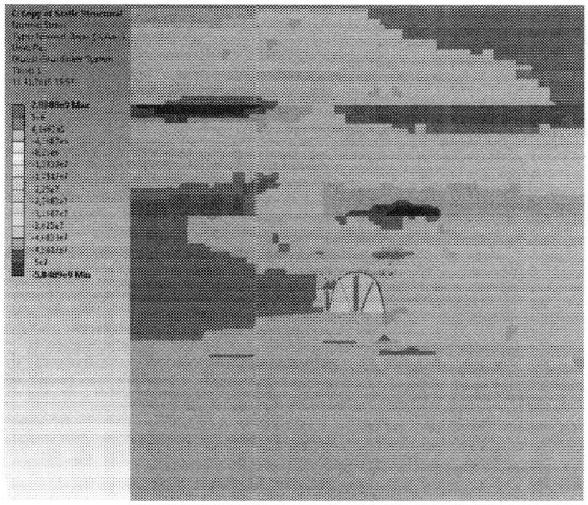

Fig. 9.2. The curve of horizontal stresses σ_x in the coal-bearing strata around the 861 belt entry

To increase the stability of sandstone, it is necessary to create an appropriate bearing from the side of underlying layers and most likely from the water-flooded argillite layers, as well as from the side of a very fractured coal seam C_8^u. These rocks are mainly in varying degrees of unloading from stresses σ_x, which indicates their low resistance to the rock pressure and the possibility of providing any significant backing for the overlying sandstone. Here also the low deformation characteristics of argillite and coal is a decisive factor, which is further reduced due to fracturing. It is possible to increase the rigidity of these lithotypes and their resistance to deformation by combining separate layers and blocks into a single load-bearing structure with the use of rope bolts. Owing to their length, the rope bolts are able to connect with each other the rock layers with a thickness of 4 – 5 m or more, and the rigidity of any plate (arch) increases in proportion to the square of its height [96]; consequently, it is possible to form a sufficiently rigid armored and rock structure, which will prevent the development of an arch of ultimate equilibrium to a considerable height into the main roof.

Thus, according to the factor of horizontal stresses action, it is advisable, in our opinion, to strengthen the roof rocks (up to the lower sandstone layer) with rope bolts; their direction in the roof (60 – 70°) is determined by the coordinates of location (relative to the drift) of σ_x concentrations area in the sandstone.

In the side of the drift from the side of mined-out space, an area of almost completely unloaded rocks is formed with occurrence of tensile stresses σ_x up to 3 – 5 MPa. Considering that a zone of uncontrolled collapse and a zone of hinged-block displacement are located here, the action of tensile stresses σ_x even more intensifies the disintegration of rocks and creates something like a rubble band of extensive sizes. Therefore, the assumption (implemented in the technical solution of the mine) has the right to exist about the collapse (caused by single-row breaker props and three rows of side prop stays of the strengthening support) of the immediate roof and lower layers of the main roof in order to create a backing when lowering the overlying roof. Moreover, there is a relatively low gradient of σ_x change in height of the roof (from the side of mined-out space) up to the coordinates of the sandstone occurrence at the boundary of the zone of layers smooth deflection without discontinuity. This relative homogeneity of σ_x distribution (with the exception of hard sandstone) indicates the low-intensity deflections of the main roof layers, and, consequently, the similarity of the rubble band of the collapsed rocks performs its functions of creating a proper backing.

In the side of the drift from the side of virgin massif against the general low concentration of σ_x (in the range 1.0 – 2.8), two local areas are of interest with a concentration coefficient σ_x up to 4.0 – 5.0. The upper area is located near the spring of the arch (yielding joists of the frame). Its strengthening (by resin-grouted roof bolts as in the basic variant of fastening) is completely substantiated, since it creates a more holistic bearing, which supports the overlying rock layers; in addition, the strengthening of this area reduces the negative impact of the oblique load on the frame. The lower area of σ_x concentration is located near the bearings of the frame prop stays and, in our opinion, it also needs the strengthening by the roof-bolts in order to: reduce the lateral rock pressure in the depth of the immediate bottom ripping and prevent (at least partial) the pressing of the frame prop stays into the bottom of the drift.

In the bottom of the drift, a sufficiently uniform field σ_x is observed, which breaks the local areas of unloading and small areas of σ_x concentration, caused by different-sized rigidity of argillite, sandstone and seam C_8^l in case of their deflection in the direction of the mine working cavity. Therefore, in general, it is possible to predict according to the factor of σ_x, quite moderate swelling of the bottom rocks, but with a predominance from the side of massif.

9.1.3. RESEARCH OF STRESSES INTENSITY

The revealed peculiarities of the massif SSS around the 861 belt entry are generalized by analyzing the distribution of the integral characteristic – the stresses intensity σ, the curve of which is shown in Fig. 9.3. The main features of the σ distribution are as follows.

In the roof of mine working the concentration of σ extends up to the sandstone, occurring at a distance of 14.0 – 14.5 m from the drift. The area of σ concentrations has an irregular shape with a predominance from the side of virgin massif in the adjacent layers of the roof and a more symmetrical shape in the remote layers of the roof. The dimensions of the area make it possible to predict the development of high rock pressure on the fastening system; the shape of the area indicates the predominance of the oblique load from the side of virgin massif. This, according to the curve of σ, is conditioned by two main factors.

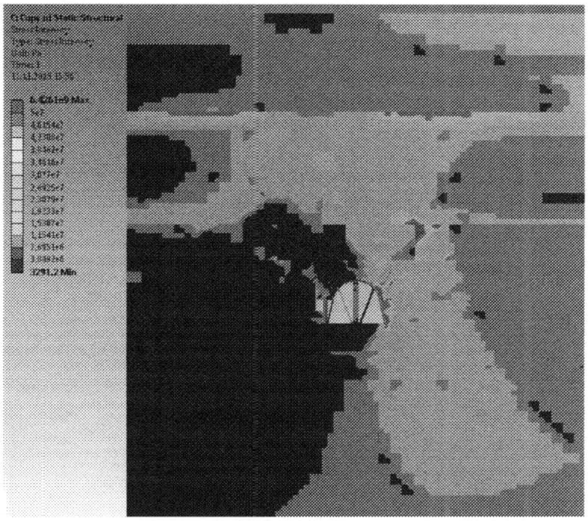

Fig. 9.3. The curve of stresses intensity σ in the coal-bearing rock strata around the 861 belt entry

The first factor – from the side of mined-out space, in the adjacent layers of the roof, an unloading area is formed up to 8.2 m, where the reduced rock pressure has been formed through the yielding mode of operation of the extensive zone of collapsed rocks. Therefore, it is fair to say about the expediency of a technical solution to create a single-row breaker props and three rows of side prop stays of the strengthening support for the protection of mine working.

The second factor is the formation from the side of virgin massif (at a distance of 4.7 – 6.5 m) of a zone with high concentrations of $\sigma = 45$ – 70 MPa, where the destruction of rocks definitely occurs. If this volume of broken rocks will have the opportunity for relatively free expansion and movement due to stratification, deformation and lowering of adjacent rock layers and blocks from the side of mine working, then in the rock layers the thrust load-bearing system is not formed, but there a gradual (step-by-step) lowering occurs and collapse of the layers onto mine work support and the border rocks. Then, the most of the volume of disturbed rocks in the roof will create a differently vectored load on the support. Therefore, there is a need for a "deep" strengthening of the roof rocks up to the zone of destructive concentrations, so that a relatively holistic armored and rock plate in the roof retains a horizontal thrust and, owing to it has such a load-bearing capacity that withstands not only its own weight, but also the pressure of the overlying roof rocks. Then the fastening system will be protected from excessive rock pressure.

The geomechanical processes described in the formation of an armored and rock plate in the roof are unattainable without creating a reliable backing from the border rocks in the side of mine working from the side of virgin massif. Here, it is clear an impact of self-destructive concentrations of σ across the width of 0.7 – 0.9 m, as well as hazardous concentrations across the width of up to 1.8 – 3.0 m, which are capable to weaken the rocks of the immediate roof and the bottom rocks of the coal seam along the height of the upper and the depth of the lower ripping of the drift. The task of strengthening the immediate roof is performed with the use of two resin-grouted roof bolts in accordance with the passport for maintaining the drift, and for strengthening the immediate bottom it is necessary to introduce an additional horizontally located roof-bolt along the depth of the lower ripping.

With regard to the rocks of the bottom (according to the stresses intensity distribution factor), the previously noted peculiarity of the predominant swelling development from the side of virgin massif is confirmed, where the bearing pressure in the side of mine working contributes to the concentration of σ and weakening of the immediate bottom to a depth of 2.1 m. In the rest of the mine working width, there is an unloading area σ and the bottom weakening is not predicted.

Summing up the results of the performed analysis, it is necessary to focus on the prediction of a sufficiently intensive development of geomechanical processes of displacement in the coal-overlaying formation and active manifestations of rock pressure against their extreme asymmetry relative to the vertical axis of mine working. Extensive and remote areas of SSS anomalies lead to the refusal of the "standard" methods of resistance to rock pressure by strengthening the fastening system, and involving the methods of combined roof-bolting strengthening of adjacent roof rocks to a distance of up to 6 m and more to create armored and rock load-bearing structures.

9.2. RESEARCH AND STRESS-STRAIN STATE ANALYSIS OF COAL-BEARING MASSIF AROUND THE 861 BELT ENTRY AT THE RECOMMENDED SCHEME OF ITS MAINTENANCE

It is generally accepted that one or another scheme for maintaining the mine working has some influence only on the border rocks, and the more distant areas of massif are not exposed to any significant impact. Nevertheless, the need to assess the SSS of the surrounding massif at the recommended fastening scheme is dictated by the fact that this scheme uses the rope bolts, which perform the so-called "deep" strengthening of rocks, and this case needs to be tested.

The analysis of the SSS of an adjacent massif has been made on three components of stresses: vertical σ_y, horizontal σ_x and stresses intensity σ.

9.2.1. ANALYSIS OF DISTRIBUTION OF VERTICAL STRESSES COMPONENTS

The curve of vertical stresses, shown in Fig. 9.4, is characterized by the following peculiarities of anomalies distribution of vertical rock pressure around the 861 belt entry.

From the side of mined-out space, a sufficiently homogeneous field of σ_y distribution is observed, which is changed mainly in the range of 2.3 – 6.7 MPa, and this indicates the unloaded state of the rocks with a value of 0.21 – 0.61 from the initial value γH of virgin massif. The most uniform distribution of σ_y occurs in the bottom rocks over the entire depth (not less than 18.4 m) of a model, and in the lateral direction extends to its vertical boundary and occupies a width of 17.1 to 20.9 m. In the side of mine working in height of up to 7.7 m, there are local foci $\sigma_y = 11.0 - 15.3$ MPa, which influence on the fastening and security systems of the drift. Here the concentration of compressive stresses σ_y is only 1.0 – 1.39 and only the upper value of this range corresponds to the compressive resistance of argillite to the rocks of the immediate roof and bottom. There is another very limited area above a single row of wooden prop stays of the security system, where $\sigma_y = 15.3 - 19.7$ MPa; in this area, weakening of soft argillite (compressive resistance range is 9 – 20 MPa) is more likely, but its small size (in height up to 1.4 m, in width – up to 0.4 m) do not have a significant impact on the overall unloaded state of the immediate roof rocks from the side of mined-out space. In the rocks of the main roof, against the unloading, mainly at the level of 0.21 – 0.61, rather limited tensile areas of up to 2 MPa are formed, along with even more local areas of compressive stresses σ_y concentration to a value of 1.39. These areas of vertical stresses of different signs have arisen due to the partitioning of rock layer of the main roof by fractures into blocks and their interaction with each other during the process of lowering into mined-out space. Here, the main factors of the different "outbursts" of σ_y are contact stresses over the surfaces of the interacting blocks and their flexure strain under the influence of overlying rock layers. In general, these local areas of the component σ_y pertur-

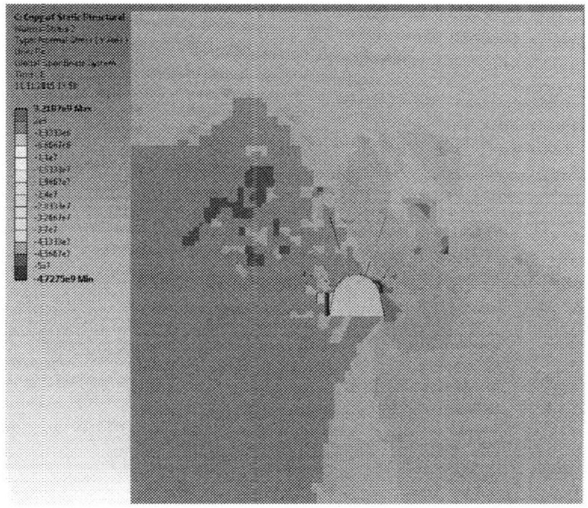

Fig. 9.4. The curve of vertical stresses σ_y in the coal-bearing strata around the 861 belt entry at the recommended scheme of its maintenance

bations do not have a significant impact on the overall picture of the unloaded state of the main roof rocks to a height of up to 16.6 m.

From the side of virgin massif, a zone of lateral bearing pressure is formed, which has the following peculiarities of concentrations distribution of the compressive stresses σ_y. The manifestations of lateral bearing pressure are distributed in the roof above the mine working to a height of up to 13.1 m. Here the concentration of σ_y is only 1.39 – 1.79; the area is extended into the roof and has an irregular shape with asymmetry in the direction of virgin massif; only in certain local areas, the concentration of vertical stresses reaches a value of 2.18, and in the absolute value it does not exceed 24 MPa. The area of σ_y concentrations is restricted by the thickness of the zone of hinged-block displacement, not reaching the sandstone, located at a height of 14.4 m from the arch of the drift. Comparison of the strength characteristics inside this area with the acting concentrations of σ_y indicates that mainly the argillite of the immediate roof is exposed to weakening, and the coal seam C_8^u, sandstone and siltstone of the main roof are in the prelimiting state, which is sufficiently stable.

The similar conclusions can be made relative to the other areas of the lateral bearing pressure, located outside of some zone near mine working, which is formed in its side and occupies a position in height of up to 9.7 m, and in width – up to 7.2 m. In this area, the concentrations of σ_y 1.79 – 2.18 level are destructive for argillite of the immediate roof and bottom, and the coal seam C_8^u, interlayer C_8^{l0}, sandstone and siltstone of bottom remain in a holistic state. Inside the studied area, a subarea is formed with a concentration of σ_y in the range of 3.75 – 4.15 to a width of up to 1.2 m and over the entire height of the drift, where definitely the intensive destruction of border rocks of the immediate roof and bottom, as well as the weakening of the coal seam occur. This subarea needs to be strengthened with resin-grouted roof bolts in both the roof and the bottom of the seam, which is implemented in the recommended scheme of fastening the drift.

An attention should be paid to another area of σ_y concentration, located in the main roof and extending in height of up to 7.0 m and in width of up to 3.1 m. This area includes the coal seam C_8^u, argillite and sandstone of the main roof and is 2.5 to 3 m away from the contour of mine working in the direction of virgin massif. Its function is as follows. *Firstly*, the area is located within limits of reach of strengthening by rope bolts or strengthening of those rock layers of the main roof that are included in its composition. *Secondly,* it is from these rock layers of the main roof the load-bearing armored and rock structure is formed (by rope bolts), designed to protect the frame support of the drift from excessive rock pressure. *Thirdly,* if there is a significant discontinuity of the presented lithotypes, then the width of this area of 3.1 m is quite enough to cause such shifts in the thrust system from the rock blocks of the main roof, at which it would lose stability; then the task of forming an armored and rock plate in the roof would not have been completed and the whole load of the weight of unstable rocks would have effect on the frame support. Such a negative situation is considered unlikely on the basis of a comparative analysis of the field σ_y distribution and the strength characteristics of lithotypes in the studied area. Indeed, the vast majority of the area is experiencing $\sigma_y = 19.7 - 24.0$ MPa and only in certain limited areas of $\sigma_y = 37 - 46$ MPa. Consequently, it can be concluded that the upper layer (sandstone) and the underlying seam C_8^u of the studied area will be in a stable state (according to the factor of σ_y action), since their compressive resistance is by $1.3 - 3.0$ times higher than the value of vertical stresses in the main volume of rocks. The argillite located between the coal seam C_8^u and sandstone, will be in the initial stage of weakening, so it is advisable to strengthen it by connecting the rope bolts into a single load-bearing structure with more stable adjacent lithotypes.

The above considerations on the analysis of the field of vertical stresses distribution sufficiently substantiate the expediency of strengthening the main roof by rope bolts. As for the resin-grouted roof bolts, their positive strengthening effect contributes to increasing the stability of the immediate roof rocks in two areas of location of relatively low concentrations of σ_y on the lower part of the end sections of the armored and rock plate and in the sides of mine working: above the security system from the side of mined-out space and from virgin massif. These areas serve as a kind of backing for the armored and rock plate and the degree of their stability depends on the efficiency of the load-bearing armored and rock structure to restrict the rock pressure manifestations in the belt entry.

9.2.2. DISTRIBUTION OF HORIZONTAL STRESSES COMPONENTS

Further on, consider the peculiarities of the distribution field of horizontal stresses σ_x (Fig. 9.5) at the recommended scheme of maintaining the 861 belt entry. It is appropriate here to recall that the curve σ_x rather clearly reflects the flexure strain of the rock layers and their direction.

From the side of mined-out space, the "deep" unloading of rocks from the action of horizontal stresses is observed along the entire thickness of the uncontrolled collapse and the hinged-block displacement. If not yet to take into account the rock cantilevers hanging above the mined-out space, then the remaining area is characterized by a uniform field of tensile stresses $\sigma_x = 0.5 - 5$ MPa; in a harder and more rigid sandstone, the tensile stresses σ_x increase to $10 - 15$ MPa and definitely divide it into rock blocks by vertical tension cracks. In the area of hanging rock cantilevers, a predominantly initial stress state is established with separate local areas of tensile stresses σ_x action of up to 5 MPa and the concentrations of compressive stresses σ_x of up to 13 MPa, which is conditioned by some bending of the rock layers in the process of their hanging above the mined-out space. Therefore, a relatively uniform field σ_x is established with limited anomalies within the immediate roof and underlying layers of the main roof (up to the sandstone); here, the value σ_x is close to the initial state of virgin massif and extends above mine working both in the direction of the hanging rock cantilevers and towards the massif at a distance of up to 4 – 5 m from the mine working contour.

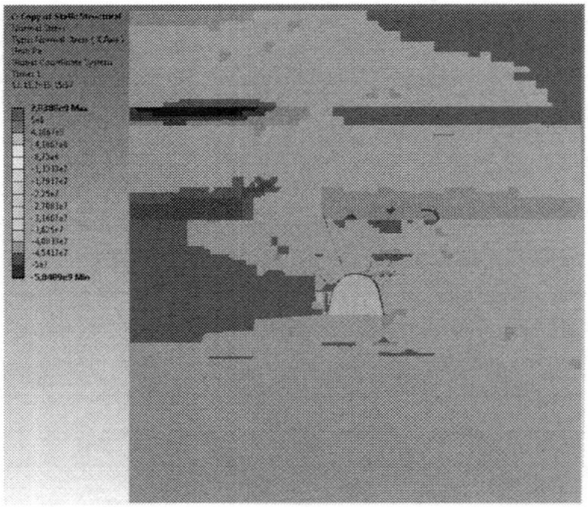

Fig. 9.5. The curve of horizontal stresses σ_x in the coal-bearing strata around the 861 belt entry at the recommended scheme of its aintenance

Such a distribution of σ_x in the immediate roof and lower layers of the main roof uniquely affects their stability, if to consider the components σ_x and σ_y in combination with each other. In the rock cantilevers hanging above the mined-out space, the absence of significant concentrations σ_x is combined with unloading σ_y, and even soft argillites do not experience destructive efforts. On the contrary, from the side of virgin massif there is a lateral bearing pressure with rather high concentrations of σ_y; under these conditions, the action of relatively low compressive stresses σ_x contributes to an increase in the difference $\sigma_y - \sigma_x$ and, correspondingly, to an increase in the probability of weakening a certain volume of border rocks.

It should be especially focused on the distribution of σ_x in the sandstone of the main roof, which actively resists to the rock pressure due to its hardness ($\sigma_{compr} = 30 - 60$ MPa) and thickness of 3 m. These parameters of the sandstone indicate its sufficient rigidity, owing to which the reduced flexure strains load the underlying rock layers to a lesser extent (a relatively homogeneous field σ_x has previously been noted) and limit the deflections of the overlying layer of siltstone: the field σ_x in the siltstone is close to the state of virgin massif.

This behavior of siltstone is also favored by the overlying sandstone with a thickness of 1.7 m; despite the fact that this sandstone belongs to the group of lithotypes of medium thickness, it also makes a certain contribution to the resistance to displacement in the coal-overlaying formation.

The sandstone layer adjacent to mine working, when resisting the rock pressure, forms a cantilever of 6.0 – 6.5 m in length from the mine working contour. This cantilever protects the underlying rock layers without significant field σ_x perturbations, which contributes to a decrease in the intensity of the oblique and lateral load action on the fastening system from the side of mined-out space.

From the border of the cantilever, in the direction of the mined-out space, tensile stresses up to 5 MPa are acting (along the most part of the sandstone thickness), and in its lower part – up to 10 – 15 MPa. Given that the resistance of sandstone to tension $\sigma_{compr} = 1.6$ MPa, it is possible to predict in this area the separation of sandstone into smaller pieces and blocks.

Along the length of cantilever the tensile stresses σ_x are practically absent, and compressive stresses σ_x by 6 – 12 times less than the compressive resistance of sandstone, – the console is in a holistic state.

Above the mine working in the lower part of sandstone thickness, a concentration of compressive stresses σ_x is formed, which indicates its some deflection into the cavity of mine working; the vast majority of the rock volume in this concentration area σ_x is in a holistic state due to the increased strength properties of the sandstone. Therefore, when sandstone is involved into a single armored and rock structure (by means of the rope bolts), its sufficiently stable state will contribute to increasing the load-bearing capacity of the entire volume of the roof rocks from the mine working contour to the sandstone location; this thick rock plate, by our predictions, is capable to take up the most of the vertical rock pressure.

A more distant into the roof and less thick layer of sandstone will obviously be exposed to partial weakening by small (up to 2 – 3 MPa) tensile stresses σ_x from the side of virgin massif and by the limited concentration area of compressive stresses σ_x from the side of mined-out space.

In vast volumes of rocks from the side of virgin massif, a very homogeneous field of $\sigma_x = 8 - 13$ MPa acts, which is not able to weaken any of the lithotypes located there.

In the bottom of the drift, against a relatively uniform distribution of σ_x over long distances into the depth of massif in its part, which is the nearest to mine working, there are local "outbursts" of low tensile stresses σ_x up to 2 – 3 MPa and compressive stresses σ_x up to 10 – 15 MPa; but, they are typical for the harder lithotypes (sandstone and interlayer $C_8^{l_0}$) and do not play any significant role in the development of swelling in the bottom of the drift.

In general, according to the factor of horizontal stresses σ_x action, it can be stated that the state of the rock massif around the drift is fairly stable when using the recommended scheme of its maintenance.

9.2.3. RESEARCH OF STRESSES INTENSITY

At the final stage of the SSS study of the massif around the 861 belt entry using the recommended scheme for its maintenance, the distribution field σ stresses intensity has been analyzed, the curve of which is shown in Fig. 9.6.

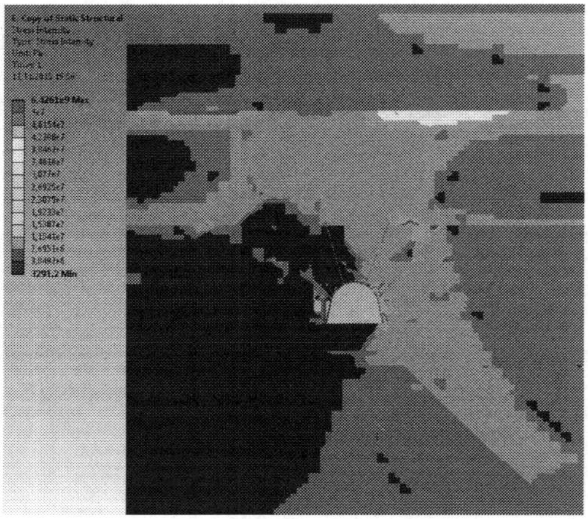

Fig. 9.6 The curve of stress intensity σ in the coal-bearing strata around the 861 belt entry at the recommended scheme of its maintenance

From the side of mined-out space, there is an extensive unloading area σ, extending in the sides and bottom to the corresponding boundaries of the model, and in the roof – mainly to the height of sandstone occurrence (about 7.5 – 8 m). This is quite natural, since the stresses intensity σ is a characteristic that generalizes all the components, and two of them (σ_y and σ_x) have the minimal values in the mined-out space.

In the rocks of the main roof, starting from sandstone with a thickness of 3 m and higher, a relatively uniform field of $\sigma = 15 - 30$ MPa is formed. In height, this area is located between two layers of sandstone, and in width it is extended to 18 – 25 m, and in the sandstones themselves – to the boundaries of the model (with a width of 50 m). The specified value (on average) is far from destructive for sandstone, and for siltstone it can be estimated either as close to the limiting state, or corresponding to initial stage of its weakening. The prospect of this siltstone state depends to a considerable extent on the degree of integrity of the lower and overlying layers of sandstone: in their holistic state, the conditions of the "constrained" siltstone deformation are preserved, with the exception (or greater restriction) of the possibility of displacing the rock volumes of siltstone in an arbitrary direction when the stage of its weakening occurs. Therefore, any movement of the rock volumes of siltstone will generate an appropriate reaction aimed at restoring its integrity.

In light of the above, it should be noted that the distribution of σ in the lower layer of the sandstone, in general, is many times lower than the value of its compressive resistance. An exception is the limited area, located from the side of virgin massif, where $\sigma = 35 - 50$ MPa approximately correspond to the hardness of the sandstone. However, the sizes of these areas of the sandstone state, which is close to the limiting, are small and do not have a decisive influence on its stability.

In the upper layer of sandstone, two areas of σ concentrations are observed. From the side of virgin massif, the concentration of σ extends across the width up to 12 m, and along the thickness of sandstone – up to 1.3 m. The value of the concentration $\sigma = 35 - 40$ MPa suggests to preserve its integrity. From the side of mined-out space, the dimensions of the concentration areas σ are somewhat less, but there is a subarea with a height of up to 0.4 m, where the value $\sigma = 45 -$ 55 MPa is close to or equal to the value of sandstone compressive resistance. Here, a certain weakening of the sandstone is possible, but its sufficient distance from mine working and location above the mined-out space does not allow predicting any significant influence on the formation of the load on the fastening system of the drift.

From the side of virgin massif, in the zone of the lateral bearing pressure, there is an extensive area of increased stresses $\sigma = 15 - 30$ MPa, affecting both the rocks of the main roof, and the main bottom of the seam C_8^l. With regard to their state, it is possible to predict the partial destruction of weak argillites, a limiting state and some weakening of siltstones and a stable prelimiting state of sandstones, coal seams and interlayers. In the zone closest to mine working, with a width of up to 1.0 m and a height of up to 4.5 m, the concentrations $\sigma = 38 -$ 42 MPa act with very local foci up to 50 MPa. This area is definitely exposed to weakening of the coal seam C_8^l and the destruction of argillites of the immediate roof and bottom, and, therefore, it is recommended to strengthen it with resin-grouted roof bolts. Approximately the same level of σ concentration acts in the previously mentioned area of the main roof, which is located in the lateral its part at a distance of 2.5 – 3 m from the mine working contour. It affects the lower layer of sandstone, which, due to its increased compressive resistance, is in a holistic state; and also it extends into the underlying argillite to most part of its thickness, and across the width of up to 2.5 m. As noted above, in this area it is necessary to restrict the horizontal movements of rocks in order to maintain the stability of the thrust system of the rock blocks. At the specified values σ, the destruction of argillite will certainly occur, but a positive moment is the relatively limited area of the destroyed state, which is deformed under constrained conditions, and also that part of the argillite's thickness is in a stable state and is capable of providing the thrust for block system of strengthened rocks of the main roof.

Summarizing the analysis results of the SSS components, it is possible to note the increase in the stability of some areas of adjacent massif when applying the recommended scheme for maintaining the 861 belt entry. To specify the changes in the massif SSS through the use of a new fastening system, a comparative analysis has been performed of the stresses components distribution relative to the variant of the basic scheme of the 861 prefabricated drift setting.

9.2.4. COMPARATIVE STRESS-STRAIN STATE ANALYSIS OF THE ENCLOSING ROCK MASSIVE WITH DIFFERENT SCHEMES OF MAINTAINING THE 861 BELT ENTRY

The degree of influence of the recommended fastening system of the 861 belt entry on the change in the SSS of adjacent massif are assessed by vertical σ_y and horizontal σ_x components, and the final conclusions have been made on the basis of differences in the stress intensity σ distribution, which are an integral parameter in the reflection of the state level (prelimiting, limiting and superlimiting) of one or another rock volume.

In the comparative analysis of the curves of vertical stresses distribution, the following points should be noted. From the side of mined-out space, no significant changes in the curve σ_y are observed. The same conclusion can be made regarding the bottom rocks of the seam C_8^l. Such a situation is quite expected, since neither from the side of mined-out space (the security system), nor from the side of the bottom rocks were made any changes (except for the lateral resin-grouted roof bolts in the depth of the bottom ripping) to the scheme for maintaining the 861 belt entry.

In the rocks of the immediate and main roof of the seam C_8^l, a number of changes is observed. Firstly, the previously continuous concentration area σ_y of 1.39 – 1.79 level is no longer integral, but consists of several fragments separated by a value of σ_y, close to the initial state of virgin massif. The volume of rocks of the main roof, which was exposed to the above concentration of σ_y, decreased at an average by 25 – 33%. Secondly, the concentration σ_y of the level 1.79 – 2.18, which weakens the argillite, at the recommended scheme of setting mainly in the immediate roof and compared with the basic variant of the drift fastening, the height of this area is reduced by 4.2 – 4.6 times if to take as the reference point the vertical coordinate of the arch keystone of a frame. Thirdly, the concentration value σ_y has changed significantly in the rocks of the main roof, located at a distance of 2.5 – 3 m from the contour of mine working from the side of virgin massif. Intensively destroying concentrations of 4.5 or more have practically disappeared here, and concentrations of the level 3.3 to 4.1 are also significantly restricted.

The revealed changes make it possible to predict a more stable state of the roof rocks in general and its lower layers, in particular, which, in our opinion, is caused by the formation of an armored and rock plate in the roof by rope bolts. This factor, in combination with resin-grouted roof bolts, made it possible to reduce the concentrations σ_y in the border rocks of the sides of the drift from the side of virgin massif. Thus, the area of deliberately destructive concentration of more than 4.5 units has disappeared at all, and with the basic fastening scheme, it occupied in the immediate bottom a width of up to 1.3 m and a height of up to 2.5 m. The same level of concentration in the spring area in the immediate roof is restricted in the sizes of distribution by 15 – 20%. Thus, according to the factor of vertical stresses distribution, there is an undeniable increase in the stability of a number of rock volumes around the 861 belt entry, at the recommended scheme of its fastening.

When comparing the fields of horizontal stresses distribution, two main changes should be distinguished against the general constancy of parameters of σ_x distribution for most part of the areas in the surrounding massif. Both differences relate to the adjacent and more distant sandstone in the space above the mine working and from the side of virgin massif.

The increase in the stability of adjacent sandstone (at the recommended scheme of the drift fastening) is conditioned by a general decrease in the concentrations of compressive stresses σ_x by 20 – 25% in the area occupying the lower half of its thickness and extending over the mine working to a distance of up to 8.9 m; here there are only three separate places of σ_x concentration up to 50 MPa, which are rather small in size and do not influence on the stable state of the sandstone.

In the more distant layer of sandstone, the concentrations of tensile stresses σ_x decrease (up to 15 – 30%), and in the underlying layer of siltstone, the tensile stresses disappear completely in the part of their thickness adjacent to the sandstone.

Thus, according to the factor of the horizontal stresses action, it should be noted a general increase in the stability of the hardest layers of the main roof. In our opinion, this was due to the formation by rope bolts of the thrust system with the rock blocks in the lower layers of the main roof.

At the final stage of the comparative analysis, the main differences in the distribution of stress intensity σ have been identified, which indicate a decrease in the intensity of rocks in the main roof, sides and bottom of the drift from the side of virgin massif.

In the rocks of the main roof, including the more distant layer of sandstone (the height of this area is up to 14.3 m) there was a decrease in the concentration of σ up to 40 – 60%, which has provided the prelimiting state of all the lithotypes located there, with the exception of argillite experiencing partial weakening.

In the remote layer of sandstone, the concentration of σ decreased by 19 – 43% and this ensured its integrity along its thickness. A similar decrease of σ occurred in the adjacent sandstone and in the underlying argillite in the lateral area (from the side of virgin massif), which was considered problematic from the point of view of the formation of a stable thrust system from the rock blocks of the lower layers of the main roof.

In the rocks of the immediate roof and bottom, the decrease of σ occurred by 10 to 12%, which somewhat has restricted the intensity of their destruction. In the part of the side rocks adjacent to mine working, the stresses intensity decreased by 30 – 40%, which changes the coal seam C_8^l from the superlimiting state to the closest to limiting state and reduces the degree of rocks destruction in the immediate roof and bottom of the coal seam.

In the rocks of the main bottom, in addition to decrease in σ by 10 – 15%, the width of their concentration area decreased by 50 – 80%.

The above results convincingly show a general decrease in the massif intensity around the 861 drift when using the recommended scheme of its maintenance. Therefore, it is reasonable to predict a decrease in the intensity of the rock pressure manifestations and the preservation of operational characteristics of mine working for its reuse.

9.2.5. STRESS-STRAIN STATE ANALYSIS OF A PARTING ROCKS OF SEAMS C_8^l AND C_9 IN THE AREA OF THE 861 BELT ENTRY LOCATION

In the Western Donbas mines, the coal seams in the series are mined jointly in two, less often – in three adjacent layers by descending order. At the same time, the stope works in the upper layer, as a rule, are ahead of the lower seam extraction by several panels for the separate mining of mining sites [94]. However, in very difficult mining and geological conditions, the descending sequence of mining of two adjacent layers should take into account the geomechanical processes of influence of the overworking on the state of parting rocks when extracting the lower layer. A certain disturbance in the bottom rocks of the upper layer during the passage of the stope face (front and lateral bearing pressure) in difficult conditions may cause the activation of rocks displacement in the coal-overlaying formation when mining the lower layer. A particularly dangerous situation arises when, in the process of mining the lower layer, the hinged-block displacement zone in the roof is connected to the area of disturbed bottom rocks of the overlying layer, which by this time has already been extracted [57]. Then, the height of unstable rocks increases many times, which includes: the zones of uncontrolled collapse and hinged-block displacement at the boundary with the mined-out space of the lower layer; the area of disturbed bottom rocks of already mined-out upper layer; the zones of uncontrolled collapse and hinged-block displacement in the mined-out space of the upper layer. In such conditions, an extensive area of stabilization of the rock pressure manifestations (around the already mined-out upper layer), which has long been in equilibrium, can again be removed from it with the corresponding vertical and horizontal movements of the massif to a considerable height (30 – 50 m and more) into the roof of the mined lower layer. The load from the displacement of stratum of this massif exceeds the load-bearing capacity of the support sections of the mechanized complex and, especially, of any fastening system of extraction mine workings; that is, an extremely dangerous emergency situation arises, the study of which has been carried out on the example of mining the two adjacent layers C_8^l and C_9 in the descending order at the West Donbas Mine of "DTEK Pavlohradvuhillia" PJSC.

For a quantitative assessment of the stress-strain state of parting rocks of the seams C_9 and C_8^l, a computational experiment has been performed using the finite element method, which simulates the conditions for mining the seam C_8^l after the completion of stope works in this area of the mine field along the seam C_9. The mining and geological situation has been studied in the area of a parting located above the 861 belt entry of the seam C_8^l. Based on the calculation results, the graphs have been constructed (Fig. 9.7) of changes in the concentration coefficients of vertical K_y and horizontal K_x stresses along the height h of a parting. The concentration coefficients are determined by expressions

$$K_y = \frac{\sigma_y}{\gamma H}; \; K_x = \frac{\sigma_x}{\lambda \gamma H},$$

where σ_y and σ_x – vertical and horizontal stresses, respectively; H – the depth of mine working location; γ – the weight-average unit specific gravity of the coal-overlaying formation to the earth's surface; $\lambda = \frac{\mu}{1-\mu}$ – coefficient of lateral thrust; μ – Poisson's ratio.

The dependence of the change in coefficient K_y of vertical stresses concentration is characterized by a number of peculiarities. Immediately, it should be noted that the above dependences $K_{y,x}(h)$ are constructed for the most intense area of the 861 belt entry located from the side of virgin massif, the vertical axis of which is spaced at a distance of 1.0 - 2.5 m from the contour of the drift.

Directly throughout the height of mine working, the concentration coefficient K_y varies in the range of 4.6 – 5.9 and certainly causes the weakening and destruction of rock, which is confirmed by comparing the compressive resistance of lithotypes occurring here and the value of stresses intensity σ according to the Mohr-Coulomb failure theory.

In the rocks of the immediate roof, because of weak compressive resistance of argillite (σ_{compr} = 14.5 MPa), its weakening develops throughout the height up to the coal seam C_8^u. It should be noted here that the height of the area of weakening and destruction is shown by the columns in the right-hand side of Fig. 9.7, but also it is necessary to consider a certain limitation of these areas along the strike of the seam. Along the thickness of the coal seam C_8^u, the weakening does not occur because its resistance ($\sigma_{compr} = 30$ MPa) exceeds the acting value of stresses intensity σ.

In the argillite of the main roof $K_y = 1.8 - 3.8$ and this concentration is sufficient for its weakening due to the weak compressive resistance ($\sigma_{compr} = 16$ MPa).

The area of weakening extends into the overlying sandstone, despite its relatively high (for the Western Donbas environment) hardness ($\sigma_{compr} = 45$ MPa), as there $K_y = 3.8 - 6.2$. But, the weakening of sandstone is not developed to its full thickness, which is caused by a decrease in the stresses intensity σ in the upper part of the sandstone thickness.

The overlying siltstone closes the hinged-block displacement zone and $K_y = 2.4 - 5.1$ acts in it. Despite the increased compressive resistance ($\sigma_{compr} = 21$ MPa) of siltstone, it is exposed to weakening along its thickness.

Above the siltstone there is a sandstone of medium thickness and, being a more rigid lithotype (in relation to adjacent lithological varieties), it takes up an increased rock pressure with a deflection towards the mine working. Therefore, in the upper part of its thickness, the weakening value of stress intensity σ develops. It should be borne in mind here, that a small concentration of $K_y = 1.4 - 25$ is considered in combination with significant horizontal tensile stresses σ_x (the coefficient K_x becomes negative to -2.6). And according to the Mohr-Coulomb failure theory, the combination of compression with tension is the most dangerous for the integrity of any rock.

The overlying siltstone is already in a stable state with a concentration coefficient K_y, decreasing from 1.4 to 1.0, that is, on vertical stresses σ_y there is a changeover to the state of virgin massif. A similar situation is for argillite, which represents the bottom of the seam C_9.

Thus, the absence of weakening in the most stress direction (vertically) is observed to a depth of 8.4 m in the bottom of the seam C_9.

The concentrations of horizontal stresses K_x play a partly subordinate role, but, nevertheless, are important in terms of formation of the destructive value of stresses intensity σ according to the Mohr-Coulomb failure theory. The value of the concentration coefficient K_x has higher values, since this dimensionless parameter is determined with respect to horizontal stresses σ_x of virgin massif, which are usually 40 – 50% of the vertical stresses σ_y. In addition, by reason of bending of the rock layers, rapid changes in the coefficient K_x occur even along the thickness of one lithotype, and the arising of tensile stresses σ_x in certain rock layers determines the negative values of K_x.

In the argillite of an immediate roof, because of its intensive bending, the coefficient K_x varies from 1.7 at the boundary with the seam C_8^l, then it increases to 3.8, and on the boundary with the seam C_8^u turns into negative values of $K_x = -1.6$. A negative value of K_x indicates a deflection of the argillite into the cavity of mine working, the maximum of bending stress σ_x is located at a distance of 1.5 – 2.0 m from the mine working contour from the side of virgin massif.

The coal seam C_8^u, despite its low thickness of 0.7 m, is experiencing an alternating concentration of K_x from -2.2 to 1.9. This is due to the increased compressive resistance of coal, which contributes to the ability to take up an increased load. The argillite, lying in the roof of the seam C_8^{ϵ}, is also bending intensively; here, the concentration coefficient K_x varies from -0.9 to 3.8.

The hardest and more rigid sandstone with a thickness of 3.0 m is most heavily loaded; therefore, the maximum value of the concentration coefficient $K_x = 11.8$, which decreases at the boundary with the underlying argillite to $K_x = 0.4$.

The thick siltstone overlying the sandstone is less prone to bending and the concentration coefficient in it varies in the range $K_x = 0.2 - 6.9$. At the upper boundary of the siltstone, the zone of hinged-block displacement is over and the interval of fluctuations of K_x is being narrowed due to less intensive deflection of layers without discontinuity. Although, in the hard and rigid sandstone, overlying the siltstone, the fluctuations of K_x remain significant: from -2.6 to 0.2. The changes of K_x along the thickness of the overlying siltstone are still significant: from -2.3 to 3.8. Then, in the argillite of the seam C_9 bottom, the concentration coefficient K_x gradually decreases from 1.8 to 1.0 as it approaches the seam C_9.

Thus, the dependences of changes in the concentration coefficients of vertical K_y and horizontal K_x stresses along the height of a parting are assessed, conditioned by the seam C_8^l mining. Here, the most interesting are not the concentrations $K_{y,x}$, as such, but the areas of weakening and destruction of rocks in the most intense area from the side of virgin massif. Next, an influence is considered of the previously mined seam C_9 and the corresponding concentration coefficients $K_{y,x}$ are shown in the form of graphs numbered 2 in Fig. 9.7.

In the argillite of the immediate and first layer of the main bottom of the seam C_9, the concentration K_y is maximum and varies from 3.5 at the boundary with the coal seam to 1.8 at the boundary with the underlying siltstone. The decrease (in depth of the bottom of the seam C_9) in the concentration coefficient K_x of horizontal stresses also occur smoothly without any perturbations from a value of $K_x = 3.9$ at the boundary with the seam C_9 to $K_x = 2.6$ along the planes of bedding with siltstone. It should be emphasized that along the entire thickness of the argillite, the intensity of stresses σ exceeds its low compressive resistance ($\sigma_c = 14.5$ MPa), which causes the weakening of the argillite along the entire thickness.

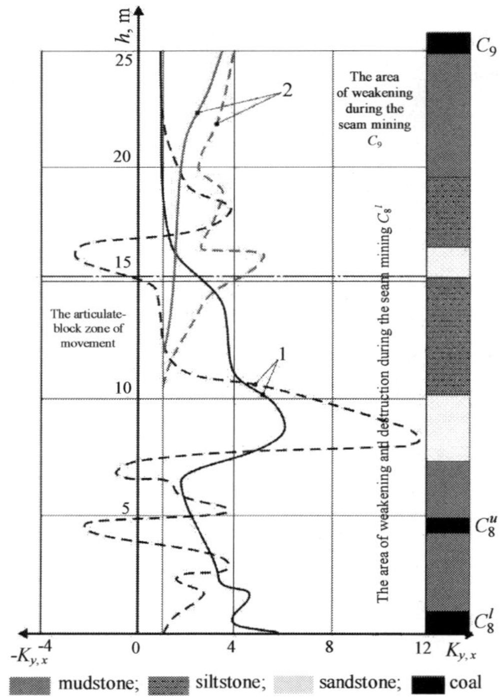

Fig. 9.7. Change in the coefficients of vertical K_y (——) and horizontal K_x (– –) stresses concentrations along the thickness h of a parting in the seams C_9 and C_8^l: 1 – when mining the seam C_8^l; 2 – when mining the seam C_9

The underlying siltstone is in a holistic state due to its increased strength, on the one hand, and on the other hand, a gradual decrease in concentration K_y from 1.8 to 1.6 at the border with a the sandstone lying deeper. The concentration coefficient of horizontal stresses along the thickness of siltstone varies in the range of $K_x = 2.5 - 3.5$; however, a combination of concentrations K_y and K_x is not capable of generating destructive stresses σ.

Further on, over the bottom depth, there is a sandstone, which is more loaded because of its increased strength and rigidity. Therefore, there is an outburst of concentrations of horizontal bending stresses: in the lower part of the sandstone thickness $K_x = 4.2$, while in the upper $K_x \geq 2.6$. The concentrations of vertical stresses gradually decrease from $K_y = 1.6$ to $K_y = 1.4$. The influence of the mined seam C_9 is such that it cannot cause the weakening of sandstone, in contrast to the impact of mining operations in the seam C_8^l.

Below the sandstone, there is a thick siltstone, which is the upper boundary of the hinged-block displacement zone. Along its thickness, the concentrations of $K_{y,x}$ gradually decrease to a value of 1.0, which characterizes the state of virgin massif according to the factor of impact of the C_9 seam mining.

As a result, the peculiarities have been revealed of the distribution (along the thickness of a parting) of the concentration coefficients of vertical K_y and horizontal K_x stresses with an appropriate assessment of the state of the constituent lithological varieties in terms of their integrity. From this point of view, an attention should be paid to the column on the right-hand side of Fig. 9.7, where the height is marked of the areas of weakening and destruction in each lithotype of a parting. These areas of disturbed rocks are not interconnected with each other, but the intervals of the entire massif are rather limited, which causes a certain probability of joining with each other of the disturbed rocks areas and the formation of a single zone of weakening along the thickness of a parting. Despite the fact that, along the strike of the seams, the area of disturbed rocks is rather limited, nevertheless, the obtained results change the concepts about the mutual influence of the jointly mined seams in the series of strata: according to the value of a parting thickness $h = 23 - 26$ m, the seams C_9 and C_8^l do not belong to the contiguous seams; nevertheless, their mutual influence is very likely and is represented in the form of the following mechanism for the geomechanical processes development.

Once again it should be noted, that the areas of weakening, shown in Fig. 9.7, are restricted along the strike to several meters, but there are also wider areas of disturbed rocks. Most of them are located in the zone of hinged-block displacement and the zone of smooth deflection of the layers without discontinuity during the C_8^l seam mining. The stable state of the rocks in these zones is determined by the horizontal forces of the thrust, especially of the rock blocks in the area of a parting which is the closest to mine working. With the destruction of a certain volume of rocks in the zone of hinged-block displacement, it becomes possible for relatively free horizontal movements of the rock layers and the thrust forces dramatically decrease; along with them, the stability of thrust rock structures decreases (or disappears at all). Their collapse is developed sequentially from the underlying layers to the overlying ones. In the process of collapse, the backing from the side of underlying layers disappears and under the influence of the vertical rock pressure, the overlying layers become unstable. According to such a scheme, there is a probability of a “chain reaction” of the collapse development along the entire thickness of a parting; then no one fastening system of the belt entry can cope with the vertical mining pressure and a serious emergency situation arises at the extraction site.

In conclusion, it should be noted that the performed studies show the possibility of significant interaction of adjacent seams (at their joint downward scheme of mining), even if they are not contiguous according to the existing classification. Therefore, in each specific complex mining and geological situation, it is recommended to assess the probability of stability loss of parting rocks under the conditions of a joint downward mining of two adjacent seams in the series of strata.

10. RESEARCH AND ANALYSIS OF THE STRESS-STRAIN STATE OF FASTENING AND PROTECTION BASIC SYSTEMS ELEMENTS OF THE 861 BELT ENTRY OF THE SEAM C_8^l

10.1. ANALYSIS OF VERTICAL STRESSES COMPONENTS

The vertical stresses distribution σ_y has been studied, the curve of which is shown in Fig. 10.1. The frame support cap board is under the action of alternating stresses σ_y, which indicates its multidirectional bending. Thus, in the central part of the frame arch with a length of up to 1.0 m, there is a concentration of σ_y, caused by the reaction of two vertically located central (wooden) prop stays of the strengthening supports, and this corresponds to the results of other studies [57, 58]. Despite the impact of rigid central prop stays, the bending of the cap board is directed toward the cavity of mine working: in the upper part of the SCP section, the distribution of σ_y is not critical (in comparison with the yield limit of St.5), but indicates a significant bending moment. When moving from the arch keystone of the drift towards the mined-out space, the stresses in the cap board decrease in absolute value and reach the minimum values ±10 – 20 MPa near the yielding joist of the frame. In the peripheral area of the cap board from the side of virgin massif, there is a change in the sign of the bending moment with the action towards the roof, and the value σ_y remains at the level like for the arch keystone. Such anomalies of σ_y in the cap board, obviously, are connected with the oblique load from the side of virgin massif and the SSS anomalies that arise in the main roof. These results indicate the predominant loading of the frame cap board from the side of virgin massif.

Fig. 10.1. The curve of vertical stresses σ_y in the elements of the basic scheme for maintaining the 861 belt entry

The previous conclusion about the dominant oblique load from the side of virgin massif has been confirmed during the study of the state of the frame prop stays. The prop stay from the side of massif is in an overloaded state for most of its length. A significant bending moment is not observed (with the exception of the area of the prop stay bearing at the height of 0.15 – 0.20 m), however, it should be considered that even a small bending moment is capable of disturbing the stable state of the prop stay. Therefore, it is expedient to increase the resistance of the frame prop stay in the lateral direction by means of its yielding mechanical connection with the lateral roof-bolts.

The frame prop stay from the side of the mined-out space is extremely unloaded: in the rectilinear part of the prop stay the values σ_y vary from 20 – 35 MPa of compression to 10 – 30 MPa of tension; a small bending moment does not constitute any danger. In the curvilinear part of the prop stay, the bending moment increases with the deflection towards the cavity of mine working, but the value of the tensile σ_y in the SCP section does not exceed 70 – 90 MPa, which is only 26 – 33% of the estimated yield limit of steel.

Thus, the asymmetry of loading the frame support is clearly observed, which makes an impact at the state of the resin-grouted roof bolts in the immediate roof. So, from the side of virgin massif, the roof-bolts are more loaded with tensile stresses σ_y, than the roof-bolts set from the side of mined-out space. This indicates more active displacements of the roof rocks from the side of virgin massif, which was previously noted in the study of the SSS of adjacent coal-overlaying formation. In detail, the relative length of areas of tension is 21 – 33% from the side of massif in comparison with 0 – 11% from the side of mined-out space. The location itself of areas of tensile stresses σ_y in the central part of the length of the roof-bolts indicates the predominance of displacements of border rocks over the displacements in more remote areas of rocks; that is, there is an active suppressing of displacement by resin-grouted roof bolts of the most weakened border part of rocks in the immediate roof. Therefore, the expediency of setting two pairs of roof-bolts in the area of the curvilinear part of the props stays is beyond doubt. Here we need, in our opinion, to make an insignificant modification of coordinates of the tail joints location of the roof-bolts, as well as their angles of gradient in accordance with the vectors of the greatest border rocks displacements.

Concerning the loading and efficiency of four central wooden prop stays of the strengthening support, the following has been determined. Two vertical prop stays and one inclined one from the side of mined-out space are exposed to a very uniform action (without noticeable bending moments) of compressive stresses $\sigma_y = 20 - 30$ MPa, which is 50 – 75% of the calculated resistance of pine to compression; that is, the prop stays are in the prelimiting state, performing their functions to resist vertical rock pressure. Similar is, at first glance, the result of a very weak (mainly up to 8% of the load-bearing capacity) stresses loading σ_y of an inclined prop stay from the side of virgin massif, because in this very direction the greatest oblique rock pressure acts. The reason for this phenomenon is in the nature of lowering the rock blocks of the lower layers of the main roof: the blocks rest on the border rocks from one side and on the central part of the frame arch (through the direct roof), from the other side. And in the middle of the span, where the load is minimal (on the immediate roof and frame), an inclined wooden prop stay is just set. In contrast, the prop stay of the frame support "collects" most of the load along the width of mine working and therefore is extremely loaded.

From the above results of vertical stresses σ_y distribution, it follows that it is not expedient to set an inclined wooden prop stay from the side of virgin massif; moreover, the remaining central wooden prop stays are not fully loaded and their resistance to rock pressure can be assessed as medium effective.

The direct opposite is the state of three rows of side wooden prop stays of the strengthening support and a breaker-prop row. All of them are overloaded, but there are differences. The side wooden prop stays of the strengthening support work either at the limit of the load-bearing capacity or beyond its limits. From the destruction, these prop stays are being "saved" by their yielding property, caused by contortion of wooden substrates (girders) and pressing into the bottom rocks of mine working. At the same time, the multiple overloading of the prop stays of the breaker-prop row unambiguously predicts their destruction after the longwall face recedes at a certain distance. The last result, obviously, is laid in the technical solution for the forced collapse of adjacent rock layers and the formation of similarity of a rubble band with extensive dimensions.

In general, the results of research on the vertical stresses σ_y distribution in the elements of the basic fastening system are as follows:

– asymmetric loading of the frame and resin-grouted roof bolts is observed with the prevailing rock pressure from the side of virgin massif; moreover, the frame prop stays need to be strengthened to resist lateral loads;

– the central wooden prop stays of the strengthening support are underloaded, and the inclined prop stay from the side of virgin massif is characterized by a very weak resistance, which calls into doubt the expediency of their usage to maintain the drift;

– the side prop stays of the strengthening support work very effectively for the forced collapse of the rocks in the immediate and main roof;

– a breaker-prop row performs its functions to induce the collapse of the immediate roof and create a 'bed' in order to bear the lowering coal-overlaying formation.

10.2. DETERMINATION OF HORIZONTAL STRESSES COMPONENTS

The second main component characterizing the state of the fastening system, is the horizontal stresses σ_x, the curve of which is shown in Fig. 10.2. Its analysis made it possible to obtain the following results.

In the frame cap board, along its length, there is a somewhat asymmetric nature of the compressive stresses σ_x distribution; the asymmetry itself is directed towards the virgin massif. In the peripheral areas of the cap board (near the yielding joists), the minimum compressive stresses $\sigma_x \leq 12$ MPa act with the occurrence of local areas of tension up to 50 MPa; here, there is a cap board deflection towards the roof rocks. When moving to the central part of the arch, the compressive stresses increase to 200 – 250 MPa, and in certain areas it reaches the estimated yield limit of SCP steel. The most important peculiarity of σ_x distribution is the action of an alternating bending moment along the entire length of a cap board: there are five areas of the bending moment maxima with a constant alternation in the direction of the cap board bending. Moreover, in its central part with a length of up to 1.3 m, the level of σ_x is such that the occurrence of plastic bending is possible. That is, despite the reaction of two vertically located central wooden prop stays of the strengthening support, the frame cap board is in a state close to the limiting one with the occurrence of plastic deformation areas according to the factor of σ_x action.

Fig. 10.2. The curve of horizontal stresses σ_x in elements of the basic scheme for maintaining the 861 belt entry

The frame prop stays are in differently loaded state depending on their location: from the side of virgin massif or from the side of mined-out space. The most loaded is the prop stay from the side of massif. In its curvilinear part, the σ_x act, approaching the estimated yield limit of steel, and form a bending moment capable of causing the plastic bending. In the area of a contact of the frame prop stay with an inclined wooden prop stay of the strengthening support, the local areas occur of concentrations of tensile stresses σ_x up to 50 – 80 MPa and compressive stresses σ_x up to 200 – 250 MPa; here, in spite of the reaction of the wooden prop stay, the bending of the frame is directed towards the mine working cavity. That is, the reaction of the inclined wooden prop stay has not given the desired results. In the rectilinear part of the frame prop stay, the component $\sigma_x = 80 - 110$ MPa is fairly uniformly distributed both along the length and in the cross section of the SCP without the occurrence of any significant bending moment. The area of the prop stay bearing to a height of up to 0.6 m is an exception, where alternating signs σ_x are observed, causing the bending into the cavity of mine working. Therefore, it is expedient to strengthen it in this area of the frame prop stay by means of a mechanical connection with a horizontally located roof-bolt in the immediate bottom of the seam C_8^l.

The frame prop stay from the side of mined-out space has significant differentials in the curve σ_x along its entire length. In its rectilinear part, the uniform compressive stresses σ_x act with minimum level (10 – 30 MPa); in the curvilinear area, a transition is observed to the tension area (30 – 80 MPa); high concentrations σ_x with different signs occur in the area of contact with the inclined wooden prop stay, the value of which approximates to the estimated yield limit of steel. From these data it follows an ambiguous conclusion about the feasibility of using inclined wooden prop stays of strengthening support from the side of mined-out space:

– on the one hand, the rectilinear part of the frame prop stay is almost completely unloaded;

– on the other hand, in the curvilinear part of the frame prop stay, a perfectly permissible bending moment acts, but at the contact with the inclined wooden prop stay, the occurrence of plastic deformations of SCP is very likely, which leads to a disturbance of the normal operation mode of the yielding joist of the frame.

In the resin-grouted roof bolts set in the immediate roof in the sides of mine working, an asymmetry of loading is manifested. The roof-bolts from the side of virgin massif are more loaded; in their reinforcement (similar to the curve σ_y), the areas of considerable expansion (30 – 80%) of tensile stresses σ_x action are observed, which indicates the active resistance of the roof-bolts to the rocks displacement of the immediate roof into the cavity of mine working. From the side of mined-out space, the roof-bolts are loaded to a lesser degree, but their reaction ensures the preservation of continuity of the strengthened rock volume, which is especially important for efficient load transfer from the armored and rock plate in the roof onto the prop stays of the breaker-prop row and in the side of mine working. Thus, according to the factor of the component σ_x action, it is necessary to note the efficiency of the resin-grouted roof bolts in the basic variant of maintaining the 861 prefabricated drift.

As for all the central wooden prop stays of the strengthening support of the basic scheme for maintaining the drift, they do not experience significant horizontal stresses. This is partly explained by their small lateral dimensions and the condition for free expansion, and partly by their vertical location. Low compressing stresses of $\sigma_x \leq 12$ MPa in inclined wooden prop stays indicate their weak efficiency in strengthening the frame support.

In general, based on the analysis results of the horizontal stresses distribution in the elements of the basic scheme for maintaining the 861 belt entry, the following conclusions were formulated:

– the strengthening of the frame cap board by two rows of vertical central wooden prop stays has not lead to the expected result: the cap board is under the action of an alternating-sign bending moment (not less than five maxima in length), some of which are capable of causing the plastic deformation of the cap board;

– the curvilinear part of the frame prop stays, despite the different degree of their loading as a whole, is in a state close to the limiting one, which is largely caused by the reaction of the inclined central wooden prop stays as concentrators of stresses;

– weak loading of inclined central wooden prop stays raises doubts about the expediency of their setting;

– lateral resin-grouted roof bolts in the immediate roof showed their high performance;

– the wooden prop stays of the breaker-prop row and in the sides of mine working cannot be assessed by the component σ_x because of their vertical arrangement.

In the final part of the assessment of the basic variant effectiveness for maintaining the 861 belt entry, the analysis results are represented of the stresses intensity σ distribution (Fig. 10.3).

Fig. 10.3. The curve of stresses intensity σ in the elements of the basic scheme for maintaining the 861 belt entry

The cap board is loaded asymmetrically and an extended area (up to 1.1 m long) of the limiting state (or close to it) of steel is formed in it; this area is shifted towards the virgin massif and represents a danger from the point of view of loss of the cap board stable form. Two rows of central wooden prop stays of the strengthening support do not improve the state of the cap board, and its plastic bending is predicted under the influence of vertical rock pressure.

The frame prop stays from the side of virgin massif along their entire length are in a limiting state with a high probability of plastic bending into the cavity of mine working. The prop stay needs to be strengthened for increasing the resistance to lateral loads. Therefore, it is recommended pliable mechanical connection of the lower part of the prop stay with the lateral roof-bolts, which also perform the functions of strengthening the side rocks and reducing, thereby, the lateral rock pressure.

The wooden central prop stay with an inclination towards the massif is loaded less than 50% of its load-bearing capacity and does not influence the limiting state of the frame prop stay and its cap board. The same level of loading relates to two vertical central prop stays, bearing the cap board. Their insufficient resistance to rock pressure calls into doubt the expediency of using them in mining and geological conditions under consideration.

The frame prop stay from the side of mined-out space is very weakly loaded (up to 8 – 10% of the load-bearing capacity) in its rectilinear part. In the curvilinear part, the intensity of stresses σ increases up to 60 – 85% of the estimated yield limit of SCP steel, and in the area of contact with the inclined central prop stay in the area of up to 0.6 m, a transition to the limiting state is observed. That is, the reaction of the inclined central prop stay of the strengthening support induces the transition of the upper part of the frame prop stay to the limiting state. For this reason, as well as according to the factor of the overall low loading of the frame support, it is doubtful whether it is expedient to use an inclined central prop stay of the strengthening support. Although, it should be noted that this particular wooden prop stay is loaded in separate areas at the level of the load-bearing capacity, that is, it actively resists to the rock pressure.

The resin-grouted roof bolts in the sides of mine working in separate areas are loaded up to 100% of their load-bearing capacity and their setting is completely substantiated. Given the complete compliance with the vector of the lateral rocks displacement, it is recommended to reduce by 10° the angle of gradient to the horizontal and slightly change the coordinates of the tail joints location of the roof-bolts.

Three rows of wooden side prop stays of the strengthening support in the side of mine workings are overloaded, but they are being saved from destruction by their yielding property, caused by contortion of wooden substrates (girders) and pressing into the bottom rocks of the drift. These maintaining elements very effectively resist to rock pressure and are recommended for use.

The single-row breaker-prop will definitely be collapsed, but at some distance behind the stope face, where an area of uncontrolled collapse with sufficient thickness and rigidity is formed to create a proper bearing for the lowering layers of the coal-overlaying formation.

Summarizing the research results, it should be noted that some of the elements of the basic fastening scheme (single-row breaker-prop, three rows of wooden side prop stays of the strengthening support in the working flank of mine working, resin-grouted roof bolts in the immediate roof) are recommended for use, and all the central wooden prop stays of the strengthening support do not work effectively to resist to rock pressure.

11. RESEARCH AND ANALYSIS OF THE STRESS-STRAIN STATE OF FASTENING AND PROTECTION RECOMMENDED SYSTEMS ELEMENTS FOT MAINTAINING THE 861 BELT ENTRY OF THE SEAM $C_8{}^l$

In accordance with the previously substantiated geomechanical model reflecting the loading condition of the fastening and protection systems of the 861 belt entry after drifting of the 861 longwall face, a set of SSS calculations for each of the constituent elements has been performed. The parameters of the SSS anomalies of the coal-bearing massif in the vicinity of mine working (the zone of bearing pressure and unloading) qualitatively correspond to the known results [55 – 58], and the features of the quantitative indicators are most evident when assessing the state of the fastening and security systems of mine working. Therefore, a detailed analysis of the SSS of each of the main elements of the recommended scheme for maintaining mine working in terms of the main task solving – to ensure the conditions for its reuse.

The SSS analysis of fastening and security systems of mine working has been performed by the components of stresses: vertical σ_y, horizontal σ_x and stresses intensity σ.

11.1. ANALYSIS OF VERTICAL STRESSES COMPONENT

The curve of vertical stresses σ_y distribution is shown in Fig. 11.1, during the analysis of which the following results have been obtained. Component σ_y in the frame support reflects the following distribution features, characteristic of the mining and geological conditions of the Western Donbas. Thus, the cap board is in an unloaded state with the action of alternating-sign stresses σ_y in the range from 30 MPa of compression to 30 MPa of tension, which in relation to the value of the estimated yield limit of SCP steel is only 11%. At the same time, the prop stays of the frame are

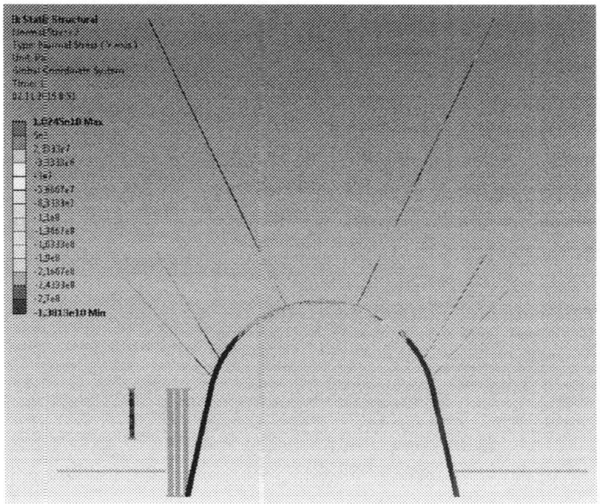

Fig. 11.1. The curve of vertical stresses σ_y distribution for the recommended scheme of maintaining the 861 belt entry

subjected to the action of a uniform compression stresses field with a value corresponding to or exceeding the estimated yield limit of St.5 steel. The uniformity of the field σ_y indicates the absence of any significant bending moment in the prop stays, but their limiting state predicts a high probability of plastic bending into the cavity of mine working. Therefore, we consider it expedient to bind the frame prop stays with the side roof-bolts by means of longitudinal pliable binders [95].

Rope bolts are characterized by different distribution of σ_y depending on place of their setting – from the side of massif or from the side of the mined-out space. Along the length of the rope bolts set from the side of massif, there is a periodic alternation of areas with high tensile stresses σ_y (about 30 – 80% of the load-bearing capacity) with the areas of compression of the similar level. This phenomenon is conditioned by the different-sized deformations of individual rock layers and blocks relative to each other along the length of the rope bolt, which indirectly confirms the significant shifts in the roof rock volumes, on the one hand, and the active resistance to the lateral bearing pressure acting from the side of massif as a result of stope works performance, on the other. The rope bolt from the side of mined-out space is exposed (by 90% of its length) to the action of compressive stresses σ_y; this indirectly indicates the attempt of the roof-bolt to connect the rock layers and blocks into a integrate formation, since there are no sharp differentials of σ_y along the length of the roof-bolt. Thus, it can be stated that rope bolts form in the roof a certain similarity to a holistic rock plate with considerable thickness (up to 5.0 – 5.3 m), which is capable of sustaining a vertical load many times exceeding the load-bearing capacity of the frame support.

The work of resin-grouted roof bolts in the immediate roof of the seam C_8^l should also be assessed positively. Such a conclusion arises from the "smooth" nature of the change in the compressive stresses σ_y along the length of the roof-bolts, which indicates the reliable binding of rock layers and blocks in the lower part of the armoured and rock plate at its end sections outside the width of mine workings. This ensures the formation in the roof of mine working of a sufficiently holistic load-bearing armoured and rock structure.

Lateral roof-bolts located in depth of the bottom drift ripping (in the immediate bottom of the seam C_8^l), are very weakly loaded with vertical stresses σ_y, since they are oriented horizontally. Their work effectiveness will be assessed when considering the curve of horizontal stresses σ_x distribution.

Three rows of lateral wooden prop stays of the strengthening support differ by a very uniform distribution of the component σ_y, and in quantitative terms, the vertical load, as a rule, exceeds the calculated resistance of pine to compression. But, in general, the stable state of the prop stays is predicted thanks to their 'deviation' from excessive rock pressure by contortion of wooden substrates (longitudinal girders) and pressing into the rocks of the immediate bottom.

In a single breaker-prop row, the level of σ_y many times exceeds the calculated resistance of pine to compression and it is possible to quite confidently predict the destruction of the prop stays in the course of development of displacement in the coal-overlaying formation behind the stope face.

Assessing in general the curve of vertical stresses σ_y distribution in the elements of the fastening and security systems, it can be concluded that it is advisable to use the recommended scheme for maintaining the 861 belt entry.

11.2. HORIZONTAL STRESSES COMPONENTS DISTRIBUTION

The curve of horizontal stresses is shown in Fig. 11.2, from the analysis of which the results were obtained, which correspond in general to the previously expressed considerations on the rather intense manifestations of the rock pressure in the vicinity of the 861 belt entry after drifting of the 861 longwall face.

The cap board is exposed to action of component σ_x with a high gradient of change along the entire length. As for instance, the peripheral areas (near the frame's yielding joists) are unloaded: here σ_x arise with different signs up to 10 – 15 MPa with a fairly uniform distribution in the cross section of the SCP, which indicates the absence of any significant bending moment. When approaching the arch keystone, the compressive stresses increase in the range of 50 – 140 MPa, and in the very central part of the arch, 1.2 – 1.3 meters long, compressive stresses σ_x reach 65 – 85% of the estimated yield limit of steel. In this case, there is an important feature of σ_x distribution – there are no significant bending moments almost along the entire length of the cap board. This indirectly confirms the expediency of creating additional bearings for the cap board by means of longitudinal flexible connections with rope bolts.

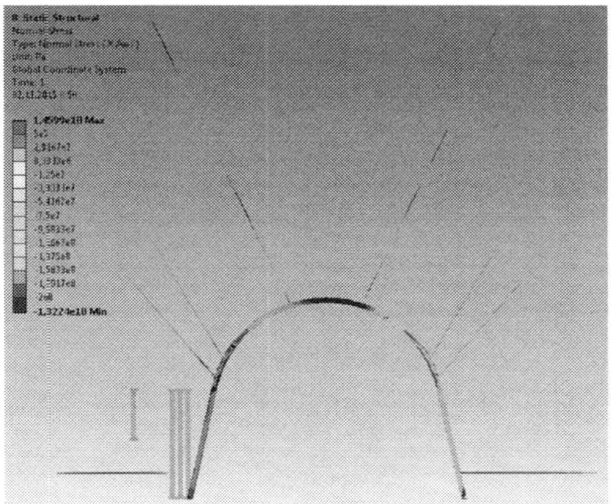

Fig. 11.2. The curve of horizontal stresses σ_x distribution for recommended scheme of maintaining the 861 belt entry

In the prop stays of the frame support from the side of massif, the compressive stresses $\sigma_x = 100 - 130$ MPa are also relatively uniformly distributed in the cross section of the SCP; the special profile bending is observed in the area of the curvilinear part of the prop stay, where two vertical lateral resin-grouted roof bolts are located. The bending is directed towards the cavity of mine working, and the spatially-yielding connection of the lower lateral roof-bolts with the frame prop stays effectively endeavors to its elimination [95]. This confirms the loading of the roof-bolt by tensile forces σ_x. In the prop stays from the side of mined-out space, there is an similar pattern of σ_x distribution but increased in the range of 140 - 220 MPa. Thus, it is expedient to bind the frame prop stays with the lower side roof-bolts set in the immediate bottom of the seam C_8^l.

The rope bolts from the side of massif actively resists the horizontal shifts of the rock layers and blocks, binding them into a relatively holistic structure. This is evidenced by a periodic change in sign σ_x along the length of the roof-bolt. The rope bolt from the side of mined-out space also prevents the horizontal shifts of the rock layers and blocks relative to each other, but here the fluctuations of σ_x occur within compressing stresses.

Resin-grouted roof bolts in the immediate roof are loaded more uniformly along entire length, which indicates their moderate resistance to horizontal displacements of massif in this area; in addition, their predominantly vertical arrangement does not allow the horizontal component σ_x to fully assess the effectiveness of their work. This cannot be said of the lower side roof-bolts in the immediate bottom: almost along the entire length they are exposed to tension and actively resist the displacements of the immediate bottom rocks into the cavity of mine working.

The rows of lateral wooden prop stays of strengthening support and a single breaker-prop row are unloaded from horizontal stresses (from 8 to 12 MPa of compression), which is conditioned by their free transverse deformation and small transverse dimensions.

In general, the results of the field σ_x analysis do not contradict the previously stated concepts about the rational operation of the proposed fastening and security systems of the 861 belt entry.

11.3. STRESSES INTENSITY RESEARCH

In the final part of the analysis, the stresses intensity curve σ (Fig. 11.3) is considered as the most informative and generalizing parameter. The revealed peculiarities of σ distribution confirm in general the previously stated concepts (by scientific assessment) on the intensity of the rock pressure manifestations and expediency of using a combined roof-bolting system consisting of resin-grouted roof bolts and rope bolts that strengthen the adjacent massif along the whole frame contour. The results are presented for each element of the recommended scheme for maintaining the 861 belt entry.

As it is known [55 – 58], the peculiarity of deformation of frame supports of mine working from the special SCP profile in the conditions of soft coal-bearing rocks of the Western Donbas is the predominant loading of the prop stays with relatively unloaded cap board. This is the main trend with a wide range of variations in the degree of loading of the cap board up to its plastic bending. An essential contribution to this pattern is the so-called supporting roof-bolting fastening widely used in the Western Donbas, which takes up a part of the rock pressure thanks to the roof rocks strengthening. On the other hand, the central wooden prop stays of the strengthening support, set in the area of the stope works influence ahead of the longwall face, induce the plastic bending of the cap board. This information will help to objectively assess the state of the frame in the proposed scheme of fastening the 861 belt entry.

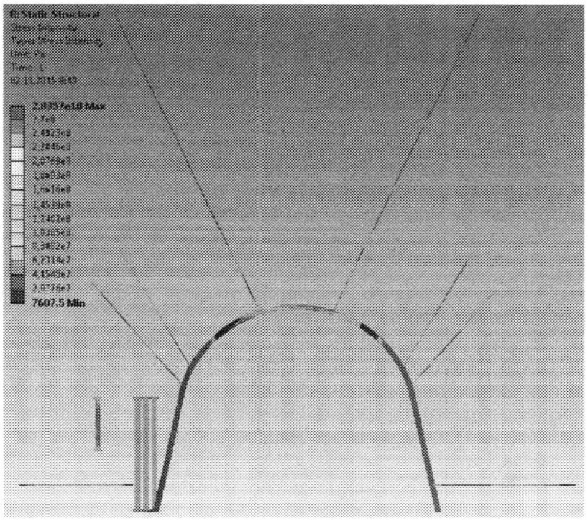

Fig. 11.3. The curve of stresses intensity σ_x distribution for the recommended scheme of maintaining the 861 belt entry

In general, the cap board is in a stable state with a different level of loading along its length. The central area near the arch keystone with a length of 0.8 – 1.0 m is exposed to action of $\sigma = 200 - 240$ MPa, which is 74 – 89% of the estimated yield limit of St.5 steel. In the peripheral areas of the cap board, the stresses intensity gradually decreases from 120 – 190 MPa to 20 – 80 MPa in the area of the yielding joists. There is a significant influence (on a decrease of σ) of additional bearings in the form of pliable binders, connected with rope bolts. Nevertheless, despite the shortened cap board, for the most part of the length of its span it is loaded by more than 50%. The active resistance to the vertical rock pressure of the roof-bolting system should also be taken into account. The combination of these factors indirectly confirms the previously stated assumption of the formation in the roof of an extensive arch of ultimate equilibrium caused by an unstable state of the main roof layers. This conclusion is supplemented by the high loading of the prop stays of the frame support along almost their entire length. The predominantly limiting state of the prop stays of the frame requires their strengthening to avoid plastic bending and deformation into the cavity of mine working. It is therefore advisable, in our opinion, to apply the well-known [95] technical solution for connecting the side roof-bolts and frame prop stays with longitudinal spatially-yielding connection.

Analysis of stresses intensity distribution in rope bolts shows their high loading, the level of which substantially exceeds the results of computational experiments conducted for other mining and geological conditions [55 – 58]. More specifically, we note that the rope bolts operate at 85 – 100% of their load-bearing capacity along the length, which is 75 – 80% of the active length of the roof-bolts located in the main roof of the seam C_8^l, consisting of soft fractured argillites and a very fractured coal seam C_8^u. The high loading of rope bolts indicates the effectiveness of their work on the formation of an armoured and rock plate in the roof and the appropriateness of the selected parameters of setting. If to assume that the rope bolts managed to fasten these lithotypes of the main roof into a single plate, then, even with account of their breaking into blocks, tension crack, the load-bearing capacity of such a thick armoured and rock structure is by 3 – 5 times higher than that of the frame support according to the most understated estimates. Then the conclusion is obvious about the expediency of using rope bolts in this complex mining and geological situation of unstable rocks of the main roof.

The immediate roof of the seam C_8^l and part of the lower layer of the main roof strengthen the resin-grouted roof bolts in the sides of mine working. They, as previously noted, are designed to create relatively holistic and stable bearings in the sides of mine working for effective resistance of the rock plate in the roof. The loading of resin-grouted roof bolts is very high, both from the side of massif and from the side of mined-out space; that is, they actively resist not only to the vertical, but also to the oblique rock pressure. The specific quantitative values of this resistance are as follows:

– the upper resin-grouted roof bolt in the immediate roof from the side of mined-out space is 80 – 100% (from the load-bearing capacity) loaded along 40% of its length adjacent to mine working; the remaining its buried part resists with forces of 30 – 75% of the maximum;

– the bottom roof-bolt in the immediate roof from the side of mined-out space is 90 – 95% of its length has a load in the range of 80 – 100% of the load-bearing capacity;

– from the side of massif, the upper lateral roof-bolt is similarly loaded at 50% of its length located in the buried part of the roof-bolt, and on the rest length adjacent to mine working, the relative resistance is 40 – 75%;

– the bottom lateral roof-bolt in the immediate roof from the side of massif is least loaded (up to 70 – 80% of the load-bearing capacity).

The data represented again focus on the rational choice of the parameters for the side resin-grouted roof bolts setting in the immediate roof of the seam C_8^l, as well as indirectly indicate a high probability of intense oblique rock pressure. Given the equal-sized load-bearing capacity of the resin-grouted roof bolts and inclined wooden prop stay of the strengthening support (in compliance with the passport of maintaining the 861 drift), a simple technological solution of assembling the longitudinal spatially-yielding couplings between these roof-bolts and the frame prop stays [95] will allow to refuse from setting the inclined wooden prop stays, which clutter up a section of a drift.

The need to set lateral roof-bolts along the bottom ripping depth of the drift (in the immediate bottom of the seam C_8^l) is substantiated by their considerable loading by stresses σ. The level of stresses intensity in the roof-bolt from the side of mined-out space makes up 30 – 55% of the load-bearing capacity, which is explained by the relative unloading of the border area of the drift berm: it is loaded only by one breaker-prop row and consolidated collapsed rocks. At the same time, the lateral rock pressure from the side of massif contributes to 100% loading of the corresponding roof-bolt at 40 – 45% of its length, adjacent to mine working. Such results confirm the expediency of roof-bolt strengthening of the immediate bottom of the drift along the depth of its ripping, especially with account that these roof-bolts actively resist to horizontal displacements of the border rocks of the immediate bottom into the cavity of mine working: this is evidenced by the analysis results of the horizontal stresses curve.

Thus, the SSS calculation data confirm the operational efficiency of the combined roof-bolting system as a whole: rope bolts (and partly the resin-grouted roof bolts) create an armoured and rock plate with high load-bearing capacity, and the rest lateral resin-grouted roof bolts provide reliable bearings for it.

In terms of reliability of the armoured and rock plate bearing from the side of mined-out space, its strengthening is assessed by three rows of lateral wooden prop stays. All of them are loaded enough uniformly both along the length and in the cross section with a high reaction of resistance, equaling to their load-bearing capacity or exceeding that. Protection of wooden prop stays from destruction is provided by pliable substrates, as well as by the phenomenon of pressing the prop stays into soft bottom rocks. The work of all three rows of lateral wooden prop stays at the level of their load-bearing capacity indicates the effectiveness of such a technical solution to counteract high loads from the side of roof rocks. It can be stated that the lateral wooden prop stays perform their function of induced collapsing the layers of the main roof and creating a reliable bearing for the armoured and rock plate in the roof.

A single breaker-prop row made of wooden prop stays is one of the most loaded elements of the scheme for maintaining the 861 drift. Here the level of acting stresses σ is many times higher than the calculated resistance of pine to compression and it quite predictably destroys the prop stays even taking into account their pliability due to pressing into the rocks of the immediate roof and bottom. As previously assumed, the breaker-prop row performs the function of breaker timbering in the restricted area behind the longwall face; further on, as the longwall face recedes, the rocks of the zone of uncontrolled collapse, consolidating, create sufficient bearing to the lowering of the overlying layers of the main roof and stabilize the process of displacements of the coal-overlaying formation.

Based on the complex of SSS studies of the fastening and security systems, it is advisable to apply this scheme for maintaining the 861 belt entry for its repeated use while mining the adjacent extraction site.

11.4. RECOMMENDATIONS FOR SELECTING THE PARAMETERS OF THE FASTENING AND PROTECTION SYSTEMS OF REUSED MINE WORKINGS THROUGH THE EXAMPLE OF MINING THE 861 LONGWALL FACE

Based on the analysis of mining and geological conditions for maintaining the 861 belt entry and the results of a computational experiment for calculating the SSS of the geomechanical system under research, recommendations have been developed on the choice of the parameters for fastening and securing of mine working for the purpose of its repeated use. The proposed technical solution is presented in Fig. 8.3 and is characterized by the following parameters.

Frame support of the TSYS series is set in accordance with the passport of maintaining the 861 belt entry.

The rope bolts with length of 6.0 m are being arranged symmetrically (in cross section) with respect to the vertical axis of mine working, with an angle of gradient 65° to the horizontal. In terms of extraction mine working, the rope bolts are set in increment of $L_{r.a} = 3.2$ m, that is, every four frames in the middle of the inter-frame gap. For the most dangerous area (PK 128+1 – PK 132+8), the step of the rope bolts setting is reduced by half ($L_{r.b} = 1.6$ m). In the cross section of mine working, the distance between the tail joints of rope bolts is 1.4 m (0.7 m from the vertical axis of mine working). This scheme of setting will allow to perform the so-called "deep-seated" strengthening of the rocks in the main roof and form a load-bearing armoured and rock plate, bearing on the strengthened (by resin-grouted roof bolts) rocks of the immediate roof and bottom, rows of lateral wooden prop stays of the strengthening support and a single breaker-prop row of prop stays from the side of mined-out space.

The connection (along the mine working) of the cap board of the frame by means of pliable binders (for example, from previously used ropes) will provide a number of technological advantages:

– strengthening of the cap board of the frame, especially when dismantling the prop stays in the area of "windows" of the longwall face, which increases the safety of work in the junction area;

– there is no need to dismantle and mount the wooden prop stays of strengthening support during the subsequent bottom ripping after drifting of the 861 longwall face;

– the "live" section of the drift increases and the resistance to the movement of the air jet reduces, which increases the efficiency of ventilation of the extraction site;

– It is easier to ensure distances and clearances in accordance with safety norms and rules.

The resin-grouted roof bolts with a length of 2.4 m, are set in the immediate roof of the seam C_8^l in the middle of the interframe gap. The setting parameters:

lower roof-bolts – at a distance of 0.3 m from the edge of the coal seam at an angle of 50° into the roof; upper roof-bolts – at a distance of 0.4 m from the lower roof-bolts at an angle of 60° into the roof. Such a scheme of roof-bolts arrangement ensures the formation of a holistic rock bearing transferring the load from the armoured and rock plate in the main roof to the three rows of lateral wooden prop stays of the strengthening support and to a single breaker-prop row. The increased resistance of the bearings constraints the lowering of the armoured and rock plate in the roof and reduces the section loss of the drift.

In the immediate bottom of the seam C_8^l, the resin-grouted roof bolts with length of 2.4 m are set horizontally in the interframe gap in the middle of the depth of the bottom ripping of the drift. Along the length of the drift, the tail joints of the resin-grouted roof bolts are connected with pliable binders to the frame prop stays. This ensures:

– the creation of a sufficiently rigid lower part of the bearing for armoured and rock plate by strengthening the soft rocks of the immediate bottom;

– constrain of bending of the frame prop stays into the cavity of mine working and reducing the intensity of swelling, which contributes to decrease in losses of the drift section.

Three rows of lateral wooden prop stays of the strengthening support are fully loaded and provide induced collapsing of the rock cantilevers of the main roof beyond the width of mine workings, which reduces the loading intensity of the fastening system.

A single-row breaker-props perform temporary functions in the junction area and in a limited area behind the longwall face; its task is to provoke a collapse of the immediate and lower layer of the main roof (up to the seam C_8^l), to create a bearing of collapsed rocks, sufficient to stabilize the rock pressure behind the stope face with a small distance of its withdrawal.

In the area of the belt entry with intensive rock pressure manifestation (ahead of the longwall face in the bearing pressure zone and behind the longwall face to the zone of its stabilization), it is recommended to set under the cap board of each frame a demountable yielding central prop stay of the strengthening support, for example, a two-segment one made of the SCP segments connected by the yielding nodes and synchronously triggered by an appropriate amount with the yielding joists of the frame.

The above recommendations, in our opinion, will ensure the repeated use of the prefabricated drift as a marginal adjacent extraction site with a minimum amount of necessary repair and restoration works.

SUMMARIZING THE RESEARCH RESULTS TO CHAPTER II

1. Analysis of mining and geological conditions of downward and joint mining of seam C_9 and C_8^l, has revealed the probability existence of transition of rocks of a parting into unstable state in separate areas of the extraction panel, which is conditioned by the emergence of a series of emergency situations both in the stope face and in the drifts when mining the underlying seam C_8^l. This mining-engineering situation is not occasional, but is stably repeated at a number of mine fields in the Western Donbas. Therefore, the relevance of the problem of assessing the stability of a parting is obvious, although the seams being mined cannot be assigned to the category of contiguous coal seams by their classification characteristics. The assumption itself of the loss of stability of a sufficiently thick (22 – 28 m) parting that have prompted a complex of computational experiments in order to study the state of the constituent rock massif, the fastening and security systems of the drifts.

2. The performed studies have substantiated the possibility of a significant mutual influence during mining the coal seams, even if they are located at a distance of more than 20 m. The essence of mutual influence is in structural transformations of a parting affecting the bottom in the process of mining the overlying layer, as well as the roof – when mining the underlying seam. When the areas of unstable rocks are being coupled, an extremely high rock pressure develops in the parting. Therefore, in each specific complex mining and geological situation, it is recommended to assess the probability of loss of stability by the rocks of a parting in conditions of a descending order of joint mining of two seams in the series of strata.

3. A geomechanical model has been developed and substantiated of the enclosing massif behavior, as well as of the fastening and security systems of mine working in the period after drifting the stope face, and which is maintained for its repeated use. The elastic-plastic formulation of the computational experiment (for each lithological variety and fastening materials) has approximated the conditions for geomechanical system deformation to real ones, as well as reflection of the geometric parameters of the ways and means of fastening and security in full compliance with the mine working passport. As a result, the results of SSS calculation can be assessed as sufficiently adequate and reliable.

4. Contrary to popular belief about the insignificant influence of the fastening structure resistance on the SSS of the surrounding massif, a complex of studies has been carried out for two variants of the fastening schemes, which are conditionally called “basic” and “recommended”. This intentional search for differences in the states of the massif is intended to clarify the degree of influence of the fastening system parameters in the previously unexplored conditions of the possible loss of stability by a parting, since the negative development of events leads to

the formation of a load many times exceeding the load-bearing capacity of existing fastening structures.

5. A comparative analysis of the SSS of the enclosing rock massif with different schemes of maintaining the drift has created an evidentiary basis for two new concepts in the conditions of probable loss of stability by a parting:

– an essential change in the fastening scheme of mine working is capable to influence significantly on the stresses components distribution not only in the border rock volumes but also in quite remote ones;

– the combination of resin-grouted roof bolts and rope bolts plays a decisive role in the stresses field transformation in the massif surrounding the drift; their rational parameters make it possible to create an armoured and rock structure capable of controlling the rock pressure in the direction of increasing the stability of a parting.

The positive impact of the combined roof-bolting system is manifested both in the rocks of the roof and sides, and the bottom of the drift: dangerous concentrations of stresses components decrease from 10 – 15% to 30 – 40%, and their areas of action – up to 50 – 80%, in some zones – up to 4.2 – 4.6 times.

Based on the obtained results, it is quite substantiated to predict a decrease in the intensity of the rock pressure manifestations and maintenance of the operational characteristics of mine working for its repeated use.

6. A comparative analysis of the state of the basic and recommended fastening systems by the main stresses components has confirmed the expediency of using a combination of resin-grouted roof bolts and rope bolts in combination with the used security method to maintain the belt entry for its repeated use.

It has been revealed that some of the elements of the basic fastening structure insufficiently effectively resist to the rock pressure and in the case of loss of stability by rocks of a parting, such a fastening system is not capable of protecting mine working.

The increased efficiency of the combined roof-bolting system is confirmed by data for calculating its SSS: an armoured and rock structure with high load-bearing capacity is created in the roof, and lateral resin-grouted roof bolts provide reliable bearings for it.

As a result, the recommendations have been developed on the choice of the parameters for the fastening and security schemes of the belt entry, which will ensure its repeated use as a marginal adjacent extraction site with a minimum amount of necessary repair and restoration works.

12. ANALYSIS OF THE PROBLEM OF PREDICTING THE BOTTOM ROCKS HEAVING OF MINE WORKINGS

12.1. TECHNICAL AND ECONOMICAL ASPECTS OF MINE WORKINGS MAINTENANCE DURING ROCKS HEAVING

Increasing the competitiveness of coal mining in Ukraine is directly associated with a decrease in the prime cost of final products by main cost objects, from which the cost of maintenance of mine workings is significant. A modern analysis of the production activity of mines in Ukraine [97] indicates that repair work in mining operations is by 60% related to the elimination of the consequences of bottom rocks heaving, where about 42 thousand of underground workers are employed. According to other estimates [98], such a geomechanical phenomenon as a heaving of bottom rocks is the determining factor in the performance of repair and renewal operations in 40 – 50% of mine workings in an unsatisfactory operational state.

At the turn of the 21st century, in the "Dobropilliavuhillia" LTD, the measures of bottom ripping accounted for about 75% of the total volume of repair and renewal operations [99]. The miners of the Vorkuta coal field, where the annual volume of mine workings being retimbered reaches 60 – 70% for the main reason – the bottom rocks heaving, face similar problems.

The situation has not changed even in the present period of time – so, it is noted that the costs of repair and renewal operations in heaving rocks are often comparable to the cost of conducting new mine workings. The processing of information from the geological and surveying services of the Donbas coal mines made it possible to conclude [100] that the probability of heaving the bottom rocks in the preparatory mine workings is 70 – 80% and will increase in the future with increasing depth of mining operations.

The problem of bottom rocks heaving is the most acute in the Western Donbas, where a thin coal-bearing massif is prone to plastic flow, and the process of bottom rocks heaving is already intensively developing at low depths. In in-seam workings, which have the greatest total length, the heaving is activated by the so-called "stamp effect" [67, 68], when the stronger and harder coal seam, under the influence of the bearing pressure presses out the rocks of the immediate bottom not only into the sides, but also into the bottom of the drift.

Bringing the drift bottom into proper operational condition causes not only substantial material and labor costs, but also the level of mechanizing the bottom ripping processes is 6.3%, which is not comparable with the high degree of mechanization of driving operations. Here there is another technological problem – the bottom ripping is associated with difficulties in the functioning of transport, which along with other factors of maintaining the exploitation state of in-seam workings, puts the task of protection from heaving into a number of priorities.

On the other hand, in different mining, geological and mining-engineering conditions, the intensity of bottom heaving has significant differences, which requires a differentiated approach to the measures to restrict this geomechanical phenomenon. After all, even within the limits of a single mine field for a specified seam, its bottom is characterized by a considerable variety of structure under local influence of weakening factors (mainly, water-cut provoked by both the tectonic of the field and performance of mining operations), which in total create conditions for a significant fluctuation in the intensity of the rock pressure manifestations in the bottom of in-seam workings. Therefore, the diversity of structure and properties of adjacent rock layers requires an assessment (monitoring) of their stability along the entire length of in-seam working; accordingly, recommendations for improving the stability of the mine working bottom in its certain areas should be developed differentially, depending on the structure and properties of the bottom rocks and, conditioned by them, heaving intensity.

For the reasons stated above, it becomes especially relevant to predict the state of bottom rocks of in-seam workings in different areas of the mine fields for each mined-out seam within the entire geological and industrial region, for example, the Western Donbas region. This will allow not only in advance (even at the design stage) to provide for a complex of measures to avoid heaving, but also to perform their systematization and zoning, depending on the degree of influence of the main geomechanical factors on the development of heaving process of the bottom rocks in in-seam workings.

12.2. ANALYSIS OF CONCEPTS OF THE MECHANISM OF OCCURRING THE BOTTOM ROCKS HEAVING AND FACTORS AFFECTING ITS DEVELOPMENT

Any prediction of the rock pressure manifestations is based on a certain amount of mine observations, laboratory and analytical studies of the geomechanical process. The history of the concepts formation of ideas about the mechanism and reasons for the occurrence of heaving of bottom rocks in mine workings dates back to the 20's – 30's of the last century, when a number of well-known scientists paid close attention to the problem of bottom rocks swelling (heaving) of underground structures for various purposes. Along with observing the heaving process and searching for ways and means to avoid it, at that time attempts were already made to explain the nature of this geomechanical phenomenon and to determine the factors that have the most significant influence on the development of heaving the bottom rocks in mine workings. Several hypotheses have been put forward on the occurrence of heaving, which can be grouped as follows:

– heaving is conditioned by swelling and increase in the volume of clay rocks lying in mine working bottom;

– heaving is a consequence of the transition of the potential energy accumulated in the border massif of the initial stressed state into the kinetic energy of its movement into the cavity of mine working including from the side of bottom;

– occurrence in the bottom of some similarity of thrust systems from the rock layers and blocks, which lose their stability under the action of horizontally directed forces;

– pressing-out the bottom rocks as a consequence of the bearing pressure acting in the sides of mine working.

These concepts were developed at the beginning of the second half of the 20th century. So, for example, in the works of V.M. Gorodnichev, an attempt was made to disclose the mechanism of bottom rocks heaving on the basis of a complex of laboratory and mine research, followed by a generalization of the results based on the techniques of continuum mechanics. However, the insufficient development (at that time) of the mathematical apparatus for solving such a complex visco-plastic problem forced the use of elasticity theory, which significantly distorted the physical sense of the phenomenon. Nevertheless, the author in his classification of the reasons for heaving the bottom rocks in mine workings was not limited only to the action of internal forces in the border massif, but also pointed to other groups of factors mentioned above. In particular, with regard to the swelling factor of watered clay rocks, a number of specialists agree that this reason is secondary, and not decisive.

Different representation and understanding of the heaving mechanism have formed (in a subsequent period of time) the two main directions describing this process:

– development, based on the static of the granular medium, of the schemes for pressing-out the rocks of the sides and the bottom of mine working under the influence of vertical rock pressure or on the basis of structural mechanics – the schemes for the stability of thrust systems from rock layers and blocks in the bottom of mine working;

– the use of continuum mechanics techniques to simulate the plastic flow of coherent mine rocks into the cavity of mine working from the side of bottom.

The first direction includes the works [101, 102], where attention is focused on the quasi-brittle fracture and stratification of rocks in the coal-bearing Donbas stratum, including argillites and siltstones. Therefore, the continuity of bottom rocks is disturbed, it acquires a block structure and in these conditions the techniques of structural mechanics are more applicable than that of the continuum medium. This view is opposed by the works [103, 104], where it is stated that heaving is a rheological process, which is developed for a long time, and the main factor here is the creeping of deformations in a coherent continuous medium. As an argument for this understanding of the heaving phenomenon, the mine observations of upheaving the bottom of mine workings over time are presented. The earlier engineering and geological surveys of the Donbas rocks [105], where three forms of behavior of bottom rocks are distinguished, combine these contradictory representations: plastic flow, brittle fracture and swelling of water-saturated rocks.

The predominant influence of one of the forms depends on the type of rock and mining-geological conditions for maintaining the mine working.

To complete the review of concepts of the directions of describing the process of heaving the bottom rocks of mine workings, it is necessary to single out a group of researches [98, 99], where a combined approach is used: first, by the continuum mechanics techniques, the area of limit equilibrium of the massif in the vicinity of mine working is determined. And then, based on schematic representations of the loss of stability by the thrust rock system in the bottom, the process of heaving itself is simulated.

As for the factors provoking the heaving of bottom rocks, one of the founders of research into this phenomenon [106] in the 20s of the last century distinguished the following:

– displacement of the mine working contour when unloading an adjacent massif;

– pressing-out the rocks of sides and bottom of mine working under the action of bearing pressure;

– increase in the volume of water-saturated rocks.

At present, the opinion of a number of specialists has not changed greatly, taking into account some details of the brittle fracturing the bottom rocks and forming the thrust systems in it. Of special note is the work [100], in which, based on the analysis and systematization of information on the 699 Donbas mining areas, the 34 mining-geological factors and phenomena have been identified. Based on the correlation-dispersion analysis, the 11 most influential factors were selected, which are recommended to be taken into account when assessing the state of the bottom rocks. Such a probabilistic approach deserves attention due to the generalization of the various mining and geological conditions of Donbas and a large number of factors, which characterizes a high degree of accuracy. Therefore, it is considered expedient to take into account in these researches a number of key indicators, such as: the depth of mining, resistance of bottom rocks to compression and their thickness; such parameters as the extraction thickness of the coal seam and its angle of incidence for conditions, for example, in the Western Donbas, are quite constant and their influence can be estimated by the averaged indicators.

12.3. ANALYSIS OF PREDICTION TECHNIQUES OF THE BOTTOM ROCKS HEAVING IN MINE WORKINGS

Based on a review of existing concepts about the factors that have a significant impact on the process of formation and development of bottom rocks heaving, it seems necessary to analyze modern techniques for predicting this geomechanical phenomenon in underground mine workings. The main criteria for assessing a particular technique accuracy is an adequate reflection of mine observations results of the rock pressure manifestations, structural and technological parameters for maintaining mine working and an appropriate application of the mechanics of deformable solids. From this point of view, the variety of approaches developed to date has been considered, which are expedient to combine into several groups according to the essence of modeling the process of heaving of the bottom rocks of mine workings.

The first group of techniques uses the main provisions of the static of the granular medium, in particular, on the formation and displacement of sliding prism into the cavity of mine working. A positive feature is an attempt to take into account the experimentally established relation between the bearing pressure in the sides of mine working and the pressing-out the bottom rocks. This interrelation is suggested and substantiated by many experts on the basis of various research techniques, both analytical and modeling on equivalent materials, as well as indirectly confirmed by mine observations. Therefore, in our opinion, the prerequisites laid down into the basis of the first group of techniques are a definite contribution to the discovery of the mechanism for the development of the process of the bottom rock heaving in mine workings. Nevertheless, a high degree of schematic representation and the use of techniques of static of the granular media predetermine a number of disadvantages in this group of prediction techniques, despite the simplicity and availability of final expressions for assessing the bottom rocks state.

The first position is a modelling by flowing medium or quasi-flowing medium of the mine rock massif adjacent to mine working, using such strength characteristics as the angle of internal friction φ and adhesion c. Of course, the border massif is exposed to various degrees of weakening and even loosening, but it only concerns certain rock volumes, while other areas in the vicinity of mine working are in a partially coherent or completely holistic state. That is, the real conditions for maintaining mine workings are characterized by the physically heterogeneous behavior of the surrounding massif and its description within the framework of any one model is not objective enough. Moreover, there are problems with determining the parameters φ and c at different stages of mine rock deformation.

The second position – a modelling of the border rocks as a homogeneous medium does not reflect the real laminal structure of the coal-bearing stratum with essentially different mechanical properties, both between lithological differences and along the planes of their laminations. Therefore, in order to increase the accuracy of the prediction one has to use a set of characteristics c and φ for different computational areas, which significantly complicates the task. Hence, the initial advantage in simplicity and accessibility of the technique is neutralized. In addition, as a rule, there is not enough information on the parameters c and φ, at least on the structure of the coal-bearing stratum adjacent to mine working, which makes it difficult to assess the bottom rocks state.

The third position is the schematic representation of the processes of rock prisms displacement into the cavity of mine working. The assumption of flat sliding surfaces of border rock volumes of an idealized geometrically regular shape does not correspond to reality. The introduction of the curvilinear sliding surfaces into the computational schemes significantly complicates the calculations, which again reduces the level of the initial simplicity and accessibility of this group of techniques, along with a sharp restriction of the domain of accurate process reflection. Attempts to use the techniques of correlation-dispersion analysis of the calculation results lead to a multiple increase in the volumes of calculations, the awkwardness of the final expressions without eliminating the above disadvantages.

Based on the above mentioned data, the first group of techniques can only be considered as a first approximation in assessing the bottom rocks state of mine workings.

The second group of techniques is based on the concept of the formation in the bottom of a certain similarity of the thrust rock system in the course of weakening and stratification (of varying degree and depth) of the originally holistic rock volume of the immediate and main bottom [99, 101, 102]. Displacement of bottom rocks in mine workings induces a contact loss between its layers, which are deformed relatively independently of each other. In this case, the rock layer is represented in the form of a beam (or plate), pinched (with varying degree of rigidity) along the sides of mine working. Under the pressure of the underlying layers, the beam bends into the cavity of mine working with the occurrence of the bending moment maximums in the central and lateral sections in accordance with the classical concepts of structural mechanics [15, 16]. In these sections, first of all, the fractures develop in the areas of horizontal tensile stresses σ_x action (x – the horizontal coordinate of the mine working width) and the total moment of resistance to the rock beam bending is decreased by two or more times, depending on the ratio of deformation moduli of the specified mine rock for tension and compression [107]. Residual resistance to the bending moment is realized only by means of the compressive stresses σ_x that create the so-called restoring moment $F\delta$, as shown in Fig. 12.1, *a* (F – the resultant thrust force, δ – the shoulder of the force F action). When the maximum σ_x exceeds the value of resistance to compression σ_{compr} of the rock layer, part of sections under consideration (in the area of compression) will collapse and the rock beam will be divided into two blocks. A thrust system will be formed, the stability of which is determined by the ratio of the bending moment from the active forces action (the pressure of the underlying rock layers) and the restoring moment from the resultant horizontal forces F action that summarize the stresses σ_x in the formed so-called quasi-plastic hinges.

Such representations have certain advantages in comparison with the previous ones, since they initially consider a continuum medium (homogeneous or laminal), and the process of its discontinuity (using such a mechanical characteristic as rock resistance to compression σ_{compr}) to some extent reflects the phenomena observed in practice of preferential swelling of bottom in the central part of mine working in comparison with the lateral parts. However, there are also significant disadvantages caused, in our opinion, by the high degree of schematic description of the actual process of the bottom rocks heaving.

Firstly, there is some uncertainty in the choice of thickness m^R of the bottom rock layer that can lose stability (according to the scheme in Fig. 12.1, *a*) and move into the cavity of mine working. Typically, the value m^R is determined from the boundary of adjacent planes of bedding or by the surface of mine working bottom, as well as by the boundary of separation between the immediate and main bottom. Here it is assumed that, within the mine working width, along the contact of the considered layer surfaces, the adhesion is lost simultaneously in all points, and the rock layer itself is endured by thickness, that is, the value m^R does not depend on the coordinate X. Such idealization does not fully correspond to the actual process, since, on the one hand, the disturbance of the contacts of the considered lithological variety occurs over certain separate areas of the surface of bedding. On the other hand, the disturbance can primarily develop somewhere inside of the layer, since both siltstones and argillites have an open laminal texture with very weakened bonds on the contact surfaces. As a result, the probability of a partial (in some areas of contact) separation of a rock layer with a stepped shape of the boundaries and a variable thickness $y_i(i=1,...,5)$ across the width of mine working is high, as shown schematically in Fig. 12.1, *b*. In this regard, it is appropriate to recall that the moment of resistance to the rock layer bending is in a quadratic dependency on its height y_i, which determines the significant influence of the parameter y_i on the stability of the bottom rock layer. A similar degree of influence is conditioned by the presence of contact areas of adjacent layers, where their coherency is not completely disturbed.

Secondly, there is no certainty that the thrust system will only be represented by two rock blocks across the width of mine working, as shown in Fig. 12.1, *a*. Indeed, in the idealized representation of bending a beam (plate) with constant thickness and having physical homogeneity [15, 16], the maximum of the bending moment appears in the central and lateral sections, independently from the degree of pinching rigidity on its ends. Therefore, it is assumed that the integrity of the beam is violated in the mentioned sections and the formation of two rock blocks occurs. The factors of changing (across the width of mine working) the height y_i of the beam, the non-uniformity of the disturbance of contacts with adjacent layers, the presence of planes of weakening (for example, in the form of natural fracturing), etc., transform both the curve of bending moment and the coordinates of the most dangerous beam cross sections. This causes the change not only in the rock blocks length, but also in their quantity (see Fig. 12.1, *b*), which is especially important in further assessment of thrust system stability. It is well known [15] that the stability of hinged beams depends on the quantity of these hinges (change in the span of a rock block) and with their growth the stability of the thrust rock system in the bottom decreases. Thus, uncertainty arises when assessing the degree of stability of the thrust system, since the number of rock blocks composing it is unknown. There is another uncertainty, conditioned by the degree of opening the fractures (contacts) between blocks in the areas of the previously acting tensile stresses σ_x; the position of the resultant force F_i depends on it (Fig. 12.1, *b* shows an example for the third rock block), as well as its value, which sets the height of the shoulder δ_i of the resultant forces F_i action and directly influences the value of the restoring moment in the thrust system. In addition, with the analyzed approach, the mechanism of cleavage of the rock prisms in the area of the blocks contacts in the areas of the compressive stresses action σ_x is not usually disclosed, when the resultant forces F_i and the shoulder δ_i of their action are changed in the course of their destruction (when $\sigma_x > \sigma_{compr}$). Thus, according to the factors considered, the degree of accuracy of the techniques for predicting the bottom rocks heaving, which are based on the hypothesis of the formation of thrust systems, also raises doubt.

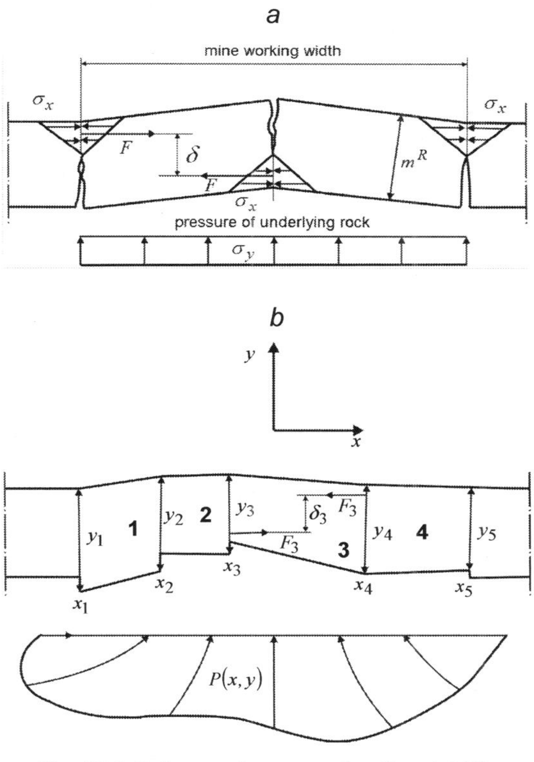

Fig. 12.1. Schemes for assessing the stability of the thrust rock system in the mine working bottom: *a* – conventional; *b* – controversial

Thirdly, the forces exciting bending of an arbitrary layer of the bottom rocks, which act from the side of the underlying layers are very schematically represented (see Fig. 12.1, *a*). Here, there are various variants from a complete lack of effort σ_y from the side of underlying rock layers (due to the supposed separation of the investigated layer with complete loss of contact) to a predominantly uniform (across the width of mine working) curve of vertical rock pressure. Obviously, none of these variants corresponds to the actual process of interaction between bottom rocks layers in mine workings. When using the techniques of structural mechanics, it would be necessary to create a procedure that takes into account the force interaction of the layers during the bottom heaving by analogy with the works [107, 108] on layer-by-layer lowering of the roof. Then, it would be possible to argue about a certain degree of a model adequacy to the real process. As regards the curve of loading the rock beam with the underlying bottom layers, it should be noted here the works [67, 68, 73] on computer modeling of the displacement processes of massif in the vicinity of in-seam working: the obtained curve of complete displacements clearly shows how the oblique displacement vector in the sides of mine working changes the direction to horizontal (coordinate X) in the immediate bottom under the bearings of prop stays of the frame support. But, as the horizontal displacements move to the central part of the mine working width, they are first transformed into oblique (into the mine working cavity) and in the middle of mine working they are in almost transformed into vertical. Thus, a complex curve of loading is observed, for example, of a layer of immediate bottom, in which not only the uneven load (from the side of the adjacent bottom areas) occurs, but also the varying vector of this load, as shown in Fig. 12.1, *b*.

Summarizing the results of the analysis of the second group of techniques for predicting the bottom heaving of mine workings, it is necessary to draw a general conclusion about the high degree of idealization of the real process, the schematic description of which in most positions predetermines a significant error in assessing the bottom rocks state.

The third group of techniques [99, 109] is based on the basic principles of the mechanics of deformable solids with the description of SSS of the massif in the vicinity of mine working as a continuous coherent medium, which enables to assign this group of techniques to exclusively analytical studies with the following advantages and disadvantages, inherent in them when predicting the rock pressure manifestations. In this regard, consideration of not the local area of border rocks, but the general field of stresses and displacements around the mine working should be regarded as a positive side – the SSS in the roof, sides and bottom is analyzed with an appropriate assessment of the state of these areas and their mutual influence. That is, the whole pattern of displacement of the rock massif enclosing mine working is analyzed as a coherent medium that obeys the conditions of continuity of displacements and deformations. Disadvantages are also conditioned by the capabilities of the existing mathematical apparatus when solving the problems of calculating SSS around mine workings, the main ones are as following.

The first factor is the reflection of the structure of the coal-bearing stratum in the vicinity of mine workings. As a rule, a homogeneous rock massif is studied with constant mechanical characteristics in the entire design area, since it is extremely difficult to describe the actual lamination at least in the form of a division into lithological varieties (microlamination) using the available mathematical apparatus. This assumption substantially distorts the fields of distribution of stresses and displacements according to the studies [55, 67, 68, 73], since the strength and deformation characteristics of the lithotypes constituting the massif differ from each other up to several times, and in existing normative documents [71, 94], the accounting of mechanical properties at a distance of up to 20 m from the central axis of mine working is compulsory. In addition, studies [55, 67] have revealed that accounting the discontinuity along the planes of laminations of adjacent lithological varieties (stratification) cardinally changes not only the stresses distribution curve in the border rocks, but also the value of their displacements into the mine working cavity. Thus, the idealization of the coal-bearing rock massif structure used in these techniques cannot be considered satisfactory when describing the heaving process, since the constructed models are not sufficiently adequate to the real object.

The second factor is the reflection of the mechanical properties of lithological varieties composing the coal-bearing stratum. In the analyzed group of predicting techniques, advantageously either an elastic or elastoplastic problem for calculating SSS in the vicinity of mine working is being solved. It is well known [50, 74] that the elastic model of mine rock behavior under load is sufficiently adequate with values of stresses much lower than the corresponding values of compressive resistance of the rock, and in weak rocks the linearity of the stress-strain relation is valid only for relatively low loads. By themselves, the elastic displacements of border rocks are not of practical interest due to their smallness compared to the actual values of bottom rocks heaving observed in mine workings, conditioned by the processes of weakening and loosening the separate volumes of adjacent massif. Consequently, the elastic model of mine rock behavior does not reflect the actual state of massif in the area of weakly metamorphosed rocks, which is of great interest, and it can be used as a starting point in the SSS analysis. The onset of the plastic state is usually reflected by the condition of the elementary volume constancy in accordance with the classical concept of an incompressible medium. The onset of the limiting state of an elementary volume is established on the basis of some theory of failure; the lineal envelope of Mohr's circles is used most often to simplify the solution. An intermediate parameter in solving this class of problems is the size of the plastic deformations zone, which is represented by either a circle or an ellipse depending on the accepted initial stress state of the massif. In accordance with the classical solutions of elastic-plastic problems, around mine working of usually a circular shape, the size of inelastic deformations $r_L(x, y)$ zone is directly proportional to the value of the bottom rocks heaving, which, nevertheless, does not reach the values observed in practice. This is conditioned by several reasons for the correspondence of the model to the real object of research; but, as regards the reflection of the mechanical properties of mine rocks, one of the reasons is to use the deformation characteristics of the holistic medium, which do not describe the complete diagram of the rock deformation, namely, the weakening and loosening stages. In this respect, there are solutions [74] that take into account the mentioned stages of mine rock deformation, however, they are not directly aimed at predicting the bottom rocks heaving, and their sketchiness and high degree of idealization of the process (homogeneous massif, round shape of mine working, no support, etc.) do not allow to recommend for practical use.

It should be noted separately the solutions [109], where the superlimiting stage of rock deformation is considered using a parameter called 'the average value of the relative increase in rock volume θ_{av} in the zone of inelastic deformations', which describes in some averaged way the phenomenon of loosening the massif. These solutions can be called combined, because they combine the classical calculation of the SSS of the elastic-plastic problem with the so-called elastic-plastic stability scheme of the border massif, which is essentially a scheme for the limiting state of the thrust system formed in the bottom from the loose rock volumes. Consequently, the disadvantages characteristic of this and the previous group of techniques are combined. In terms of the accuracy of the mechanical properties reflection of the massif enclosing mine working, the expediency of averaging the phenomenon of weakening and loosening by a single parameter θ_{av} is questionable, since the zone of inelastic deformations has a wide distribution around mine working (as indicated by the solutions being analyzed – the value r_L usually amounts to 2 – 3 radii of mine working) and in each of its points the superlimiting state differs from that at neighboring points. Figuratively speaking, the state of the massif on the outer boundary of the inelastic deformation zone is fundamentally different from its state on the mine working contour.

The third factor is the reflection of the shape and dimensions of mine working, its fastening system, the mechanical properties of the materials used and the structural features of the support elements. As already noted above, the solution of the plane problem for mine working with a round shape is characteristic for this group of techniques, because of serious mathematical difficulties in reflecting its real shape. It is well known that even in a homogeneous massif, the fields of distribution and displacements of the stresses components for a real shape of mine working and for a shape of a circle are fundamentally different [67, 73]. And it is precisely the parameters of the massif's SSS generate the development of the bottom rocks heaving. The round shape of mine working, together with the assumption of homogeneity of the massif, predetermines the closed and smooth shape (circle or ellipse) of the inelastic deformations zone (with the distance r_L from the mine working contour). While, the more rigorous solutions [67] indicate essentially different forms of the inelastic deformations zone and its dimensions, including discontinuous (local) along the mine working contour. The parameter r_L has been paid close attention due to its decisive influence on the value of the predicted heaving of bottom rocks in the group of techniques under consideration. In addition, these techniques do not simulate the support and, accordingly, do not consider its effect on the processes of heaving – mainly unfastened mine working is studied and only in some cases a support reaction is introduced, which is uniform along the mine working contour, independent of the deformation processes of the enclosing rocks, and therefore not reflecting the real nature of the interaction between the elements of the "massif – support'" system. Modeling of structural peculiarities of the mine workings support is not considered at all. In this regard, it should be recalled such phenomenon as the pressing into of the frame prop stays into soft rocks of the bottom. According to the results of mine observations [102] and computational experiments [68, 73], the value of pressing the frame prop stays into can reach several hundreds of millimeters and significantly affect the sizes of the section loss of mine working and its operational state in general. In analytical techniques, the reflection of the phenomenon of pressing the prop stays of frame support into the bottom rocks is practically impossible.

From the above, there is an obvious conclusion about the low description adequacy of the real conditions of the bottom rocks heaving process in mine workings when predicting the phenomenon using the third group of techniques.

The ways to increase the accuracy of techniques for predicting the bottom rocks heaving, as well as other geomechanical phenomena, are seen in the application of a multifactor computational experiment on the basis of modern software for calculating the geomechanical systems state; such studies are combined into a fourth group of techniques. The potential opportunities of the computational experiment are extremely large, the most widely used finite element method (FEM) eliminates the majority of disadvantages in the previously described predicting techniques:

– representation of the real structure of the coal-bearing massif with modeling the disturbance of contacts between lithological varieties and other weakening surfaces (for example, fracturing);

– modeling of all stages of the complete diagram of deformation of each lithological variety composing the investigated geomechanical system;

– a reflection of the actual form of mine working, the structural peculiarities of its fastening system using mechanical characteristics within the framework of the elastic-plastic fastening materials behavior.

In this direction, a number of techniques have been developed [99, 109], which for some factors or another have not fully realized the above advantages of FEM. So, many solutions are made in an elastic setting, often for mine workings of round shape, and with most of them without support. Also, the possibility of reflecting the real geometry of the frames, the cross section of the SCP special profile, the structural peculiarities of yielding joists, the supporting plates under the prop stays, the discrete setting of the frames along mine working (i.e., the solution of the spatial problem), the parameters of the inter-frame fencing, and the like are have not used. These solutions can be described as “halved”, since not always the reasonable assumptions and idealizations on individual positions give in the total significant discrepancies with the practically observed value of bottom rocks heaving, as the authors themselves sometimes notice.

Another side of the FEM study is the “discreteness” of the computational experiment conditions with a certain set of specific parameter values included in the calculation. To generalize the recommendations for a certain range of mining-geological and mining-engineering conditions, it is necessary to identify a set of patterns of each parameter influence, which necessitates conducting the large-scale multifactor computational experiments, and then processing the obtained results by the techniques of correlation-dispersion analysis. This requires considerable expenditure of time and careful planning of the algorithm for conducting the research, but in return the degree of the geomechanical model objectivity and the accuracy of the subsequent predicting the bottom rocks heaving in mine workings are being increased. This direction for solving the geomechanical problems seems to be the most promising, taking into account the need to reflect and generalize the peculiarities of the bottom rocks heaving of mine workings in, for example, the Western Donbas mines, where the characteristics of the coal-bearing stratum differ significantly from other Donbas regions. The technique of generalizing the results of multifactor computational experiments is expedient to use in combination with a constant comparative analysis of data of existing experimental studies of the bottom rocks heaving manifestations, analytical-experimental and statistical techniques, as well as recommendations of regulatory documents [71, 94].

12.4. STRUCTURE, PURPOSE AND OBJECTIVES OF RESEARCH IN THE GEOLOGICAL AND INDUSTRIAL WESTERN DONBAS REGION

The successful implementation of the chosen direction in the development of a technique for predicting the bottom rocks heaving in mine workings is based on the substantiation of the strategy for performing the research, which is more clearly represented by the structural-logic scheme in Fig. 12.2. It consists of separate blocks and tasks, the relationship of which is subordinated to a single goal – to create a base for planning science-based technical solutions for mines in the Western Donbas by monitoring the state of the bottom rocks on the basis of a technique development for predicting the processes of its heaving development.

The first group of tasks, combined in block I, is aimed at substantiating and developing such a geomechanical model of the process of bottom rocks heaving that would most fully reflect the peculiarities of the phenomenon that are typical for the conditions of maintaining the in-seam workings in the Western Donbas mines, with the most adequate modeling of the main structural elements of the fastening system. This formulation has required a very balanced approach to the choice of research strategy using a set of techniques:

– the basic technique is a multivariate computational experiment using modern FEM software;

– multivariance and multifactority of calculations are substantiated by detailed analysis and systematization of structural features and properties of coal-bearing stratum in the area of in-seam workings, as well as the need for accurate reflection of the fastening system structural elements;

– the degree of the results accuracy is tested by means of a comparative analysis with the data of existing mine studies of bottom rocks heaving in mine workings of the Western Donbas mines;

– the results of multivariate calculations serve as the initial database for constructing the regression equations describing the patterns of the development of the heaving process.

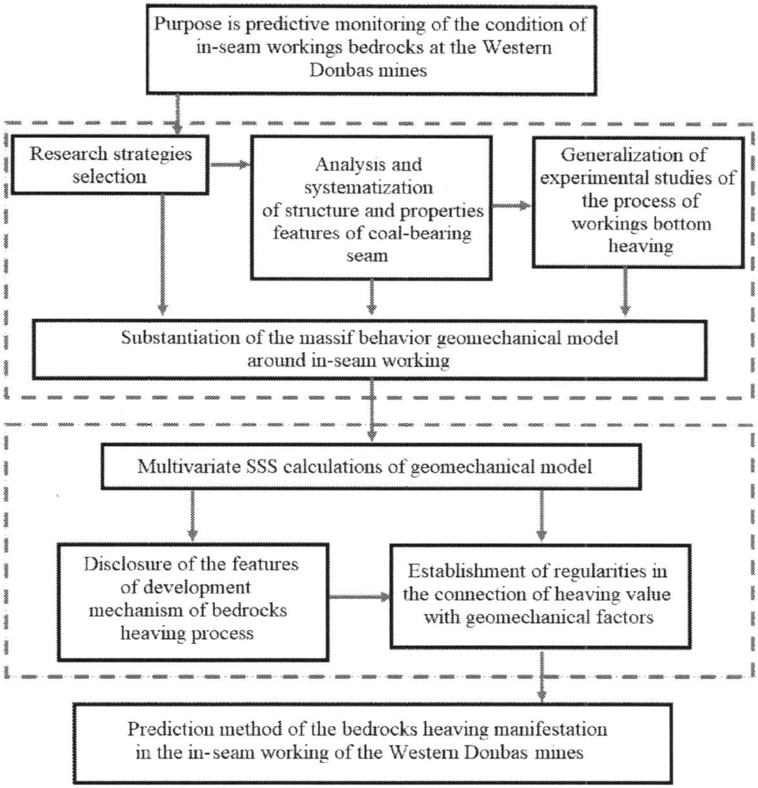

Fig. 12.2. Structural-logic scheme for the research purpose realization at the Western Donbas mines

The final result of the block I research is a comprehensive substantiation of the geomechanical model of behavior of the soft rocks laminal massif around in-seam working.

Block II includes a range of tasks to create a database for the development of a technique for predicting the bottom rocks heaving in in-seam working at the Western Donbas mines:

– compiling a set of variants for the values of the initial model parameters to comprehensively reflect the geomechanical conditions for maintaining the in-seam workings;

– conducting the multivariate SSS calculations of the massif (based on the computational experiments results) and the disclosure of the mechanism of the process of the bottom rocks heaving, depending on the structure and properties of the coal-bearing massif;

– the establishment of patterns of connection between the value of the bottom rocks heaving and the geomechanical parameters of maintaining the mine working and their separation according to the nature of the process development.

The last of the block II tasks is the final research result and is directly related to the ultimate goal – to develop a technique for predicting the bottom rocks heaving at the Western Donbas mines (see Fig. 12.2), which should be completed as an independent document that allows an early assessment of the bottom rocks state over the entire period of mine working under the changing mining-geological situation. In addition, the achieved commonality of decisions and recommendations makes it possible to monitor the manifestations of the bottom rocks heaving in a wide range of mining-geological conditions at the Western Donbas mines, which contributes to the timely planning and implementation of scientifically based measures to neutralize the negative consequences of this geomechanical phenomenon.

13. CONSTRUCTION OF GEOMECHANICAL MODEL OF THE "MASSIF – IN-SEAM WORKING" SYSTEM

13.1. STRUCTURE OF ADJACENT COAL-BEARING ROCK STRATUM OF THE MINED-OUT COAL SEAMS OF THE WESTERN DONBAS

13.1.1. GENERAL PROVISIONS

A substantiated and adequate to real conditions modeling of the structure and properties of the coal-bearing massif in the vicinity of mine working is a basis for obtaining accurate results for assessing the state of the "massif – in-seam working" system.

According to large-scale geological surveys of the past century, the Donbas coal deposits are mainly represented by terrigenous rocks – sandstone, siltstone and argillite, among which coal seams and limestone layers are enclosed in the form of thin layers. The terrigenous rocks in the coal-bearing stratum compose 90 – 95%, in some series argillites and siltstones are sharply predominant, which compose up to 80%. Carbonate rocks are mainly represented by limestones, which among coal-bearing deposits of the Donbas play a subordinate role, amounting to 0.2 – 5% in the sections of separate series. Coals and carbonaceous-argillaceous rocks usually make up 0.5 – 2% of the total series thickness. From the above data follows a sufficient objectivity of modeling the coal-bearing stratum, consisting of coal seams and rock layers of argillite, siltstone and sandstone.

The coal-bearing deposits of the Western Donbas correspond to general patterns, but when reflecting the structure of the coal-bearing massif of the model, the question arises first of all of the dimensions (in height) of the simulated stratigraphic column. In other words, what height into the roof and how deep into the bottom of the coal seam, and most importantly – what structure of the coal-bearing rock stratum is viable to model? Here one should also turn to the extensive and long-lasting (half a century as minimum) studies of the last century, which unambiguously confirm that outside the zone of the stope works influence, the stability of in-seam working is determined by the state of adjacent rocks of the roof and bottom: analytical and experimental studies show that the SSS perturbations extend to a height (depth) of 10 – 15 m; in industry normative documents, this area is limited from one and a half widths of mine working when drifting to a distance of 20 m from its central part, which correlates well with analytical and experimental studies. Finally, the decision on the rational coal-bearing massif model dimensions will be made based on the results of the forthcoming test calculations. But at the stage of its construction (as a first approximation) the maximum dimensions (15 m in the roof and bottom and 10 m on each side in the sides) are accepted and for them, an analysis of an adjacent coal-bearing massive structure of coal seams of the Western Donbas has been made, the most characteristic parameters of which are reflected in the computer model.

A number of researchers [102, 110] agree, that the development of rock heaving in mine working is influenced not only by the mechanical rocks properties, but also the structure (thickness ratio) of the adjacent rock layers of the bottom. In addition, many experts believe that the heaving intensity also depends on the thickness and properties of rocks in the immediate roof and adjacent layers of the main roof in in-seam working; certain patterns of this connection for the Western Donbas are presented in [111]. Thus, it is necessary to analyze the information on the most accurate variants of the roof and bottom rocks structure to a depth of up to a maximum of 15 m (in details up to 10 m). Such studies for the Western Donbas were conducted on the basis of processing information on 553 geological exploration wells of the Pavlohrad-Petropavlivsk region and are particularized in [110]. Our task is to develop, based on the data [110], the variants of the most characteristic structures of the rock layers of the roof and bottom adjacent to the coal seams, which will be adopted for modeling the SSS of the "massif – in-seam working" system.

13.1.2. LITHOTYPE AND THICKNESS OF ADJACENT ROOF ROCK LAYERS OF COAL SEAMS

The lithological composition of the rocks of the immediate and adjacent main roof of the mined coal seams of the Western Donbas is not very diversified and according to the studies of [110] is represented by the following varieties:

– immediate roof: argillite – 52%, siltstone – 36%, sandstone – 12%;

– main roof: siltstone – 45%, argillite – 33%, sandstone – 22%.

The probability of occurrence of one or another lithological variety in the main roof of the coal seam is not stable and depends in some degree on the lithotype of the immediate roof rocks, which will be considered when modeling the properties of the in-seam working roof rocks.

For analyzing the thicknesses m_i^r of the immediate and main roof rock layers, a breakdown into the following intervals was proposed [110]:

– immediate roof: less than 1 m, from 1 to 2 m and from 2 to 4 m;

– main roof: less than 4 m, from 4 to 8 m and more than 8 m.

Thus, the total thickness of the prospected roof layers completely reflects the possible height of the SSS perturbations in the roof caused by conducting mine working. The probability distribution of the rocks structure of the immediate and adjacent main roof is as follows:

– immediate roof: less than 1 m – 25%, from 1 to 2 m – 31%, from 2 to 4 m – 44%;

– main roof: less than 4 m – 26%, from 4 to 8 m – 41%, more than 8 m – 33%.

In addition, a harmonic law of a change in the roof rock layers' thicknesses has been established [110], with the removal from the coal seam, which is characterized by a period that is a multiple of two layers. This means an increased probability of such combinations:

– reduced thickness of the first and third layers with increased thickness of the second and fourth;

– increased thickness of the first and third layers with reduced thickness of the second and fourth. But, as the distance from the coal seam increases, the harmonic law of the thickness fluctuations of adjacent layers fades.

Based on the available material sample, the variants have been constructed (Table 13.1) of the structure of adjacent roof rocks, which simultaneously satisfy the harmonic character of changing the layers' thicknesses throughout the roof height and the probability of thickness distribution within a single layer. In addition, the set of variants should ensure sufficient reliability of the subsequent correlation-dispersion analysis.

Table 13.1

VARIANTS FOR THE STRUCTURE OF ADJACENT ROCK LAYERS OF THE ROOF

variant	m_1^r, m	m_2^r, m	m_3^r, m	m_4^r, m	variant	m_1^r, m	m_2^r, m	m_3^r, m	m_4^r, m	variant	m_1^r, m	m_2^r, m	m_3^r, m	m_4^r, m
1	0.5	1.5	3	8	8	1.5	1.5	8	4	15	2	1	6	4
2	0.5	2	4	4	9	1.5	0.5	4	8	16	2	6	2	6
3	0.5	3	2	6	10	1.5	8	1.5	2	17	3	0.5	6	3
4	1	3	2	8	11	1.5	4	4	4	18	3	3	3	3
5	1	1	8	2	12	2	4	4	2	19	3	2	4	6
6	1	2	4	6	13	2	1	6	3	20	4	2	8	2
7	1	4	1	4	14	2	2	2	8	21	4	8	1	2

Table 13.1 shows the possibility of dividing: the immediate roof – into two rock layers, where lower is often a ramble; the adjacent main roof – into two or three rock layers, which is often found in the structure of the coal-bearing rocks stratum of the Western Donbas.

Also, in Table 13.1, the designation of the roof rocks thickness m_i^r with the numbering of layers ($i = 1$, 2, 3, 4), starting from the coal seam are introduced.

13.1.3. STRUCTURE AND THICKNESS OF MINED COAL SEAMS

To simulate the most accurate structure of the coal seam, the structure of all the mined seams in the Western Donbas has been analyzed on the basis of processing the data of the works [112, 113], which are summarized in Table 13.2.

The analysis of the coal seams structure was carried out in three directions:

– the probability of distributing the total thickness of the seams to identify the most characteristic interval of its change;

– thickness and lithotype of interlayers in the seams with complex structure;

– probability of distributing the thickness coal bands and their number in seams with complex structure.

Table 13.2

PARAMETERS OF COAL SEAMS DEVELOPED AT THE WESTERN DONBAS MINES

Mine	Index of the seam	Total thickness, m		Structure of the seam			
		range of change	medium	upper (middle) coal band, m	lower coal band, m	thickness of rock interlayers, m	lithotype of rock interlayers
Ternivska	C_5^l	0.70 – 1.01	0.86	–	–	–	–
	C_6^l	0.84 – 1.15	1.00	–	–	–	–
	C_8^l	1.59 – 1.69	1.64	0.75 – 0.85	0.80 – 0.87	0.02 – 0.03	Argillite
Yuvileina	C_5	0.96 – 1.00	0.98	–	–	–	–
	C_6^1	0.73 – 0.86	0.80	0.13 – 0.18	0.58 – 0.66	0.02 – 0.03	Argillite
Stepova	C_6	0.75 – 1.00	0.88	–	–	–	–
	C_6^1	0.55 – 0.74	0.65	0.02 – 0.08	0.47 – 0.64	0.03 – 0.04	Sandstone Argillite
Pavlohradska	C_5	0.85 – 1.19	1.02	–	–	–	–
	C_6	0.95 – 1.26	1.11	–	–	–	–
Samarska	C_1	0.81 – 0.93	0.87	–	–	–	–
	C_4	0.79 – 0.92	0.86	–	–	–	–
	C_5	0.91 – 1.10	1.01	–	–	–	–
Heroiv Kosmosu	C_9	1.02 – 1.02	1.02	–	–	–	–
	C_{10}^u	0.80 – 1.00	0.90	0.60 – 1.00	0.00 – 0.26	0.00 – 0.02	Kaolin
	C_{11}	0.80 – 0.84	0.82	0.06 – 0.13 0.26 – 0.28	0.37 – 0.42	0.02 – 0.04 0.03 – 0.04	Kaolin
Dniprovska	C_8^l	0.74 – 0.86	0.80	0.07 – 0.12	0.63 – 0.71	0.02 – 0.03	Sandstone
	C_8^u	0.70 – 1.17	0.94	0.02 – 0.82	0.00 – 0.88	0.00 – 0.26	Argillite
	C_{10}^u	1.05 – 1.25	1.15	–	–	–	–
Zakhidno-Donbaska	C_8^u	0.97 – 1.05	1.01	0.13 – 0.14	0.77 – 0.87	0.03 – 0.03	Sandstone
	C_8^u	0.56 – 0.67	0.62	–	–	–	–
	C_{10}^u	0.75 – 0.75	0.75	–	–	–	–
M.I. Stashkova	C_5	0.93 – 1.05	0.99	0.084 – 1.05	–	0.00 – 0.14	Coal slate
	C_6	0.60 – 0.86	0.73	0.52 – 0.86	0.00 – 0.22	0.00 – 0.10	Argillite
	C_{10}^u	1.05 – 1.10	1.08	–	–	–	–
Blahodatna	C_4	0.90 – 1.22	1.06	0.00 – 0.37	0.45 – 1.02	0.00 – 0.28	Argillite
	C_5	1.10 – 1.27	1.19	–	–	–	–

The total thickness of coal seams mined in the Western Donbas varies from 0.56 to 5.54 ft, and its average values (for each seam within a certain mine field) have an interval of 0.62 m$\leq m^c \leq$1.64 m. To assess the most characteristic range of the total thickness fluctuations of coal seams, the intervals have been distinguished according to the existing classification: less than 0.7 m, from 0.7 to 1.2 m and more than 1.2 m.

The probability of distribution of the total thickness of coal seams mined in the Western Donbas is the following: very thin – 7%, thin – 89% and layers with medium thickness – 4%. Thus, modeling of only thin layers (in the interval 0.7 m$\leq m^c \leq$1.2 m) is represented as quite substantiated, within which more narrow interval 0.8 m$\leq m^c \leq$1.1 m is the most probable (83%). This is taken into account in the compilation of variants for calculating the SSS of the "massif – in-seam working" system.

The second direction of analysis, concerning the lithotype and rock interlayers thickness in coal seams with a complex structure, provides the following results. The number of seams with a complex structure is up to 37% in the total number of mined coal seams. Mainly (77%), the rock interlayers are represented by soft rocks (argillite – 54%, kaolin – 18% and carbonaceous-argillaceous shale – 5%) and to a lesser extent (23%) – by sandstones. The rock interlayers of sandstone are very consistent in thickness in the range from 0.02 to 0.04 m with a sufficiently stable average value of 0.03 m (67%). Obviously, an interlayer of sandstone with a such low thickness will be weakened (in the zones of high rock pressure) simultaneously with the coal seam disintegration since their strength properties are close to each other, and the increased deformation characteristics of the sandstone cause the formation of an increased load on it. Soft rock interlayers are not stresses concentrators due to their reduced deformation characteristics (at one level with a coal seam). Presumably, the influence of the interlayer will depend on its thickness, which varies from 0 to 0.28 m. The main range of fluctuations of the rock interlayer thickness from 0.02 to 0.04 m occurs in 73% of cases, and in other conditions of increased rock interlayer thickness, it is not constant within the mine field, but stably degenerates until complete disappearance. From the analysis results it follows that the most objective will be modeling in the test calculations of the most probable rock interlayer thickness of 0.03 m, represented by argillite.

As for the third direction of analysis, the coal seams with a complex structure, mined in the Western Donbas, are characterized by 91% with double coal band structure. The exception is the seam C_{11} within the "Heroiv Kosmosu" Mine field, where the upper coal band has a thickness of just from 0.06 to 0.13 m with an underlying interlayer of kaolin with a thickness from 0.02 to 0.4 m with deformation characteristics close to coal characteristics. In the remaining coal seams with a complex structure, the thickness of the upper coal bands varies from 0.00 to 1.05 m, of the lower ones – from 0.00 to 1.02 m. And there is no distinct prevailing range of thickness changes of the upper and lower coal bands. Therefore, the decision has been made to distribute the rock interlayer into three variants – in the upper, middle and lower parts of the coal seam and, based on the test calculations of the SSS of "massif – in-seam working" system, to finally form the coal seam model.

Based on the results of the structure analysis of coal seams being mined at the Western Donbas mines, it is considered expedient for previous test calculations to form the following variants of coal seam structures:

– a simple structure with a thickness of 0.7, 1.0 and 1.2 m;

– a complex structure with a constant total thickness of 1.0 m, a rock interlayer of argillite with a thickness of 0.03 m and three variants of its location (lower surface of an interlayer relative to the seam bottom): 0.30, 0.50 and 0.70 m.

As a result of subsequent test calculations of the system's SSS, the structure of the coal seam will be finally substantiated to study the rock pressure manifestations in in-seam workings of the Western Donbas mines.

13.1.4. LITHOTYPE AND THICKNESS OF ADJACENT ROCK LAYERS OF THE COAL SEAM BOTTOM

The rocks of the bottom of the coal seams developed in the Western Donbas are represented mainly by the same terrigenous rocks as the roof – argillite, siltstones and sandstones. The structure of the bottom rocks, of interest to us, up to a depth of 10 – 15 m includes, as a rule, no more than four layers, which have been studied in detail in the work [110]. The common pattern of the periodic change (similar to the roof of the coal seam) of the lithologic composition with increasing of distance from the bottom of a seam has been determined. This is represented by the fact that the increased probability of argillites occurrence is observed in the odd rock layers (the first and the third), and the harder siltstone and sandstone are in the even-numbered rock layers. More specifically, the following probability of lithotypes distribution has been revealed [110] for four adjacent rock layers of the bottom, which should be taken into account when modeling the variants of bottom rocks structures of in-seam working:

– the first rock layer: argillite – 76%, siltstone – 15%, sandstone – 8%;

– the second rock layer: siltstone – 54%, sandstone – 23%, argillite – 13%;

– the third rock layer: argillite – 46%, siltstone – 21%, sandstone – 15%;

– the fourth rock layer: siltstone – 45%, argillite – 23%, sandstone – 21%.

This specified lithological composition of the bottom rocks substantiates the expediency of forming the following modeling positions:

– *the first* rock layer is composed of 91% argillite and siltstone, which can be characterized as soft rocks prone to slaking with very evident rheological properties; therefore, in the future, the first layer of bottom should be modeled with reduced strength and deformation properties;

– *the second* rock layer is composed of 90% siltstone, sandstone and argillite, which at an average have higher strength and deformation characteristics. This is facilitated by the increased probability of a harder sandstone occurrence and the fact that, when removed from the coal seam, the mechanical characteristics of the softest argillite, as a rule, increases; weakening factors of humidification and rheology act less intensively because of the increased probability of the sandstone occurrence and a reduced probability of the argillite occurrence;

– *the third* rock layer is represented by terrigenous rocks by 82% and the increased probability of the argillite occurrence predetermines a decrease in the mechanical characteristics according to the above factors; but the probability of the argillite occurrence is much less than in the first layer, and for this reason the third layer occupies an intermediate position by the average properties between the first and second layers;

– *the fourth* rock layer of the bottom is composed of 89% terrigenous rocks, the probability of occurrence of each of the lithotypes is somewhat leveled, and in terms of mechanical properties, the fourth layer is similar to the second one.

Describing in general the lithologic composition of adjacent bottom rocks, it should be noted two conclusions. *Firstly*, the modeling of only terrigenous rocks is substantiated, since the probability of their occurrence is from 82% (third layer) to 91% (the first layer without sandstone – 8%). *Secondly,* the coal interlayers (the second layer – 10%, the third – 18%, the fourth – 11%) by their strength characteristics occupy an intermediate value between siltstone and sandstone, and by deformation properties – between argillite and siltstone; therefore, they do not make a significant heterogeneity in the nature of mechanical characteristics distribution of the bottom rocks, and their accounting when calculating the SSS of the "massif – in-seam working" system will bring the adequacy of modeling the second, third and fourth bottom layers to 100%.

The next step is to assess the most probable variants of the adjacent bottom rocks structure on the basis of a data analysis of the work [110], where it is proposed to split the entire range of variation in the thicknesses of the rock layers into five intervals: less than 1 m, from 1 to 2 m, from 2.1 to 4 m, from 4.1 to 6 m and more than 6 m. The probability of their occurrence in each of the four studied rock layers of the bottom was determined by these intervals (Table 13.3).

Table 13.3

DISTRIBUTION BY THE THICKNESS OF ADJACENT ROCK LAYERS OF THE COAL SEAM BOTTOM

Number of a layer	Thickness of a layer, m				
	1<	1 – 2	2.1 – 4	4.1 – 6	>6
m_1^R	43%	25%	21%	9%	2%
m_2^R	16%	32%	27%	14%	11%
m_3^R	22%	29%	24%	10%	15%
m_4^R	20%	26%	28%	11%	15%

Table 13.3 introduces the designations of the bottom rocks thickness m_i^R with the numbering of layers ($i = 1, 2, 3, 4$), starting from the coal seam.

From Table 13.3 it follows that if to consider when modeling the first four intervals of variation m_i^R (from less than 1 m to 4.1 – 6 m), a very significant range of variation in the thickness of adjacent bottom rocks of the coal seam will be covered: the first rock layer is 98%, the second rock layer – 89%, the third and fourth rock layers – 85%. Such a solution provides a high degree of accuracy in the modeling of the bottom rocks structure of the coal seam. The second principle of selecting the structure variants is in reflection of the probability of the particular interval distribution of m_i^R variation for each particular layer of bottom rocks. The third principle of improving the modeling accuracy is the inclusion of coal interlayers into the variants of changing the thickness of the second, third and fourth rock layers of the bottom – then, in these layers the lithotype of the bottom rocks will be reflected 100%.

In addition to ensuring a high adequacy of the models structure to the real object, it is necessary to find a compromise in the multivariate combinations of thicknesses of the adjacent rock layers of the bottom:

– on the one hand, it is required to provide an opportunity to identify the patterns of the thicknesses m_i^R influence of adjacent rock layers of the bottom on the SSS of the "massif – in-seam working" system;

– on the other hand, a certain limitedness of the calculation variants is necessary to provide a "reasonable" time frames for conducting the calculation experiment and the subsequent analysis of its results.

The formulated principles have become the basis for drawing up the variants of the bottom rocks structure of the coal seam, adopted for the SSS study (Table 13.4).

Table 13.4

THICKNESSES m_i^R VARIANTS OF ADJACENT LITHOLOGICAL VARIETIES IN THE BOTTOM OF THE COAL SEAM

variant	m_1^R, m	m_2^R, m	m_3^R, m	m_4^R, m	variant	m_1^R, m	m_2^R, m	m_3^R, m	m_4^R, m	variant	m_1^R, m	m_2^R, m	m_3^R, m	m_4^R, m
1	0.5	1.5	2	5	11	1	3	1.5	3	21	2	4	0.5	1.5
2	0.5	2	4	2	12	1	4	2	2	22	3	0.5	4	1
3	0.5	3	1.5	4	13	1	6	1.5	1.5	23	3	1	1	4
4	0.5	4	6	0.5	14	1.5	0.5	5	4	24	3	3	3	3
5	0.5	5	3	1	15	1.5	1.5	1.5	4	25	4	0.5	2	2
6	0.5	6	1	0.5	16	1.5	3	0.5	3	26	4	1.5	3	0.5
7	0.5	0.5	1	6	17	1.5	5	1	1.5	27	4	1	1	4
8	1	0.5	3	5	18	2	0.5	4	2	28	5	1	1.5	1
9	1	1	1	6	19	2	2	2	2	29	5	4	4	0.5
10	1	2	0.5	6	20	2	1.5	5	1.5	30	6	1.5	6	1

The final moment when substantiating the structure of an adjacent coal-bearing stratum is the establishment of the most probable conditions for contact of adjacent lithological varieties along the planes of laminations. It is known [102, 110] that the coherence of lithological varieties in the coal-bearing stratum is characterized as weak, especially in the conditions of the Western Donbas. Mining operations induce perturbations in the initial field of the massif SSS, which leads to deformation of the rock layers (in the vicinity of mine workings) and their shifts, including in the horizontal direction. Thus, there is a high probability of coherence disruption between lithological varieties near mine workings, and it is generally known that deformed rock layers with abnormal contacts intensify the rock pressure manifestations. Therefore, at the stage of preliminary testing the models of the “massif – in-seam working” system, a research has been performed comparing the adhesion forces and the value of shearing stresses over the contact surfaces of layers near mine working, on the basis of which an adequate character of coherence of the lithological varieties along the planes of laminations was modeled.

13.2. MODELING THE SOFT MINE ROCKS DEFORMATION IN THE VICINITY OF IN-SEAM WORKING

Having assessed the structure of the adjacent coal-bearing stratum of coal seams being mined in the Western Donbas, the most characteristic mechanical properties for the region should be set for each lithotype. The determining factors of this task are:

– lithotypes distribution of terrigenous rocks over the layers of the adjacent roof and the bottom of the coal seam;

– the choice of the most adequate model of rock deformation: elastic, elastic-plastic, limiting and superlimiting state;

– setting the ranges of changes in the mechanical properties of argillites, siltstones, sandstones, and coals, characteristic of the Western Donbas, within the chosen model of lithotypes behavioral within coal-bearing stratum.

The first factor concerning the lithotypes distribution is partly described in §13.1, which is expanded in terms of substantiating the ranges of change in the rock layers’ hardness with increasing distance from the coal seam. Here, the data are used of direct geological surveys on a number of mine fields, as well as studies [102, 113], the analysis of which allowed to formulate the following provisions:

– the coal seams mined in the Western Donbas have a fairly narrow range of hardness changes – uniaxial compressive strength of coal consists $\sigma^{c}_{compr} = 30 -$ 40 MPa;

– the rocks of the immediate roof are represented predominantly (88%) by soft argillite and siltstones, for which the most adequate interval for changing the compressive resistance is $5 \le \sigma^{r}_{compr_i} \le 20$ MPa;

– in the main roof, siltstones are prevailing (45%), which together with sandstones (22%) slightly increase (at an average) the strength characteristics; however, due to the occurrence of soft argillites (33%), the most objective interval of change in the compressive resistance will be $10 \le \sigma^{r}_{compr_i} \le 40$ MPa;

– the first rock layer of the bottom is represented by 76% soft argillite, which together with soft siltstone (15%) causes its low compressive resistance and the most adequately reflecting interval is 520 MPa;

– the second, third, and fourth rock layers of the bottom are characterized by a decrease in the probability of the soft argillite occurrence (13 – 46%) and an increase in the probability of composition by harder sandstones and coal interlayers (32 – 33%), which is accompanied by an increase (at an average) in strength properties with the most expedient interval change in the compressive resistance $10 \le \sigma^{R}_{compr_{2,3,4}} \le 40$ MPa.

The formed intervals of change in the compressive resistance of adjacent rocks of the coal-bearing stratum unambiguously indicate a high probability of forming the areas of the limiting and superlimiting state in the vicinity of in-seam working with the corresponding depth of its location. Therefore, the study of the second factor is significant – the substantiation of the model of the soft rocks behavior in the Western Donbas.

The half-century history of studies of various mine rocks lithotypes deformation, conducted both in Ukraine and abroad, has led to a fairly unambiguous interpretation of the rock behavior under load. Typically, a graphical illustration of mine rock deformation is given for uniaxial loading [74, 114], and the parameters of the diagram in the coordinates "stress σ_1 – relative deformation ε_1" are considered to be the main rock characteristics, which are used for studies of three-dimensional SSS in conjunction with a theory of mine rock strength. The line describing the state of the rock in the coordinates "$\sigma_1 - \varepsilon_1$" from the beginning of the sample loading to its complete destruction is usually called the total deformation diagram. Most experts consider that the reflection of the real diagram of the rock behavior by means of three linear sections is quite adequate (Fig. 13.1): OA – stage of elastic-plastic deformation (prelimiting state); point A – is the limiting state under uniaxial compression; AB – is the area of reduction of compressive resistance of the rock (the stage of weakening); BC – an area of the so-called "ruinous" destruction of the rock, ac-

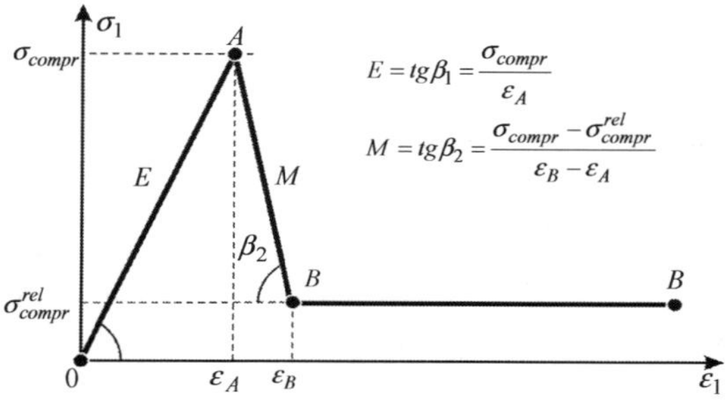

Fig. 13.1. Approximation of the complete diagram for mine rock deformation: *OA* – the prelimiting state; *AB* – the stage of weakening; *BC* – the stage of fragmentation

companied by its increase in volume (the stage of fragmentation). The piecewise line approximation of the complete deformation diagram is characterized by four parameters: the deformation module E, defined as the quotient of the compressive resistance division σ_{compr} by the value of elastic-plastic deformation ε_A, accumulated before the beginning of the rock destruction; the uniaxial compressive strength σ_{compr} of the rock, characterizing the beginning of its limiting state; the decreasing module M, defined as the quotient from division of value ($\sigma_{compr} - \sigma_{compr}^{rel}$) by the value of the relative deformation ε_B of the beginning of the rock fragmentation after deformation ε_A; residual strength σ_{compr}^{rel} of the destroyed rock, which is slightly dependent on the value of deformation at the stage of fragmentation, that is, the increase of the rock in volume. Usually, the superlimiting stage of rock deformation is characterized by relative indicators: $\frac{\sigma_{compr}^{rel}}{\sigma_{compr}}$ – the relative residual strength and $\frac{M}{E}$ – the relative module of decreasing, which are used in further studies.

Another peculiarity of the soft rocks of the Western Donbas is their tendency to creeping, expressed in very long processes of development of rock pressure manifestations in underground mine workings. The most striking rheological properties are observed in argillites and siltstones, to a lesser extent – in sandstones and coal. Therefore, to increase the adequacy of modeling the geomechanical processes in the vicinity of in-seam workings, the rheological formulation of problem has been analyzed with account of the time factor.

In many modern computer programs, the rheological model of the body is described by a generalized creep equation of the form

$$\varepsilon_1 = \varepsilon_e + C_1 \sigma_1^{C_2} t^{C_3}, \qquad (2.1)$$

where ε_e – the elastic component of the relative deformation; for the approximation used (see Fig. 13.1) $\varepsilon_e = \varepsilon_A$; t – the time of loading the rock sample; C_1, C_2 and C_3 – the coefficients of approximation of the real creep diagram obtained experimentally.

The form of equation (13.1) is different from the approximation of M.I. Rozovskyi widely used for rocks and the equation of hereditary plasticity of Yu.M. Rabotnov. However, by selecting the coefficients C_1 and C_3, it is possible to describe quite accurately the experimental diagram of a particular rock lithotype creeping according to the corresponding Fisher similarity criteria.

Thus, an attempt was made to simulate the main peculiarities of the laminal massif behavior of soft rocks in the vicinity of in-seam working.

The third component of the problem is in selecting the most adequate ranges of changing the mechanical characteristics of the Western Donbas rocks and constructing the necessary quantity of variants for mechanical properties combinations of the adjacent coal-bearing strata to assess accurately. All the variety of variants is divided into three positions: coal seam, rock layers of the roof and bottom.

For coal seams, the characteristics of prelimiting state are sufficiently stable: $\sigma^{c}_{compr} = 30 - 40$ MPa, $E^{c} = (0.30 - 0.35)\cdot 10^{4}$ MPa; the characteristics of superlimiting state have been studied to a much lesser extent, but, taking into account the results of certain experiments and the increased brittleness of coal, it is possible to assess the variation intervals in this way: $\frac{\left(\sigma^{c}_{compr}\right)^{rel}}{\sigma^{c}_{compr}} = 0.05 - 0.10$; $\frac{M^{c}}{E^{c}} = 2 - 3.$ Three variants (Table 13.5) of the mechanical coal characteristics have been constructed according to these intervals, which were used for the preliminary testing of the model "massif – in-seam working" system.

The adjacent rock layers of the roof and the bottoms of the coal seam are represented by the same lithological varieties, so the ranges of fluctuations of mechanical characteristics will be the same for the roof and bottom. Analysis of the data from a large number of tests of the rock samples made it possible to determine the most adequate intervals for changing the mechanical characteristics of the prelimiting and superlimiting state of terrigenous rocks of the Western Donbas (Table 13.6).

Table 13.5

VARIANTS OF THE COAL SEAM MECHANICAL CHARACTERISTICS

Variants	Mechanical characteristics			
	σ^{c}_{compr}, MPa	$E^{c}\cdot 10^{4}$, MPa	$\left(\sigma^{c}_{compr}\right)^{rel} / \sigma^{c}_{compr}$	M^{c} / E^{c}
1	30	0.30	0.10	2.0
2	40	0.35	0.05	3.0
3	35	0.33	0.08	2.5

Table 13.6

RANGES OF CHANGING MECHANICAL CHARACTERISTICS OF COAL-BEARING ROCKS

Type of rock	Mechanical characteristics			
	σ^{c}_{compr}, MPa	$E^{c}\cdot 10^{4}$, MPa	$\sigma^{rel}_{compr} / \sigma_{compr}$	M / E
Argillite	5 – 25	0.2 – 1.0	0.10 – 0.25	0.5 – 1.7
Siltstone	10 – 35	0.4 – 1.5	0.08 – 0.20	1.0 – 3.0
Sandstone	30 – 45	0.8 – 2.0	0.05 – 0.15	2.0 – 5.0

Variants of the mechanical characteristics combination in the roof and bottom (Tables 13.7 and 13.8), intended for calculating the SSS of the "massif – in-seam working" system, in addition to involving the main range of variation, correspond to the probability of the lithotypes distribution along the adjacent rock layers an provide the possibility of subsequent correlation-dispersive analysis.

Table 13.7

VARIANTS OF MECHANICAL CHARACTERISTICS OF ADJACENT ROCK LAYERS OF THE COAL SEAM ROOF

Variants	Rock layers of the roof							
	First				Second			
	σ^r_{compr1}, MPa	$E^r_1 \cdot 10^4$, MPa	$\frac{\left(\sigma^r_{compr1}\right)^{rel}}{\sigma^r_{compr1}}$	$\frac{M^r_1}{E^r_1}$	σ^r_{compr2}, MPa	$E^r_2 \cdot 1C^4$, MPa	$\frac{\left(\sigma^r_{compr2}\right)^{rel}}{\sigma^r_{compr2}}$	$\frac{M^r_2}{E^r_2}$
1	5	0.3	0.2	0.5	5	0.3	0.2	0.5
2	5	0.6	0.15	1	10	0.6	0.15	1
3	5	0.3	0.25	0.5	40	2	0.15	2
4	10	0.6	0.2	1	10	0.6	0.2	1
5	10	0.6	0.2	1	20	1	0.15	2
6	10	0.6	0.2	1.5	30	1.5	0.1	3
7	10	0.3	0.25	0.5	5	0.3	0.25	0.5
8	10	0.6	0.2	1	10	0.6	0.2	1
9	20	1	0.15	2	5	0.3	0.2	0.5
10	20	1	0.15	1.5	10	0.6	0.2	1
11	20	1	0.15	2	20	1	0.15	2
12	30	1	0.05	3	10	0.6	0.2	1
13	30	1.5	0.1	3	20	1	0.15	2
14	40	2	0.05	5	40	2	0.05	5

Continuation Table 13.7

Variants	Rock layers of the roof							
	Third				Fourth			
	σ^r_{compr3}, , MPa	$E^r_3 \cdot 10^4$, , MPa	$\frac{\left(\sigma^r_{compr3}\right)^{rel}}{\sigma^r_{compr3}}$	$\frac{M^r_3}{E^r_3}$	σ^r_{compr4}, MPa	$E^r_4 \cdot 10^4$, MPa	$\frac{\left(\sigma^r_{compr4}\right)^{rel}}{\sigma^r_{compr4}}$	$\frac{M^r_4}{E^r_4}$
1	10	0.6	0.15	1.5	10	0.6	0.15	1.5
2	20	1	0.1	2	40	2	0.05	5
3	10	0.6	0.2	1	20	1	0.15	1.5
4	30	1.5	0.1	3	30	1.5	0.1	3
5	20	0.6	0.2	1.5	10	0.3	0.25	0.5
6	40	2	0.05	5	10	0.6	0.15	2
7	30	1	0.15	3	30	1	0.10	3
8	10	0.6	0.2	1	10	0.6	0.2	1
9	20	1	0.15	2	20	0.6	0.15	1
10	30	1.5	0.15	3	20	1	0.15	2
11	20	1	0.15	2	20	1	0.15	2
12	10	0.6	0.15	1.5	40	2	0.05	3
13	10	0.6	0.2	1	10	0.3	0.2	0.5
14	40	2	0.05	5	40	2	0.05	5

Table 13.8

VARIANTS OF MECHANICAL CHARACTERISTICS OF ADJACENT ROCK LAYERS OF THE COAL SEAM BOTTOM

Variants	Rock layers of the bottom							
	First				Second			
	σ^R_{compr1}, MPa	$E_1^R \cdot 10^4$, MPa	$\frac{\left(\sigma^R_{compr1}\right)^{rel}}{\sigma^R_{compr1}}$	$\frac{M_1^R}{E_1^R}$	σ^R_{compr2}, MPa	$E_2^R \cdot 10^4$, MPa	$\frac{\left(\sigma^R_{compr2}\right)^{rel}}{\sigma^R_{compr2}}$	$\frac{M_2^R}{E_2^R}$
1	5	0.3	0.2	0.5	10	0.6	0.15	2
2	5	0.3	0.25	0.5	20	1.5	0.1	3
3	5	0.6	0.15	1	20	1	0.15	2
4	5	0.3	0.2	0.5	10	0.6	0.15	2
5	5	0.3	0.25	0.5	40	2	0.05	5
6	5	0.3	0.2	0.5	40	2	0.1	3
7	10	0.6	0.15	1	10	0.6	0.15	1
8	10	0.3	0.2	0.5	10	0.6	0.15	1
9	10	0.3	0.2	1	10	0.3	0.2	1
10	10	0.6	0.15	1.5	20	1	0.15	2
11	10	1	0.15	1.5	20	1.5	0.15	3
12	10	0.3	0.25	0.5	40	2	0.05	5
13	10	0.6	0.15	1	40	1.5	0.1	3
14	20	1	0.15	2	40	2	0.1	3
15	20	1.5	0.1	3	10	0.3	0.25	0,5
16	20	1	0.2	1	40	2	0.1	3
17	20	1	0.15	1	10	1	0.15	1
18	20	1	0.2	1	40	2	0.1	3

Continuation Table 13.8

Variants	Rock layers of the bottom							
	Third				Fourth			
	σ^R_{compr3}, , MPa	$E_3^R \cdot 10^4$, , MPa	$\frac{\left(\sigma^R_{compr3}\right)^{rel}}{\sigma^R_{compr3}}$	$\frac{M_3^R}{E_3^R}$	σ^R_{compr4}, MPa	$E_4^R \cdot 10^4$, MPa	$\frac{\left(\sigma^R_{compr4}\right)^{rel}}{\sigma^R_{compr4}}$	$\frac{M_4^R}{E_4^R}$
1	10	0.6	0.2	1	10	0.6	0.15	2
2	10	0.3	0.2	0.5	20	1	0.15	2
3	20	1	0.15	2	20	1	0.15	2
4	40	2	0.1	3	40	2	0.1	3
5	20	0.6	0.2	1.5	10	0.3	0.25	0.5
6	10	0.6	0.2	1	40	2	0.1	3
7	10	0.6	0.15	1	10	0.6	0.15	1
8	10	0.3	0.2	0.5	40	2	0.05	5
9	40	2	0.05	5	40	2	0.05	5
10	40	1	0.15	3	20	1	0.15	2
11	10	0.3	0.2	0.5	20	1.5	0.15	3
12	10	0.3	0.25	0.5	40	2	0.05	5
13	40	1.5	0.1	3	10	0.6	0.15	1
14	40	2	0.1	3	20	1	0.15	2
15	20	1.5	0.1	3	40	2	0.05	5
16	10	0.6	0.2	1	10	0.6	0.15	2
17	10	1	0.15	1	20	1	0.15	1
18	20	1	0.2	1	40	2	0.1	3

As a result, the structure and properties of the adjacent coal-bearing stratum have been substantiated, which most adequately reflect the mining-and-geological conditions of the coal seams mined in the Western Donbas. The variety of variants for calculating the SSS of the system, generated by a large number of considered parameters, is preliminarily tested for the detection of factors that do not significantly affect the rock pressure manifestations in in-seam working, and, thus, having limited the set of calculation variants to the technically feasible volume.

13.3. SUBSTANTIATING PARAMETERS OF IN-SEAM WORKINGS AND THEIR SUPPORT

Object of research – preparatory in-seam workings have a number of general parameters, conditioned by their purpose in the technological scheme of the coal mine functioning.

At first, all mine workings are conducted (to the dip, to the rise and along the strike) along the coal seam with overbreaking (roof) or more often with a combined roof and bottom ripping. This means that in the sides of mine working there is a coal seam with a gradient angle in its cross section from 0º (mine workings, having conducted to the dip and to the rise of the seam) to 6º – the maximum angle of incidence of the seams in the Western Donbas (mine workings, having conducted along the strike). The value of the bottom rocks ripping of the main preparatory mine workings usually does not exceed 1 m, (extraction drifts – 1.5 m). Thus, the orientation of in-seam working relative to coal seam is characterized by the following ranges of changes: the gradient angle of the seam in the cross section of mine working $\alpha = 0 - 6$º, the depth of the bottom ripping $h_{b.r.} = 0 -$ 1.5 m. In the work [115], the influence of these parameters was analyzed by means of test SSS calculations of the "massif – in-seam working" system and it has been concluded that they insignificantly impact within the marked ranges of variation. Therefore, it seems expedient to fix the gradient angle of the seam $\alpha = 0°$ in the cross section and the depth of the bottom ripping $h_{b.r.} = 1.0$ m for further studies.

Secondly, the shape of the preparatory in-seam working and the corresponding types of supports in the Western Donbass mines are not diverse. The metal three-link arched support of demountable pliable series with elongated prop stays TSYS series has been widely distributed (up to 82.7%). Typical sections of preparatory in-seam workings for various purposes (from extraction drift to main drifts, cundies, gravity inclines and dipping drifts) as a rule vary in the range from TSYS-9.5 to TSYS-20.2, accepted to study for adequate modeling of mining-engineering conditions. In Fig. 13.2 and in Table 13.9, the dimensions are given of the typical sections, which will be reflected in all the details of the demountable form. The requirement to accurately reflect the real design features of the support and inter-frame fencing consists in accurate modelling:

– geometry of the SCP special profile, from which the TSYS support is made; the SCP-19, 22, 27, 33 is used depending on the purpose of the preparatory mine working and the degree of the enclosing rocks stability in the Western Donbas mines;

– the construction of the yielding joist, which provides a rational mode of the support interaction with soft enclosing rocks;

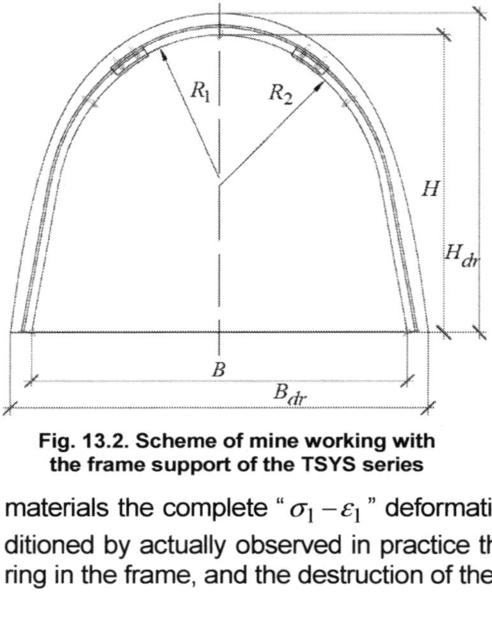

Fig. 13.2. Scheme of mine working with the frame support of the TSYS series

– the design of the backing plate (step bearing) under the prop stays of the frame, limiting their pressing into soft bottom rocks;

– incremental step of the frame support along mine working, which is usually 0.5, 0.8 and 1.0 m;

– the construction of interframe fencing.

Thirdly, the general principle of modelling the properties of the elements of the “massif – in-seam working” system and, in particular, the support with interframe fencing is the description of all the stages of the materials behavior of (from the elastic state to the complete destruction). Thus, for the fastening materials the complete “ $\sigma_1 - \varepsilon_1$ ” deformation diagrams are also used, which is conditioned by actually observed in practice the processes of plastic state areas occurring in the frame, and the destruction of the interframe fencing.

Table 13.9

PARAMETERS OF TYPICAL SECTIONS ACCEPTED FOR STUDY

TSYS size	H_{dr}, mm	H, mm	B_{dr}, mm	B, mm	Cross-sectional area, m^2		R_1, mm	R_2, mm
					project rock section	inside		
9.5	3240	3040	4350	3880	11.2	8.8	1650	2110
10.5	3500	3300	4520	4050	11.8	10.5	1580	2060
11.0	3290	3090	4940	4470	12.7	11.2	2030	2445
11.7	3660	3460	4780	4310	13.1	11.8	1775	2110
12.1	3640	3440	4880	4410	13.4	12.1	1800	2990
13.6	3660	3460	5410	4940	15.6	13.8	2280	2700
14.4	4093	3880	5226	4730	16.0	14.4	1850	2445
15.0	3853	3640	5676	5180	17.1	15.3	2380	2700
17.7	4403	4190	5856	5360	19.6	17.7	2240	2700
20.2	4633	4420	6196	5700	22.7	20.7	2530	3100

For the frame support, a real diagram of St.5 steel loading (from which a special SCP profile is made) is modeled using the data [16, 114]: the elastic characteristics of the prelimiting stage – elastic modulus $E^F = 21 \cdot 10^4$ MPa, Poisson's ratio $\mu^F = 0.3$; the stage of almost perfect plasticity (the yield segment) occurs when the stresses intensity σ increases to level of the yield limit $\sigma_{yield} =$ MPa; with further increase of σ, St.5 steel passes into the stage of strengthening to the value of the temporary tear resistance $\sigma^t = 500 - 620$ MPa.

The deformation peculiarity of the interframe fencing headboard is in the fact that when testing under load a free deflection into the cavity of mine working (between adjacent frames), when the limiting state happens ($\sigma = \sigma^p_{compr,\,t}$), its brittle fracture occurs in the area of the median span section. Then, as a result of it, the headboard practically loses its load-bearing capacity (here denoted by $\sigma^p_{compr,\,t}$ – resistance to compression of the reinforced-concrete headboard stone and resistance to tension of wooden lagging). Therefore, for the headboard materials, it is advisable to use the model of ideally brittle body characterized in the superlimiting state by zero residual strength and a slump modulus tending to infinity. The remaining mechanical characteristics of the headboard are as follows: reinforced-concrete – deformation modulus $E^p = 2 \cdot 10^4$ MPa, Poisson's ratio $\mu^p = 0.17$, uniaxial compressive strength $\sigma^p_{compr} = 20$ MPa; wooden (pine) – deformation modulus $E^p = 1.4 \cdot 10^4$ MPa, Poisson's ratio $\mu^p = 0.20$ ultimate tensile strength $\sigma^p_t = 80$ MPa.

The fastened area is modeled with an average width of 0.17 m (as it is accepted in the project documentation) and has mechanical characteristics in accordance with the materials used:

– when filling the fastened area with hardening mixtures, the mechanical characteristics of the hardened stone are determined depending on the mixture formulation according to the recommendations [51];

– when filling the fastened area with a quarry stone, the "$\sigma_1 - \varepsilon_1$" diagram is directly used of compacting the destroyed rock in cramped conditions [73], which is constructed on the basis of the experimental studies results [78] of the deformation of the protective rubble bands at the Western Donbas mines.

Thus, the substantiated parameters of in-seam workings and their fastening are the final stage of the preliminary construction of a general geomechanical model of the "massif – in-seam working" system, which is a subject to further modernization on the basis of the forthcoming testing.

CONCLUSIONS

1. The mathematical model has been substantiated, which reflects the geomechanics of the coal-bearing massif behavior of soft rocks in the Western Donbas

in the vicinity of preparatory in-seam working, fastened by a frame yield support of the TSYS series. The adequacy and accuracy of the subsequent SSS calculations of the "massif – in-seam working" system is provided by a phased and detailed development of the entire set of peculiarities that significantly influence on the force interaction process of component elements: ranges of changes in the structure of adjacent rocks of roof and bottom, conditions at lithological varieties contacts, ranges of changes in the structure of coal seams, the most accurate reflection of rock and coal properties along the full diagram of their deformation with the corresponding ranges of changes in mechanical characteristics, modelling of the real construction of the support and interframe fencing with substantiation of necessary idealizations to provide the sustainability of the procedure for the FEM calculation and other positions.

2. Adjacent rock layers of the coal seams mined in the Western Donbas (to the height of the roof and to the depth into the bottom up to 10 – 15 m) are mainly composed of soft argillites and siltstones, and to a lesser extent of sandstones with medium strength, the patterns of structural changes of which are considered in the developed model. For this purpose, a massif of structures calculation variants has been constructed that simultaneously reflects the probability of lithotypes distribution and their thickness over the rock layers of the immediate and main roof and bottom, which maximally approximates the geomechanical model to the real mining and geological conditions for maintaining in-seam workings in the Western Donbas mines.

3. In accordance with the probability of lithological varieties distribution along the adjacent rock layers of the coal seams roof and the bottom, a massif of calculation variants of mechanical characteristics has been developed, describing the complete diagram of the Western Donbas rocks deformation, most adequately reflecting all stages of the rock state from elastic-plastic to loosening.

14. TESTING THE MODEL OF THE GEOMECHANICAL SYSTEM "MASSIVE – IN-SEAM WORKING"

The substantiation of the system parameters that most adequately reflect the real mining-geological and mining-engineering conditions of the in-seam workings operation has led to the appearance of an extremely large number of variants of various parameters combinations due to an attempt of their simultaneous accounting. Therefore, the first task of preliminary model testing was to identify those parameters which (at a given degree of accuracy of mining-engineering calculations) can be averaged. The criterion for assessing the degree of influence of a particular parameter is the subject of research – the process of heaving development of the in-seam working bottom. The second task is to substantiate the objective parameters of the force interaction of elements of the "massif – in-seam working" system, for example, the degree of contact disturbances between adjacent layers, the mode of the yielding support operation and its interaction with the boundary rocks through the interframe fencing, etc. The third task is a qualitative assessment of the conformity of test SSS calculations of the system to the current concepts of massif deformation in the vicinity of mine working. The formulated tasks are divided into two main system components – the enclosing massif and in-seam working with a support and the interframe fencing.

14.1. THE SUBSTANTIATION OF PARAMETERS OF THE COAL-BEARING ROCK MASSIF

The first stage of testing the model of the "massif – in-seam working" system is the substantiation of its minimum permissible sizes in space, which, on the one hand, allows obtaining reliable calculation results, and on the other hand, saving the computing resource for the required detail of reflecting the system elements peculiarities. The condition for choosing the minimum model dimensions – the fluctuations of the stresses components on its surfaces, conditioned by the edge effects from the boundary conditions of loading, should not introduce a significant error into the SSS of the border rocks and mine working support. When performing the mining-engineering calculations, an error of 10% is considered quite permissible, which is caused by some uncertainty due to mechanical properties variation of massif in the space. In this case, it is necessary to take into account quite a stable pattern: the fluctuations of stresses components on the surfaces of a model affect the SSS of the border rocks and support rocks, as a rule, by an order less, which in many ways ensures a minimum error in the computational experiment.

For an example of the foregoing provisions implementation, the curves of the specific stresses σ intensity (Fig. 14.1, *a*) are chosen as an integral parameter, most fully reflecting the state of the massif, which is determined in accordance with the Mohr-Coulomb failure theory of testing the rocks. *The first task* is to estimate the distance in the plane XY, to which the perturbations σ are distributed, caused by conducting the mine working. At the same time, the SSS system calculations are performed in an elastic setting, which generates, as it is known, maximum distribution and value of stresses σ concentration. Total exception of mine working influence is observed: at a height of 12.1 m in the roof, at a depth of 16.0 m in the bottom and across the width of 10.4 m in sides. If to limit the dimensions of the model across the width to $X = 13$ m and at a height to $Y = 18$ m (with account of mine working dimensions), then the fluctuations σ along the side vertical surfaces reach 14%, and over horizontal surfaces – up to 20%; however, these fluctuations introduce an error into the SSS of the border rocks no more than 1.7%, which is quite acceptable. Nevertheless, taking into account the variation in the structure and properties of the adjacent coal-bearing stratum (an increase in the area of mine working influence is possible) with a certain reserve of accuracy and reliability of calculations, the model dimensions throughout the height $Y = 30$ m and across the width $X = 20$ m have been chosen for further investigations. The model thickness (according to the axial coordinate Z of the mine working length) is selected according to the same criteria, and here a decisive role is played by the frame support discretely set along mine working, which generates perturbations of the SSS field in itself and in the border rocks. The calculations have shown that when modeling five frames along the mine working axis (the thickness of the model is up to 4 m), three central frames have fluctuations σ (relative to each other), not exceeding 7.2%, and in the border rocks along the length of the three central frames, the differences σ are not more than 3%.

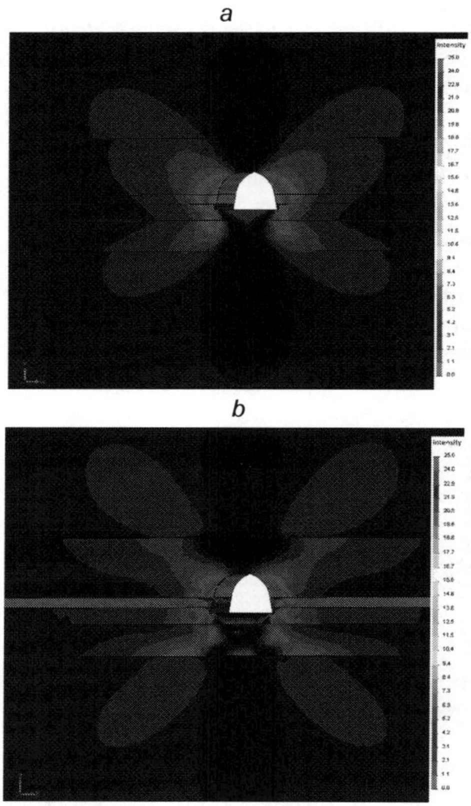

Fig. 14.1. Curves of stresses σ intensity in case of undisturbed *(a)* and disturbed (*b*) contacts between lithological varieties

Thus, based on the test SSS calculations of the "massif – in-seam working" system, it is proved that sufficient accuracy of calculations is ensured by the following model dimensions: $Y = 30$ m, $X = 20$ m, $Z = 4$ m.

The next stage of testing was to study the contact conditions over the surfaces of lithological varieties and substantiation of modeling the parameters of force interaction along the planes of laminations in the vicinity of mine working. This question for the mining-geological conditions of the Western Donbas has been studied in detail in [115], where the distribution field of tangential stresses τ_{xy}, including the planes of laminations, was analyzed in three variants of the structure and various combinations of the deformation characteristics of adjacent rock layers of the coal-bearing stratum. It is shown that already at the depth of 200 m only the action of some tangential stresses leads to disturbance of coherence and slippage of the layers relative to each other at the distance of 6 – 10 m from the mine working contour.

To the tangential stresses τ_{xy}, the action (in series of areas) of tensile vertical σ_y and horizontal σ_x stresses is added, to which the contacts of the rock layers with reduced strength properties practically do not resist, and with increasing depth of mine working location, these processes are intensified. As a result, the authors of [115] came to the conclusion that for the coal-bearing rock thickness of the Western Donbas strata, the modeling of disturbed contacts between the rock layers in the vicinity of mine working is the most objective, although this formulation of the problem substantially increases the required computing resources.

The adequacy of modeling the geomechanical processes in the vicinity of in-seam working is of interest, first of all, from the point of view of the degree of influence of the coherence disturbance between rock layers along the planes of laminations on the system SSS. To assess this influence, the same massif structure in two variants has been calculated: with disturbed and undisturbed contacts between layers. The comparative analysis of the field σ in the rock massif enclosing in-seam working, has shown the following results. First of all, the differentials of σ are clearly observed (which are mainly generated by horizontal stresses σ_x), when changing from one lithological variety to another, which is conditioned by the partially free deflection of each of the layers in the case of disturbance of contact between them. This, on the one hand, confirms the correspondence of the model to the classical provisions of the mechanics of deformable solids, and, on the other hand, indicates a general increase in the intensity of perturbations and the degree of their non-uniformity in the vicinity of in-seam working.

The deflection of the immediate roof, which has lost its adhesion to the main roof, expands the discharge zone in the roof of mine working by 1.4 – 1.6 times, and geostatic pressure is transferred to the rocks in the sides, which is especially evident in the coal seam and rock layers of the bottom: at a distance of more than the width of mine working, the specific stresses σ in the coal seam are increased by 1.5 – 2.5 times, and the concentration $\sigma = 11.5 - 12.5$ MPa is propagated in the second rock layer of the bottom to a width up to 3 times and to a depth up to 2.7 time greater than that with undisturbed contacts between layers. The concentration of stresses $\sigma > 20$ MPa in the sides of mine working is propagated to a width of up to 0.6 m in the immediate bottom and up to 1.5 m in the coal seam, which is not observed with contacts cohesiveness between the layers; and the previously unloaded bottom of mine working ($\sigma < 3$ MPa) is exposed to up to σ 12 MPa, which, in soft rocks, intensifies its heaving. This result of an increase in the stresses intensity in the sides and in the bottom of mine working is explained (in accordance with the energy conservation law) by redistributing of a part of geostatic pressure from the unloading zone in the roof to the side rocks, and from them to the mine working bottom. Another conclusion of this phenomenon is that one cannot ignore a significant increase in SSS in the sides and bottom of mine working, and for this it is necessary to model the contacts disturbance between lithotypes in the vicinity of mine working.

In general, the nature of the stresses components distribution corresponds to the existing concepts of the processes of the coal-bearing strata displacement in the vicinity of mine working: the zones of unloading are formed in the roof and bottom, with occurrence of tensile stresses σ_y and σ_x; the zones of increased rock pressure are formed in the sides; the changes in the sign of tangential stresses τ_{xy} relative to the vertical axis of mine working and mirror reflection of the field τ_{xy} relative to the axis inclined at an angle of 45°, which is conditioned by the horizontal bedding of the seam close to the horizontal, and the coincidence of the vectors of the principal stresses σ_1 and σ_3 with the vectors σ_y and σ_x.

The next stage of testing is an assessment of the influence degree of the thickness and structure of coal seams mined in the Western Donbas. The influence (in the range of $m^c = 0.7 - 1.7$ m) of coal seams thickness on the SSS of the rocks enclosing in-seam working, as well as the frame support was investigated in [115], where the change in each stresses component in the most significant areas has been studied: in the zones of unloading in the roof and bottom, increased rock pressure in the sides of mine working, frame support, etc. The most representative range of changes, 0.7 m $\leq m^c \leq$ 1.2 m, more specific than the one in [115], which is substantiated, is about the permissibility of averaging the seam thickness with a value of $m^c = 1$ m, all the more is fair in the study of problem set.

The second part of the stage is an assessment of the influence of the complex coal seam structure (37% of cases in the Western Donbas). To do this, a geomechanical model has been constructed with the most "extreme" combination of parameters of the coal seam structure (an interlayer of sandstone with a thickness of 0.12 m is located in the upper third part of the seam thickness) to identify the maximum possible SSS perturbations. On the curve of these stresses σ, an increased rock layer intensity is clearly observed, which is caused by: a significant difference (up to six times) of the deformation modulus of sandstone and coal, when a more rigid element (sandstone interlayer) takes up an increased load; the low interlayer thickness, which backs onto the base (coal) with reduced rigidity, hence, the significant bending stresses occur. However, SSS perturbations have a very limited distribution and are localized practically in the very rock interlayer: there have not been any significant changes in SSS within the border rocks and in the frame support. Taking into account that the coal seams (see Table 13.2) with the complex structure in the Western Donbas are not predominant, and the rock interlayer is most often represented by argillite (coal-like deformation characteristics) with the most probable thickness of 0.03 m, it has been decided to ignore the rock interlayers when modeling the coal seam parameters.

The performed test calculations have revealed the fact of insignificant influence of mine working on the SSS of the extreme (fourth) layers of the roof and bottom, the parameters of which can be averaged and excluded from the set of calculation variants in further studies. On the other hand, the collected statistics of calculations do not yet allow us to finally conclude about the insignificant influence of the extreme layers of bottom and roof on the heaving process and the frame support stability. Therefore, the question of averaging the parameters of the extreme layers of the model will be solved based on the results of the subsequent group of computational experiments when modeling the complete diagram of the elements deformation of the "massif – in-seam working" system with a different mechanical characteristics combination. Then, the confirmation of the assumptive fact will significantly reduce the number of calculated variants of combining the system parameters.

The final stage of testing the enclosing massif SSS concerns consideration of its rheological properties. Here, a series of SSS calculations of the "massif – in-seam working" system has been carried out using the Solid Works 2009 Simulation program. It takes into account the rheological properties of mine rocks according to the equation (13.1), thus, the following facts have been revealed. Firstly, consideration of the time factor significantly complicates the task, which is difficult due to the detailed reflection of the design parameters of the system elements, the description of their mechanical properties according to the complete deformation diagram with the condition of contacts disturbances between lithological varieties. This setting requires a significant increase in computing resources, and the time of calculating one variant increases to 100 – 150 hours or more. Secondly, there is a low reliability of the calculation process itself, when the program fails at any time interval and it is necessary to repeat again and again the calculation of the same variant of original data. Together, these factors have led to a situation in which several "successful" calculations required time expenditures comparable to the period of creep development in the Western Donbas. Despite the sufficient adequacy of the obtained modeling results to the in-situ observations of the creep process development in the vicinity of mine working, this situation cannot be considered as satisfactory, since to determine the geomechanical patterns a set of design variants is required, at least two orders of magnitude greater than obtained as a result of long test calculations. The way out of this situation has been found on the basis of a compromise solution of accounting the rheological factor by decrease in values of mechanical characteristics of rock during long loading on the basis of studies [62, 110]:

– long-term strength limit of rock

$$\left(\sigma^{r,R}_{compr_i}\right)_{\infty} = \sigma^{r,R}_{compr_i}\sqrt{1-\left(\frac{x}{\beta}\right)^{r,R}_{i}}; \tag{14.1}$$

– long-term deformation modulus

$$\left(E^{r,R}_{i}\right)_{\infty} = E^{r,R}_{i}\left[1-\left(\frac{x}{\beta}\right)^{r,R}_{i}\right], \tag{14.2}$$

where $\left(\frac{x}{\beta}\right)^{r,R}_{i}$ – rheological index, for which a correlation relationship has been set with the uniaxial compressive strength of rock

$$\left(\frac{x}{\beta}\right)^{r,R}_{i} = 0.8-0.326\lg\sigma^{r,R}_{compr_i}. \tag{14.3}$$

The reduced mechanical characteristics are set into calculation, and, thus, in an integrated form a final impact of a rheological factor on a condition of “massif – in-seam working” system is considered.

14.2. PARAMETERS SUBSTANTIATION OF THE IN-SEAM WORKING AND SUPPORT

The tasks of this stage of testing the geomechanical model of “massif – in-seam working” system include determination of a level of influence of the following factors on heaving of mine working bottom and its support state: angle of seam gradient in a cross section of mine working, depth of seam bottom ripping, models of the yielding joist of a frame and interframe fencing.

Influence of the first factor was studied on two models – with parallel seam arrangement relative to the mine working bottom (the preparatory mine workings driven to the rise and to the dip of the seam) and with angle of gradient 6º to the corresponding maximum angle of seam incidence in the Western Donbas (the preparatory mine workings driven along the strike of the seam). An insignificant increase in the specified stresses σ (up to 7 – 12%) has been set in the border rocks of mine working sides to a width of 0.4 – 0.7 m from the side of a seam rise. Some asymmetry arises in the form of unloading zones in the roof and bottom of mine working (relative to its vertical axis), but practically without change in their sizes. In the frame support, the influence of the angle of seam gradient is manifested only in the areas of prop stays bearings, but is insignificant. This result is partly consistent with works [73, 111, 115] in that more essential (up to 15 – 18%) asymmetry of mine working bottom heaving and deforming the prop stays of frames is manifested only under the conditions of extremely unstable enclosing rocks, when the section loss of mine working reaches 40 – 50% and more. Given the analysis results of the system SSS in case of the maximum angle of seam incidence 6°, which in the Western Donbas is not widespread, the conclusion has been drawn about sufficient relevancy of mine working modeling with parallel arrangement of coal seams relative to its bottom.

The influence of value of the bottom ripping was analyzed in the range of 0 – 1.5 m and the following patterns have been determined. The process of bottom rocks heaving practically in the least depends on the ripping value, and is conditioned mainly by a structure and properties of adjacent rock layers. The curves of complete border rocks displacements show that their vector changes the direction by 180° from the predominantly vertical in the roof, horizontal in the sides to vertical in the bottom with the opposite sign. That is, the process of heaving affects the considerable rock volume not only in the bottom, but also in the sides of mine working and partly in the immediate roof. Hence, some repositioning of mine working (connected with the ripping value) relative to coal seam does not change the essence of heaving phenomenon. The same reasons influence on the process of frames prop stays deformation, especially their lower part, therefore, the ripping value does not significantly influence on the SSS of the frame support in the studied range.

At the first stage of the yielding joist modeling, an attempt was made to accurately reflect its design with respect to the TSYS-15.0 support. As a result of the very time-consuming and lengthy process of the software adaptation to the peculiarities of the real joist design, initial and boundary conditions of its modeling, the SSS of the "massif – in-seam working" system has been calculated, but only in the elastic approach until the support is settled (yielding joist actuation), which is characterized by small displacements of mine working contour. An attempt to model complete diagrams of all system elements deformation has revealed the instability of the procedure for performing the calculations by the FEM, because of, as it seems to us, the static imbalance of the frame in the area of the joist joint. The analysis of this situation has revealed that due to the "point" end sections contact of the profiles of the prop stay and cap board with each other in the joist, as well as the small area of this contact, it arises an ability of the frame deformation (displacements) by a larger amount (along the X, Y and Z coordinates) with a very small load increment, which causes malfunctions in the program and termination of the SSS calculation. The situation is even more aggravate by modeling the process of joist yield, when with the cap board sliding with respect to the prop stay, due to the existing frame geometry, the area of their contact in the joist area decreases and the degree of freedom of the frame displacement (static imbalance) increases in an arbitrary direction. It should also take into account the prospects of this geomechanical model development in terms of reflecting the possibility of plastic metal flow, the occurrence of not only areas of weakening, but also of loosening in the adjacent rock massif. These factors lead to the geomechanical model complication, which will objectively enhance the instability of the procedure for performing the SSS calculations and uncertainty in time to overcome these modeling difficulties.

In a situation where there is no possibility to quickly and accurately model a real object, substantiated constructive idealizations have been introduced, which allow stable continuation of the SSS calculations of the "massif – in-seam working" system. However, the simplification of the joist design should not distort its deformation-strength characteristics when working in the yielding mode. This condition, in our opinion, made it possible to adequately model the work of the yielding joist.

Modeling of the idealized design of the yielding joist was carried out in two stages. *At the first stage* – in the area of the joist location, a gasket from easily deformable material was laid, which provided its yielding property under the action of compressive forces in the frame. At the same time, the stability of the calculation process of the system SSS according to complete diagram of its elements deformation was achieved, and analysis of stresses curves showed their compliance with the classical provisions of the underground structures mechanics: modeling of driven joints with easily deformable material does not significantly distort the qualitative pattern of the frame support loading.

The second stage of the joist modeling was performed to achieve the adequacy to the real process of yielding in quantitative terms. In this connection, the deformation-strength joist characteristic is the dependence of its resistance reaction P on the yielding value U_y (as a rule, in the vertical direction along the Y coordinate). A number of existing laboratory tests of a frame three-tiered yielding support indicates the operation nature of the joist from the SCP profile, which approaches to the constant resistance mode. The same experiments have revealed that if the joist is correctly assembled, the frame support changes into a yielding mode with resistance $P_{yield} = (0.7 - 0.8)P_{\max}$, approaching to the maximum load-bearing capacity $P_{\max}$ of the frame. The described mode of quasi-permanent resistance of the yielding joist can be represented rather adequately by the diagram of ideal plasticity of the material located in the frame in coordinates of joists setting. The material parameters should reflect the real deformation-strength characteristic of the joist for a specific typical section and the number of SCP profile. Therefore, the yield point σ_{yield}^{joist} of the joist simulator is picked up as such that at an average normal stresses on cross-sectional area of SCP (in the middle of the joist length), when the load on the frame is 80% of its load-bearing capacity, the material of model passed into a state of ideal plasticity and provided yielding for the support.

The joist simulator in the form of the cross-section with the corresponding number of SCP profile is located in coordinates of the frame joists with a real length of 400 mm. It has enabled to model a frame as integral (from the left support to the right one) along the support contour with two inserts from material of the joist simulator, which is different from mechanical characteristics of the St.5 steel only by the lowered yield point. Such idealization of the yielding joist has provided stability of the procedure of the SSS calculation of "massif – in-seam working" system according to the complete diagram of deformation of its elements materials.

Along with other factors, the influence of interframe fencing type (a cyclone fencing, a wooden or reinforced concrete lagging) was studied, which can be assessed as very insignificant (within the accuracy of the FEM calculations) during the development of a bottom heaving process and deformation of the frame support.

Summing up the results of testing "massif – in-seam working" system model, it is possible to state the fact of its comprehensive substantiation on ensuring adequacy of the description to real geomechanical process in the vicinity of in-seam workings.

CONCLUSIONS

1. Based on the analysis of coal seams structures, a number of test models of SSS calculation of the "massif – in-seam working" system has been developed, the results of which have testified the insignificance of rock interlayers influence (37% of cases of coal seams with complex structure) in the stresses field and sufficient adequacy of modeling the coal seam with averaged thickness ($m^C = 1$ m) – an error from such averaging (when assessing the bottom rocks heaving and support prop stays deformation) is not beyond the allowed accuracy of FEM calculations.

2. Test calculations of the SSS of "massif – in-seam working" system have proved emergence of contact disturbances of lithological varieties along the planes of beddings within the studied areas of massif in the vicinity of mine working. This is reflected in the development of geomechanical models of a system.

3. The parameters of the preparatory in-seam workings and their support have been substantiated, which are most adequately reflecting mining-engineering conditions of their maintenance. Test calculations of SSS of the system have substantiated an insignificance of influence of the angle of seam gradient relative to the mine working bottom and the depth of its ripping under the conditions of the Western Donbas mines. Therefore, a decision was made quite reasonably to model the parallelism of the seam and the bottom of mine working with an averaged ripping value of 1.0 m, that does not reduce the reliability of calculations for assessing the rock pressure manifestations in in-seam working. The adequacy

of a yielding joist model of the frame support to its real operating mode has been proved in the qualitative and quantitative terms.

4. The complex of test SSS calculations of the "massif – in-seam working" system made it possible to work out the geomechanical model for all the key parameters to the level of its adequate reflection of geomechanical processes in the vicinity of mine working that leads to a high degree of studies reliability to predict the rock pressure manifestations.

15. RESEARCH AND ANALYSIS OF INFLUENCE OF STRUCTURE AND PROPERTIES OF THE COAL-BEARING ROCK STRATUM ON THE STRESS-STRAIN STATE OF THE "MASSIF – IN-SEAM WORKING" SYSTEM

In chapter 13, the base of structure and properties variants of adjacent coal-bearing rock stratum of the Western Donbas has been developed and substantiated, which most adequately and accurately reflects the real mining-and-geological conditions. The primary test SSS calculations of the massif – in-seam working' geomechanical system have completely proved all its geometrical, mechanical and strength parameters. However, the obtained diversity of the calculations has required further studies in terms of identification and averaging the parameters insignificantly influencing the general patterns of development of the bottom rocks heaving of mine working. This task was solved during the second stage of test calculations in accordance with the variants of Table 13.7 and 13.8.

15.1 SUBSTANTIATION OF VARIANTS OF STRESS-STRAIN STATE CALCULATION OF "MASSIF – IN-SEAM WORKING" SYSTEMS AND THE METHODOLOGY OF CONDUCTING THE COMPUTATIONAL EXPERIMENT

The testing (through a complex of SSS test calculations) of the geomechanical system was begun with an analysis of the structure influence of the immediate and main roof of the coal seams. In the works [73, 111], besides the main patterns of the rock pressure manifestations in in-seam workings of the Western Donbass mines, the so called "stamping effect" has been revealed, when the rock seam of an immediate roof with an increased rigidity and hardness "presses-out" the softer bottom rocks into mine working cavity due to formation of the increased rock pressure (IRP) zones in its sides. This influence of rigidity and hardness of the immediate roof is not dominating in the mechanism of heaving development. However according to data [111], the value of a bottom heaving can fluctuate up to 19.2% depending on the deformation modulus of the immediate roof rocks and up to 72.9% depending on the value of compressive resistance, which is very essential factor. The data provided characterize the influencing maximum when sandstone with a thickness about 4 m occurs in the immediate roof, which, because of its rigidity and hardness, without collapsing acts as a stamp, pressing-out the bottom rocks into a mine working cavity.

From such positions, the structures of adjacent rock seams of the roof in the Western Donbas mines have been analyzed. The case of bedding in the immediate roof of sandstone with thickness (up to 5 m) takes place in Yuvileina Mine.

However, it is an exception of the general patterns of coal-bearing stratum structures, is not widely spread and is not characteristic. This is confirmed by the studies [83] of structure in general across the Western Donbas which assess the probability of sandstone occurrence in the immediate roof up to 12%, and the probability of bedding the thick sandstone is even less. Thus, there is a low probability of an intensification of the rocks heaving process due to the influence of thick sandstone in the immediate roof. The following structure of the immediate roof rocks and adjacent rocks of the main roof is the most probable. In the immediate roof, the probability of bedding the argillite prevails, thickness of which has the main range of fluctuations from 1 to 6 m with predominant interval of 2 – 3 m. These data do not contradict the studies [110], where the probability is set of 52% argillite occurrence with a thickness of $m_1^r = 1 - 4$ m (76%) in the Western Donbas. Here it should be noted that argillite by its deformation and strength characteristics is not capable to create the rigid and strong rock plate in the immediate roof acting as a stamp on the bottom rocks. The stated facts substantiate the expediency of the immediate roof structure "unification", having presented it by argillite with a constant thickness of $m_1^r = 2$ m that will significantly reduce the volume of variants of further structures calculation without loss of objectivity of real mining-and-geological conditions reflection.

The structure of the main roof, from the point of view of influence on the process of bottom heaving is not of interest, as according to the data [73, 110], an essential influence of the main roof is not observed. However, in order to adequately represent the overall process of displacement of coal-bearing stratum into mine working cavity, it is necessary to substantiate also the structure of adjacent rock layers of the main roof. Based on the second stage results of test calculations of the system SSS, it is expedient to model two lower layers of the main roof from argillite and siltstone with a thickness of 4 m. Then, the roof rocks in general are modelled with a total thickness of 10 m which is enough [73, 110] for total disturbance reflection of the massif SSS caused by mine working operations.

The coal seam parameters are modelled on the basis of substantiation performed earlier.

The analysis of the degree of influence of the bottom structure on process of its heaving has allowed to make the following decision regarding the geomechanical model of the "massif – in-seam working" system: the first and fourth rock layers of the bottom are characterized by the range of changing the thickness $m_{1,4}^R = 0.5 - 4$ m; the second and third – $m_{2,3}^R = 1 - 4$ m.

Thus, the most characteristic structure has been revealed of an adjacent coal-bearing stratum of coal layers which will be reproduced in a complex of the models variants of the "massif – in-seam working" systems.

The degree of adequacy of the computational experiment results directly depends on reliable reflection of the mechanical properties of the adjacent coal-bearing stratum. First of all, the objectivity of research depends on the chosen model of mine rock behavior, which (as it has been specified earlier) describes all stages of its deformation: elastic-plastic, weakening and loosening. However, as for a real object, there is a lack of information here about its mechanical characteristics of a complete deformation diagram. In fact, in geological surveys, only compressive resistance of this or that lithotype is defined, other characteristics are determined when conducting special enough labor-consuming laboratory testing. Such studies were conducted by experts [62, 74] for all lithotypes of the weakly metamorphosed Western Donbas rocks, the results of which are used in a computational experiment. For this purpose, the most characteristic ranges of mechanical characteristics fluctuations of specific lithotypes have been analyzed and reduced in Table 15.1.

Table 15.1

MECHANICAL CHARACTERISTICS OF ADJACENT ROCK LAYERS OF ROOF AND BOTTOM OF THE COAL SEAMS

Type of rock	σ_{compr}^{c}, MPa	$E \cdot 10^4$, MPa	$\sigma_{compr}^{res} / \sigma_{compr}$	M / E
Argillite	6 – 11	0.2 – 0.4	0.20 – 0.25	0.5 – 1
Siltstone	18.2 – 20.6	0.6 – 1.2	0.12 – 0.15	1.5 – 2.5
Argillite	13 – 20	0.3 – 0.6	0.15 – 0.20	0.5 – 1
Coal seam	30 – 40	0.30 – 0.35	0.10	3
Siltstone	13.8 – 30	0.4 – 1	0.15 – 0.20	1 – 2
Sandstone	40 – 51	1.5 – 2	0.08 – 0.10	3 – 5
Argillite	4.9 – 21	0.2 – 0.4	0.20 – 0.25	0.5 – 1
Low-thickness coal seam and interlayers	30 – 40	0.30 – 0.35	0.10	3
Siltstone	8 – 30	0.8 – 1.5	0.15 – 0.20	1 – 2

Note: some lithotypes can be absent or the order of their location in vertical coordinate is changed.

In the conclusion of the analysis of mining-and-geological conditions of preparatory mine workings arrangement, an angle of the seam incidence $\alpha = 3.5°$ was accepted and the depth of bedding is changed within $H = 300 - 500$ m. Concerning the parameter H, it is also necessary to take into account that at the used step-by-step task solution of the complete diagram of rock deformation, the depth changes smoothly to a maximum, so that it is not difficult to fix calculations at any value of H and determination of the SSS of the "massif – in-seam working" system within and outside the specified range of changes in the depth of mine working location.

The following main task is to construct an algorithm for a computational experiment to set the patterns of the bottom heaving depending on its structure and properties. In this regard, variants of the system models are substantiated, which should include the most representative ranges of changes both in the structure and properties only of the seam bottom rocks, since the structure and properties of the roof rocks have already been substantiated in the previous subchapter. It has been set that argillite and siltstone bedding almost everywhere in the roof cannot have any significant influence on the development of the heaving process in terms of its intensification or, conversely, its limitation. Therefore, the roof rocks of a seam are modelled with constant parameters for all calculation variants (see Table 15.1):

– the first rock layer of the roof – $m_1^r = 2$ m

$$\sigma^r_{compr_1} = 15 \text{ MPa}, \ E_1^r = 0.4 \cdot 10^4 \text{ MPa}, \ \frac{\left(\sigma^r_{compr_1}\right)^{rel}}{\sigma^r_{compr_1}} = 0.2, \ \frac{M_1^r}{E_1^r} = 0.8;$$

– the second rock layer of the roof – $m_2^r = 4$ m

$$\sigma^r_{compr_2} = 19 \text{ MPa}, \ E_2^r = 0.9 \cdot 10^4 \text{ MPa}, \ \frac{\left(\sigma^r_{compr_2}\right)^{rel}}{\sigma^r_{compr_2}} = 0.15, \ \frac{M_2^r}{E_2^r} = 2;$$

– the third rock layer of the roof – $m_3^r = 4$ m

$$\sigma^r_{compr_3} = 8 \text{ MPa}, \ E_3^r = 0.3 \cdot 10^4 \text{ MPa}, \ \frac{\left(\sigma^r_{compr_3}\right)^{rel}}{\sigma^r_{compr_3}} = 0.20, \ \frac{M_3^r}{E_3^r} = 0.5.$$

The coal seam of a simple structure is modelled with thickness of $m^c = 1.0$ m with the mechanical characteristics specified in Table 15.1.

To adequately reflect the bottom rocks structures, the following factors are considered:

– *first,* the amount of variants of the bottom rocks structures is limited, on creation and processing of which a lot of time is spent;

– *secondly,* in already modelled structures, a technique has been introduced for elimination (if necessary) of a slipping condition of adjacent layers relative to each other, then with their identical mechanical characteristics a thicker rock layer is formed (as the sum of two adjacent layers); thus it is possible to transform one basic structure into several with the minimum time expenditures;

– *thirdly,* for the same structure model, a group of the mechanical characteristics variants for each rock layer is calculated, that reflects not only the range of changes in the lithotype properties, but also the variation of lithotypes alternation along the depth of the coal seam bottom (this operation also does not require a significant time expenditures).

The stated methodological technique has allowed to significantly reduce the time expenditures for performing the complex of computational experiments without compromising the reliability and adequacy of the patterns determination of the bottom heaving development in in-seam workings. As a result, it has been created a minimum sufficient number of variants of the models of the "massif – in-seam working" system, the bottom structure of which is reflected in Table 15.2, and mechanical characteristics – in Table 15.3.

Table 15.2

CALCULATION VARIANTS OF ADJACENT ROCKS STRUCTURE OF THE COAL SEAMS BOTTOM

Thickness of bottom rock layers, m	Numbers of the structure variants				
	1	2	3	4	5
First, m_1^R	2	0.5	4	1	3
Second, m_2^R	2	3	1.5	1	4
Third, m_3^R	2	1.5	3	4	1
Fourth, m_4^R	2	4	0.5	3	1

In Table 15.2, the basic variants of the bottom structure are presented, which are selected in such a way that, firstly, the structure in general is changed from uniform (according to layer thickness) to extremely non-uniform with layer-by-layer and group alternation of thin and thick layers; secondly, in each rock layer there are five variants of various thickness values. Paired and tripled combining of adjacent layers gives six more bottom structures inside each basic one, then, if necessary, one can obtain a total of 35 structures of adjacent rocks of the seam bottom, which in total quite adequately reflect the mining-and-geological conditions for maintaining the preparatory mine workings in the Western Donbas mines.

Given the calculation variants of combining the mechanical characteristics of the adjacent rock layers of the bottom (Table 15.3) for setting the patterns of the heaving development, it is required to carry out 35 SSS calculations of the system according to basic structure variants and at most 75 variants (if it is necessary), of derivatives of the basic structures. The patterns are determined on the basis of techniques of correlative-dispersion analysis [116] of curves of the bottom displacements, which are obtained by each separate calculation of the system SSS and in total develop an initial database for creating the regression equations.

Research is conducted with the fixed angle of seam incidence $\alpha = 3.5^{\circ}$ and depth of mine working location $H = 400$ m; SSS of the system at $H \neq 400$ m is calculated automatically with the used step-by-step technology of calculations. Thus, for accounting the parameter H influence, additional calculations are not required, and the construction of the regression equations, including the parameter H, is carried out by the same techniques of correlative-dispersion analysis.

Table 15.3

CALCULATION VARIANTS OF COMBINING MECHANICAL CHARACTERISTICS OF ADJACENT ROCK LAYERS OF THE COAL SEAMS BOTTOM

Number of variant	m_1^R				m_2^R			
	σ_{compr}, MPa	$\frac{\sigma_{compr}^{rel}}{\sigma_{compr}}$	$E \cdot 10^4$, MPa	$\frac{M}{E}$	σ_{compr}, MPa	$\frac{\sigma_{compr}^{rel}}{\sigma_{compr}}$	$E \cdot 10^4$, MPa	$\frac{M}{E}$
1	5	0.25	0.3	0.5	5	0.2	0.4	1
2	5	0.25	0.3	0.5	10	0.15	0.6	2
3	20	0.2	1	1	40	0.1	2	3
4	5	0.25	0.3	0.5	10	0.15	0.6	2
5	20	0.2	1	1	40	0.1	2	3
6	5	0.25	0.3	0.5	40	0.1	2	3
7	20	0.2	1	1.5	20	0.15	1.5	2

Continuation of Table 15.3

Number of variant	m_3^R				m_4^R			
	σ_{compr}, MPa	$\frac{\sigma_{compr}^{rel}}{\sigma_{compr}}$	$E \cdot 10^4$, MPa	$\frac{M}{E}$	σ_{compr}, MPa	$\frac{\sigma_{compr}^{rel}}{\sigma_{compr}}$	$E \cdot 10^4$, MPa	$\frac{M}{E}$
1	20	0.15	1	2	20	0.2	0.6	1.5
2	10	0.2	0.6	1	10	0.15	0.6	2
3	20	0.2	1	1	30	0.1	0.3	3
4	40	0.1	2	3	40	0.1	2	3
5	10	0.2	0.6	1	10	0.15	0.6	2
6	10	0.2	0.6	1	30	0.1	0.3	3
7	5	0.25	0.3	0.5	5	0.2	0.4	1

In the methodological term, the following research algorithm has been developed:

– calculation of the SSS of the basic models "massif – in-seam working" system from Table 15.2;

– analysis of curves of stresses and displacements of mine working bottom according to the variants of mechanical characteristics distribution from Table 15.3;

– construction at first approximation of patterns of relationship of bottom displacements with geomechanical parameters of the system;

– establishment of mining-and-geological conditions for the ambiguous interpretation of patterns of the bottom heaving manifestation;

– performing additional calculations for other variants of the bottom structure, which are derivatives from basic variants;

– detailing of patterns influence of the bottom structure and properties on the process of its heaving;

– construction of regression equations for predicting the bottom heaving of preparatory mine workings.

15.2. STRESS-STRAIN STATE ANALYSIS OF THE BOTTOM ROCKS IN IN-SEAM WORKING

According to the developed algorithm of conducting research, at the first stage the calculation of basic models has been made for five bottom structures (see Table 15.2) with seven variants of their mechanical characteristics distribution over rock layers (see Table 15.3). The preliminary analysis has revealed (with the fixed depth of mine working location) some ambiguity of influence of structure and properties of adjacent bottom rocks on the heaving process: on the one hand, the well-known tendency is confirmed of decreasing the bottom displacements with increase in the rock strength characteristics; on the other hand, the process is exposed to the influence of considerable inhomogeneity of the properties (in a number of mining – and – geological conditions) of adjacent rock layers (for example, the moistened siltstone and sandstone), which is amplified with a wide range of fluctuations of their thickness and the possibility of destruction of thin sandstone. Such situation requires a detailed analysis of the curves of stresses components and, on its basis, disclosing the features of the mechanism of the bottom rocks displacement depending on its structure and properties.

15.2.1. ANALYSIS OF VERTICAL STRESSES COMPONENTS

The SSS analysis of the bottom rocks of in-seam working has been made beginning with a curve of vertical stresses σ_y with the identical thicknesses of layers (variant No. 1 of the bottom structure according to Table 15.2). Regardless of mechanical characteristics of rock layers of the bottom, the curve σ_y has common feature (Fig. 15.1): the area of the reduced compressive stresses σ_y, which change into in tensile stresses (is closer to the surface of the mine working bottom), considerably exceeds that in the roof. This is typical for the Western Donbas conditions where manifestations of heaving are, as a rule, more intensive, than manifestations of rock pressure in the roof of mine working.

Tensile stresses σ_y form the area of weakening in the bottom (as the Western Donbas rocks almost do not resist the tensile forces taking into account the weakening factors of watering, stratification, fracturing and rheology), the depth of which in variants No. 2 and No. 6 (see Fig. 15.1, *a*, *c*) of very soft rocks bedding in the immediate bottom reaches borders of the third layer, that is makes about 4 m. The feature here is that in variant No. 6 the second rock layer is presented by hard sandstone, but this factor insignificantly promotes the limiting of a weakening zone depth. Obviously, the influence of others stresses components is shown here, and in some cases – significant inhomogeneity of mechanical properties of adjacent rock layers. In variant No. 3 (see Fig. 15.1, *b*), the stronger rocks limit the depth of the weakening area to 2.5 m, which completely fits into the framework of existing ideas of manifestations of the bottom heaving though this tendency is not so evident: to reduce the depth of weakening to 40%, a four-fold increase in the value of compressive resistance is required. On the other hand, if to allow some small resistance of sandstone to the tensile forces, then within its thickness, the exception of weakening from tensile stresses is possible σ_y and this area will be limited only by the first rock layer of the bottom.

Among other variants of ratios of mechanical characteristics it should be noted that the bedding of more soft rocks in the third and fourth layer in comparison with harder rocks of the first and second layers practically does not change the curve σ_y (as in the qualitative, and quantitative plan), though a part of more soft remote rocks is in the limiting state. Obviously, the first harder layers constrain the process of the heaving development. In the opposite variant No. 4, the bedding of harder rocks in the third and fourth layers also insignificantly influences the curve σ_y in comparison with variant No. 2 (see Table 15.3) of bedding in the bottom of only soft rocks. Thus, for structure No. 1 of the bottom rocks, the ambiguity of influence of mechanical characteristics of its layers on the curve σ_y is in that, on the one hand, the properties of the remote layers insignificantly influence on the curve σ_y; on the other hand, – approximation of sandstone to the immediate bottom also does not limit the depth of the weakening zone along with the fact that the bedding of harder rocks in the first and second layers significantly limits the depth of the weakening zone.

The following variant No. 2 (see Table 15.2) of the bottom structures is characterized by alternation of less thick (odd) and more thick (even-numbered) rock layers and the analyzed variant is essentially different from the previous, that allows to assess the influence of the bottom rocks structure (other things equal). In the soft rocks of the bottom (for all four layers – variant No. 2 from Table 15.3), the depth of acting the tensile stresses decreases (in comparison with the previous structure) and is about 3.2 m (Fig. 15.2, a) that corresponds to the depth of the bottom weakening zone. When the second layer is presented by sandstone with the increased thickness of 3 m (variant No. 6 from Table 15.3), there is a decrease in the depth of weakening zone to 2 m (see Fig. 15.2, *c*). Therefore, an increased rigidity and strength of the second layer increase the stability of the bottom that is caused also by decrease in the thickness of the first (soft) rock layer in comparison with the previous structure where sandstone's influence was practically absent. Such tendency is quite natural and is confirmed by the curve σ_y for variant No. 3 from Table 15.3 (see Fig. 15.2, *b*) when with the same sandstone in the second layer, the first is presented by harder siltstone; but because of its small thickness, the influence is insignificant. Thus, the general tendency of rocks hardness influence remains, but its intensity increases to 60% that can be explained by an increased thickness up to 3 m of the second rock layer, when it is presented by sandstone; other combinations of mechanical characteristics only confirm the revealed pattern.

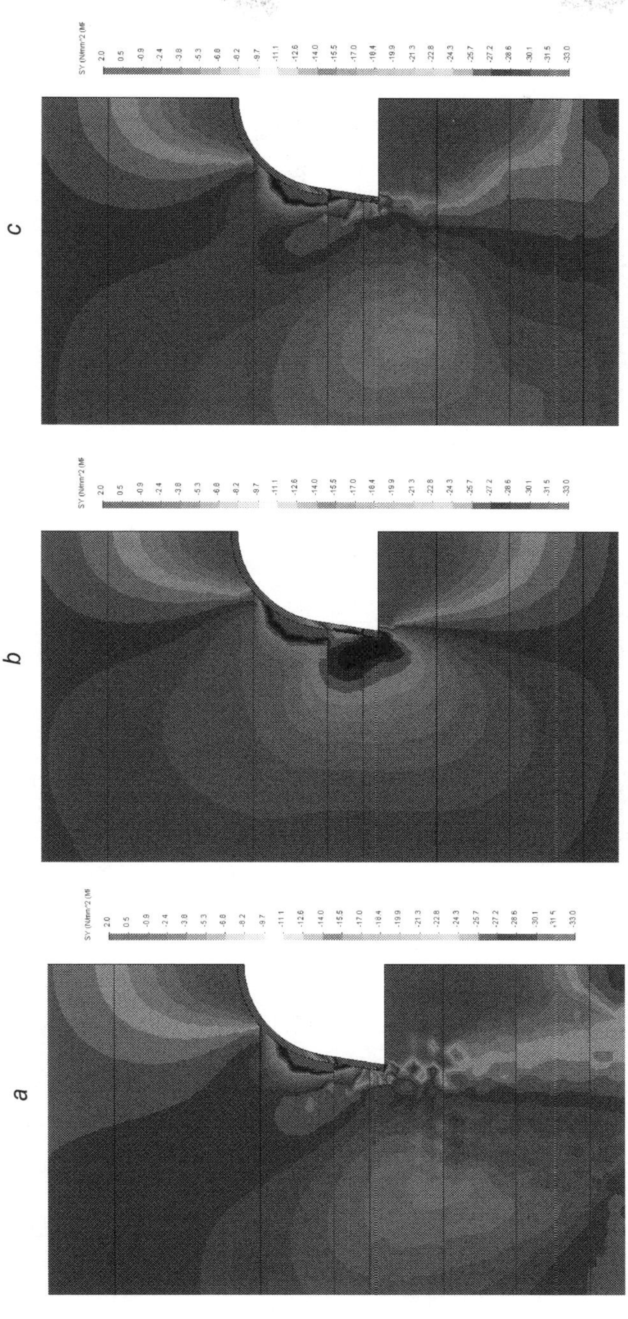

Fig. 15.1. Curves of vertical stresses σ_y at variant No. 1 of the bottom structure (from Table 15.2) and variants of mechanical characteristics distribution (from Table 15.3): *a* – No. 2; *b* – No. 3; *c* – No. 6

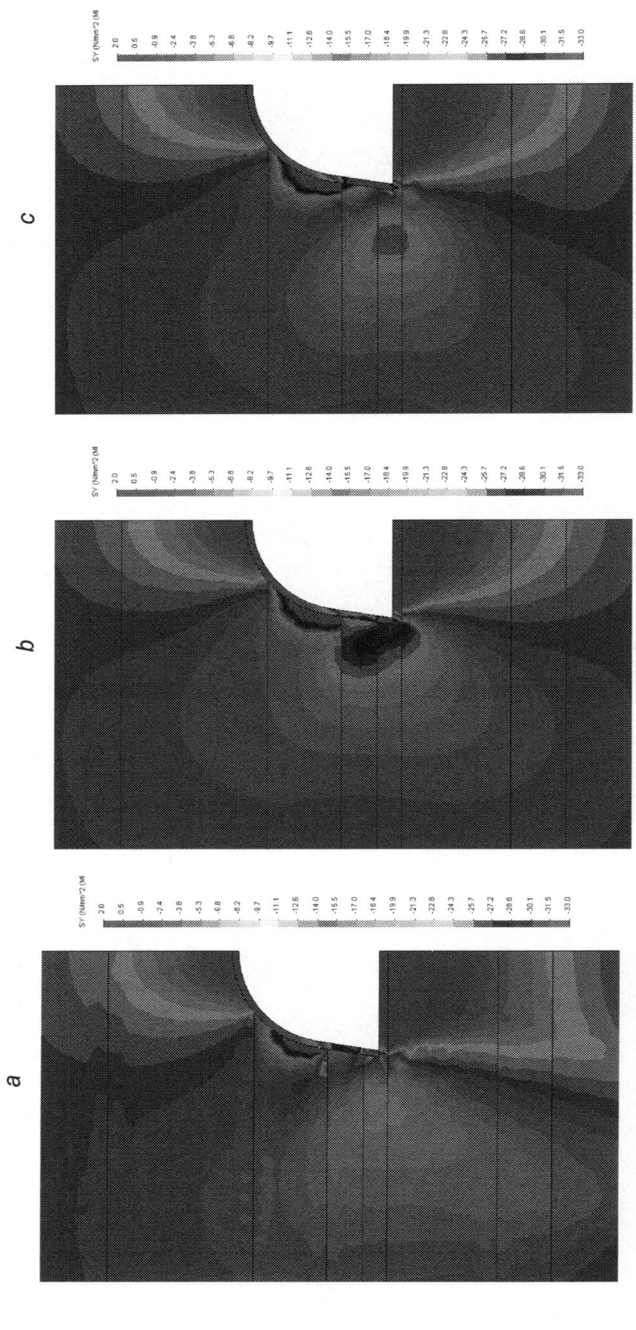

Fig. 15.2. Curves of vertical stresses σ_y at variant No. 2 of the bottom structure (from Table 15.2) and variants of mechanical characteristics distribution (from Table 15.3): *a* – No. 2; *b* – No. 3; *c* – No. 6

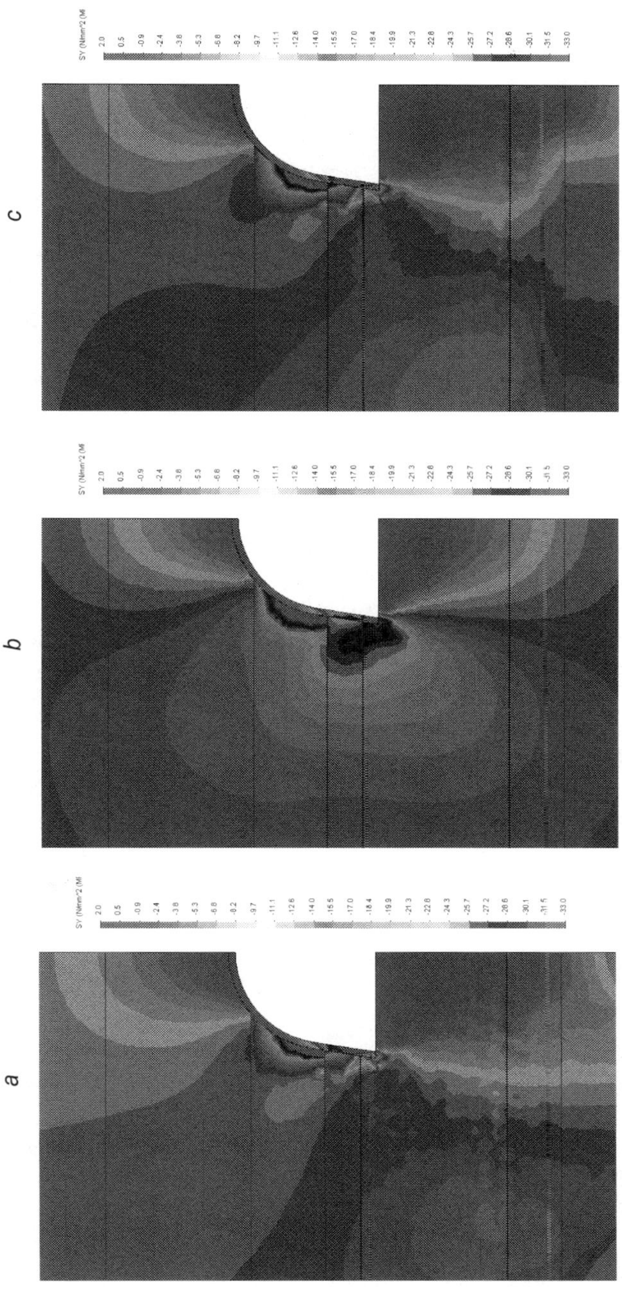

Fig. 15.3. Curves of vertical stresses σ_y at variant No. 3 of the bottom structure (from Table 15.2) and variants of mechanical characteristics distribution (from Table 15.3): a) No. 2; b) No. 3; c) No. 6

The third analyzed structure (variant No. 3 from Table 15.2) of the bottom rocks (Fig. 15.3) is characterized by alternation of layers with decreased (even-numbered) and increased (odd) thickness, but, in reverse order, if compare with the previous structure. Here, in terms of a depth fluctuations of the weakening zone, the following features stand out:

– at soft rocks in all layers (see Fig. 15.3, and – variant No. 2 from Table 15.3), the depth of weakening reaches 4.9 m that exceeds that for the previous structures;

– the depth of weakening is further increased for variants No. 6 from Table 15.3 (see Fig. 15.3, *c*) which extends to 5.6 m;

– in stronger rock layers according to variant No. 3 of Table 15.3 (see Fig. 15.3, *b*), the depth of weakening decreases to 2.5 m that corresponds to structure No. 1, but exceeds the depth of weakening for structure No. 2.

Such result is explained by action of the following factors:

– firstly, it is obvious that in soft rocks of the first layer with an increase in its thickness, the volume of the rocks weakening increases which are loaded by underlying layers and the depth of weakening in general increases;

– at the increased thickness of the soft first rock layer and bedding in the second layer of sandstone with the lowered thickness, a considerable loading is formed on it, and the load-bearing capacity is insufficient and it is partly weakened, delivering a part of load to the third softer rock layer;

– an increased rigidity of sandstone in the second layer (with reduced thickness) to a lesser extent compensates the deformation of the first soft layer and partially deformation is delivered to the third layer that causes some increase in a depth of weakening in variant No. 6 if compare to variant No. 2 (see Table 15.3);

– with an increased rocks hardness in variant No. 3, the first rock layer from siltstone has nevertheless smaller hardness if compare to the sandstone, hence, an increased thickness of siltstone promotes some increase in a depth of weakening.

Thus, it is established that the field of vertical stresses σ_y is influenced not only by mechanical characteristics, but also structure of the bottom rocks of in-seam working.

15.2.2. ANALYSIS OF HORIZONTAL STRESSES COMPONENTS

The curves of horizontal stresses σ_x distribution in adjacent rock layers of in-seam working bottom has been analyzed beginning with No. 1 structure (see Table 15.2), where all four rock layers have the same thickness of 2 m (Fig. 15.4). When bedding the soft rocks in all layers (see Fig. 15.4, and – variant No. 2 of Table 15.3), a rather uniform field σ_x in the bottom is observed, although it is the horizontal stresses, most evidently reflecting a bending of layers, are exposed to considerable differentials of σ_x (from positive value to negative one) within each of the layers and the entire thickness in general. Obviously, the reduced gradient of σ_x changing is caused by formation of zones, considerable in sizes, of limiting and superlimiting state, when plastic deformations smooth the stresses concentrations from the preceding elastic state. This conclusion confirms the almost all-round action of compressive stresses in all rock layers except for very limited areas with tensile stresses σ_x. The compressive horizontal stresses in all four layers exceeds, as a rule, resistance of rocks to compression and induces the process of their weakening that, in turn, intensifies heaving of mine working bottom (as it will be shown further).

A diametrically opposite pattern is observed in variant No. 3 of bedding in the bottom of the rocks with an increased strength (see Fig. 15.4, *b*) where the following peculiarities of horizontal stresses distribution are manifested. To some extent, the field σ_x is stable without significant perturbations because of bedding rather uniform rock layers in the bottom; however, the bending stresses are still manifested, especially in sandstone (as a stronger and rigid lithotype), where the increased compressive stresses σ_x in the lower part of a layer decrease up to zero in the upper part of a layer, that is more evidently expressed around a vertical axis of mine working. In the first rock layer, where the influence of mine working is the strongest, in its upper part the tensile stresses σ_x develop, however, the depth of this area is insignificant – to 1.2 m. Except for the limited area of tensile stresses σ_x, in the rest volume of rocks (over all bottom layers) the value of compressive stresses σ_x is less than resistance to compression of the bottom rocks, hence, they are in prelimiting state and their heaving is insignificant.

The intermediate variant No. 6 (see Fig. 15.4, *c*) is of interest when sandstone occurs in the bottom below the first soft layer: here, the peculiarities of σ_x distribution are combined of the previous variants of mechanical rock layers ratio in the bottom of in-seam working. In the first layer (with low hardness), the gradient of σ_x changing over its thickness is rather insignificant, and the value of compressive stresses σ_x, as a rule, exceeds resistance of rocks to compression, and the overwhelming volume of rock is in limiting and superlimiting states. Underlying sandstone has a significantly increased strength and rigidity, therefore, according to classical concepts, it takes up an increased load from the deformable softer bottom rocks. This is evidently manifested in curves σ_x, where an intensive sandstone layer bending generates considerable concentration of stresses σ_x with differential within the layer thickness from -20 MPa to +20MPa. In the curve σ_x, the sandstone bending into the cavity of mine working is observed (generally across its width), and beyond its limits (in the sides), the axis of a bending changes a sign of curvature and tensile stresses σ_x arise in the lower part of sandstone, and compressive stresses σ_x – in its upper part. If the value of compressive stresses σ_x is much less than the resistance to compression, then tensile stresses σ_x exceed considerably the sandstone resistance to tension σ_x and in the area of their action there is a sandstone weakening. Therefore, despite its increased hardness, it nevertheless is exposed to partial destruction. Obviously, this process induces not only a soft first rock layer, but also the soft third layer of bottom which is bedding below sandstone. The field σ_x is even more uniform in it, and their value either corresponds, or exceeds the resistance of rock to compression. Therefore, the partially limiting state of the third layer also forms an increased load on sandstone that leads to its local weakening.

Obviously, under such conditions the heaving of the bottom will be moderately manifested that is caused, on the one hand, by a limiting (superlimiting) state of soft rocks of the first and third layers and partially by holistic state of sandstone. On the other hand, it is caused, to a certain extent, limiting displacement of rocks into the cavity of mine working. The similar qualitative pattern is observed also for variant No. 4 (see Table 15.3), when the first two soft layers are in partially limiting and superlimiting states, and the third layer from sandstone is locally weakened from action of the tensile stresses σ_x in its upper part.

In the structure No. 2 (see Table 15.2) which is characterized by alternation of low-thickness layers (odd) and thick (even-numbered) in the bottom, similar tendencies with the previous structure are partly observed in terms of influence of resistance to compression of separate rock layers and their arrangement relative to the bottom of mine working (Fig. 15.5). However, there are also some peculiarities, connected with changing the thickness of rock layers, composing an adjacent bottom. So, in variant No. 2 (see Table 15.3) of soft rocks bedding over all studied thickness (see Fig. 15.5, *a*), rather uniform distribution of compressive stresses $\sigma_x = 10 - 16$ MPa takes place throughout all the area, except for the limited zones adjoining directly to the surface of mine working bottom – here the rock is more unloaded and the curve σ_x characterizes a bending of the second rock layer into mine working cavity. Increased compressive stresses σ_x (related to the initial non-hydrostatic state) points to a changeover of a part of rocks to limiting and superlimiting states; however both the value σ_x and the sizes of these zones are less, than in the previous structure that is caused by the low thickness of the softest first layer. In general, even when bedding of soft rocks, they are more stable in connection with reduction of the first rock layer thickness.

When the mechanical characteristics are distributed according to variant No. 3 (see Table 15.3) of bedding the rocks with increased hardness, their stable state throughout all studied depth of the bottom is provided, except for a bordering to mine working part of the first and second rock layers to a depth of 1 m (see Fig. 15.5, *b*) where tensile stresses σ_x act. Here the influence of the structure is shown in two positions: firstly, in the top part of the second layer, presented by sandstone, the tensile stresses σ_x increase; secondly, in the sandstone of the second and the fourth layers the compressive stresses σ_x decrease in their lower part.

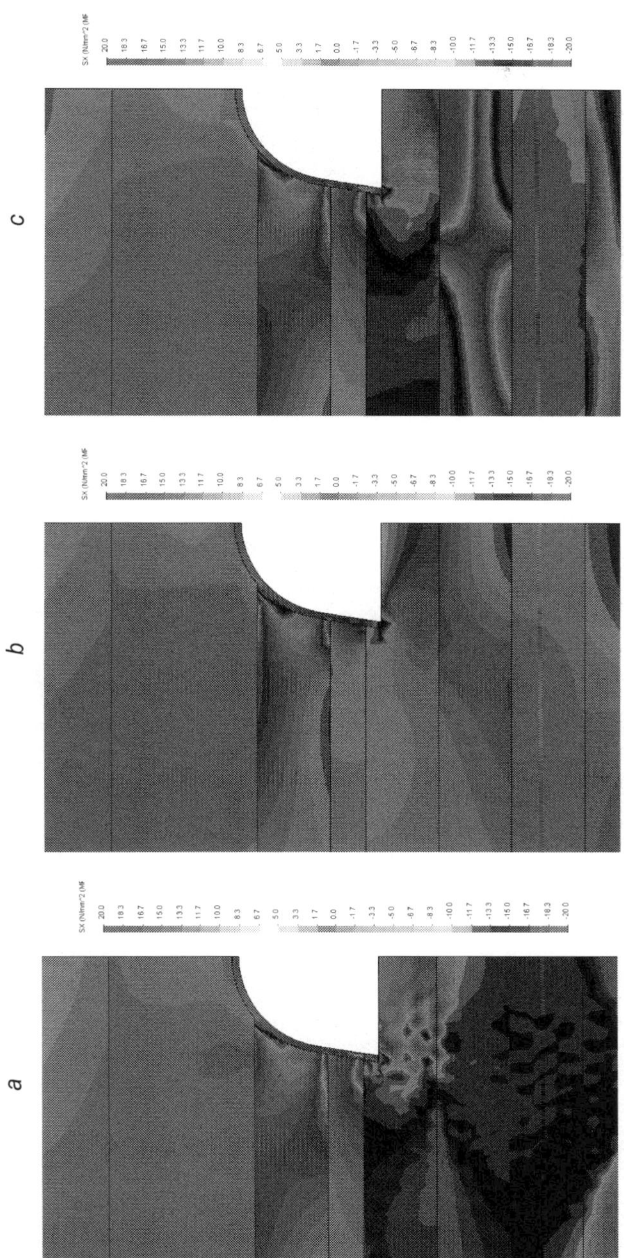

Fig. 15.4. Curves of horizontal stresses σ_x at variant No. 1 of the bottom structure (from Table 15.2) and variants of mechanical characteristics distribution (from Table 15.3): *a* – No. 2; *b* – No. 3; *c* – No. 6

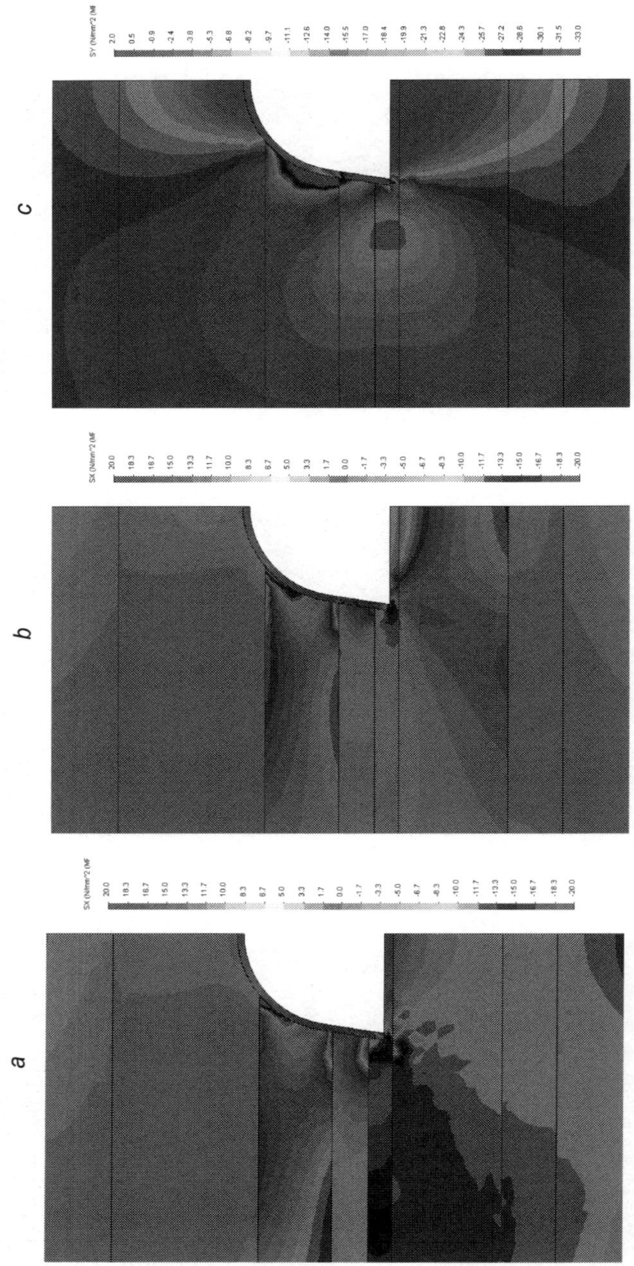

Fig. 15.5. Curves of horizontal stresses σ_x at variant No. 2 of the bottom structure (from Table 15.2) and variants of mechanical characteristics distribution (from Table 15.3): *a* – No. 2; *b* – No. 3; *c* – No. 6

However, these features are not essential for the field of horizontal stresses σ_x and their influence on the bottom stability.

More considerable changes of the curve σ_x are observed in variant No. 6 (see Table 15.3) of bedding in the second layer of the sandstone with the increased thickness (3 m) at the soft low-thickness immediate bottom (0.5 m). A decrease in the thickness of the first rock layer radically changes the curve σ_x in all underlying rock layers (see Fig. 15.5, *c*):

– stresses σ_x decreases under the sandstone deflection, and outside the mine working width, the more uniform field σ_x is observed, that reduces the area of sandstone weakening from tensile stresses σ_x and increases its stability; this is conditioned by a sudden reduction in the thickness of the soft first rock layer (which is in the superlimiting state) and, consequently, the sandstone loading;

– the third rock layer from weak argillite has, nevertheless, almost uniform distribution of σ_x, and is characterized also by a prelimiting state that is caused by an increased thickness of the overlying sandstone protecting, to some extent, from excessive loadings;

– for the same reason, the fourth rock layer is also in the more unloaded state.

Thus, except for the soft and low-thickness immediate bottom, and also a local zone of the sandstone upper part heaving, the rest volume of the bottom rocks is in a steady state that predetermines the sharp restriction of the heaving manifestation in in-seam working. Other variants of the mechanical characteristics distribution over the bottom rock layers (see Table 15.3), confirm a positive tendency of increasing the bottom stability with a decrease in the thickness of the soft immediate bottom and the sandstone bedding in the main bottom.

The following structure of the bottom rocks according to the variant No. 3 (see Table 15.2), when low-thick even-numbered layers alternate with thick odd layers, even more expands and reveals the emerging tendencies of influence on the field of the horizontal stresses σ_x distribution (Fig. 15.6) and the process of heaving development in general.

In the variant No. 2 (see Table 15.3) of the mechanical properties distributions over the bottom rock layers, their low hardness conditions the limiting and superlimiting state almost everywhere (see Fig. 15.6, *a*) to all the studied depth, however, the level of acting σ_x was slightly lower, than in the structure No. 1, that is explained by an increase in rigidity of the first layer due to an increase in its thickness (from 2 to 4 m). At the same time, relative to the structure No. 2, the value of σ_x is overestimated that is conditioned by the increased thickness of the most soft rocks of the immediate bottom and the lowered thickness of the second rock layer, the hardness of which is twice higher in relation to the first layer.

With an increased hardness of rocks in all bottom layers (variant No. 3 according to Table 15.3), their state is stable (see Fig. 15.6, b) at rather uniform field of σ_x stresses and differs insignificantly in the qualitative and quantitative plan from the previous structures.

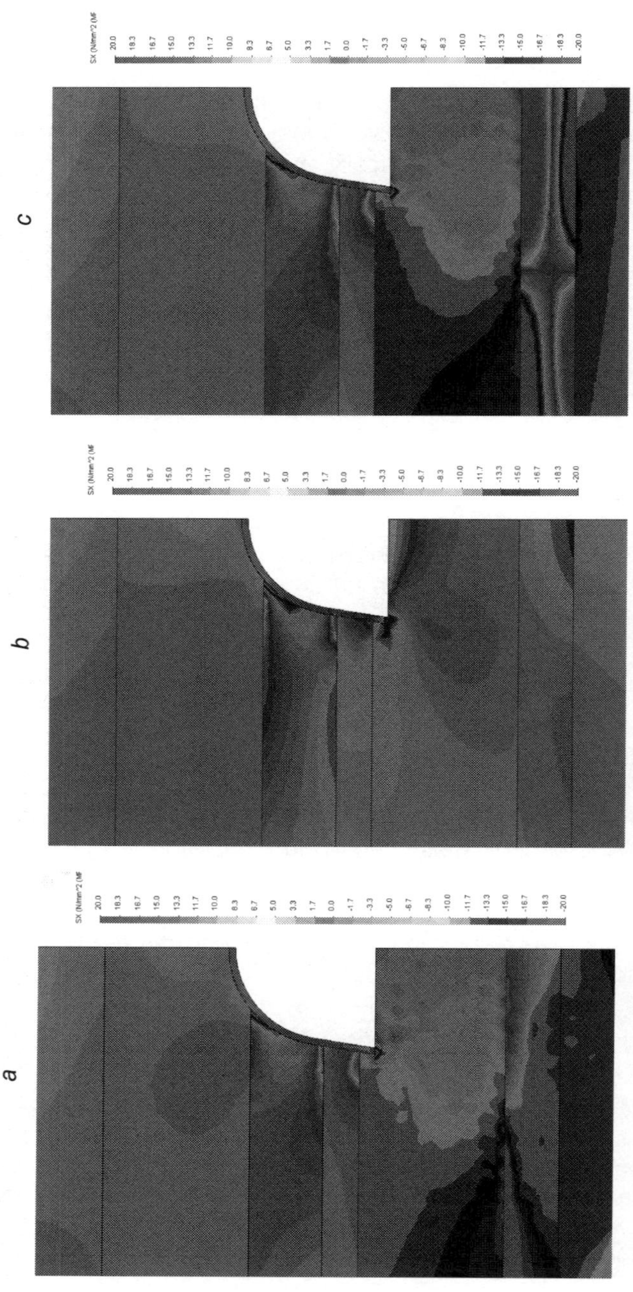

Fig. 15.6. Curves of horizontal stresses σ_x at variant No. 3 of the bottom structure (from Table 15.2) and variants of mechanical characteristics distribution (from Table 15.3): ***a*** **– No. 2;** ***b*** **– No. 3;** ***c*** **– No. 6**

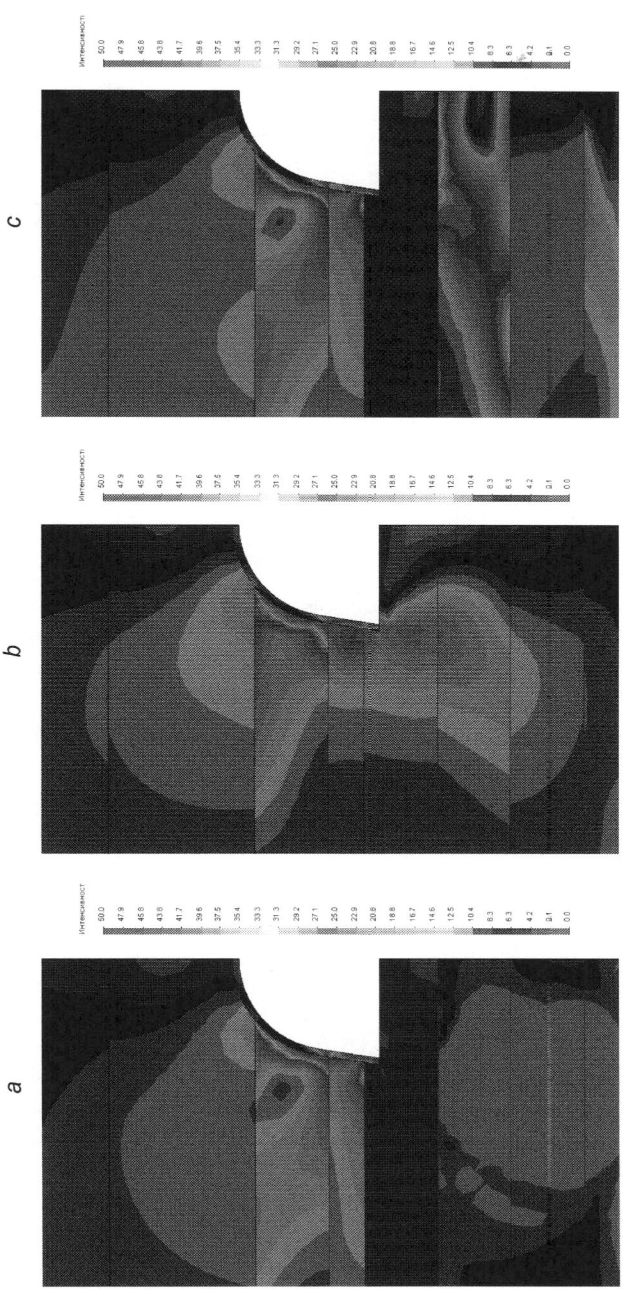

Fig. 15.7. Curves of the stresses σ intensity at variant No. 1 of the bottom structure (from Table 15.2) and variants of mechanical characteristics distribution (from Table 15.3): *a* – No. 2; *b* – No. 3; *c* – No. 6

The variant No. 6 of the sandstone bedding in the second layer at the soft immediate bottom, has both the common features and some features connected with the structure – the thick and soft immediate bottom and the sandstone with the lowered thickness (1.5 m). The immediate bottom through all the thickness (4 m) is in the limiting and superlimiting states that intensifies the process of rocks displacement into the mine working cavity, on the one hand, and, on the other hand, – increases the load on the sandstone of the second layer. The less thick (than in the previous structures) sandstone is exposed to an intensive bending and the high gradient σ_x of stress changing is observed in it.

The area of tensile is developed both under the mine working, and in the side parts of a seam almost to a half of its thickness that generates the development of extensive areas of the sandstone weakening. Therefore, the second layer is not capable to tale up all the load and transfers its part to the third weaker layer which changes into a limiting state: here, to a certain extent, an analogy is observed with the structure No. 1 and significant differences arise in relation to structure No. 3. In general, a limiting (superlimiting) state of low-thick and thick first and third rock layers should intensify the bottom heaving, as well as a partial sandstone weakening in the second layer; nevertheless, the sandstone provides some resistance to the heaving development that will be analyzed later.

As a result, the analysis of the curve σ_x has revealed a number of features of the SSS formation of the bottom rocks, which should be considered when predicting its heaving in in-seam workings.

15.2.3. ANALYSIS OF THE SPECIFIC STRESSES COMPONENTS

It is expedient to devote the final part of the SSS analysis of the laminal bottom of in-seam working to the study of the field of the specific stresses σ distribution, the value of which (according to the Mohr-Coulomb failure theory) makes possible to assess the rock state from the total action of the stresses components in the area of the compressive efforts. By analogy with the previous studies, the curves σ have been analyzed beginning with the structure No. 1 (see Table 15.2) of bedding in the bottom of the rock layers with the same thickness (see Fig. 15.7).

In variant No. 2 (see Table 15.3), extremely soft rocks occur over the entire studied depth of the bottom (see Fig. 15.7, *a*). The softest first layer is characterized by very uniform field of $\sigma = 4.2 - 6.3$ MPa distribution, which only in the local zones corresponds to the prelimiting state, and the main volume of rocks is in the limiting and superlimiting states. The second, third and fourth layers are characterized by approximately equivalent curve σ: near the vertical axis of mine working, the zone of relative unloading is located, where the value of the specific stresses either lower or corresponds to the resistance of the rocks to compression; when moving along the axis X to the lateral parts of the bottom, the limiting and superlimiting states are observed in all layers. Such a mining-and-geological situation of the predominant weakening of all four rock layers of the bottom leads to an intensification of heaving manifestation.

In variant No. 3 (see Table 15.3), of bedding in all layers of the harder rocks, the curve σ is significantly transformed (see Fig. 15.7, *b*):

– the unloading zone increases, both in size and in reduction of σ, which affects all rock layers;

– within the mine working width, the value of acting σ is several times less than resistance of rock to compression that indicates its stable state;

– in the sides (outside the mine working width), only in the first rock layer under the bearings of the support prop stays, the local areas are formed of a limiting state.

Thus, in variant No. 3, it is possible to characterize a state of the bottom as stable, which predetermines insignificant manifestations of heaving in in-seam working.

The option No. 6 (see Table 15.3) of alternating the soft and harder bottom rock layers has its distinctive features (see Fig. 15.7, *c*) from the above described curves σ:

– the first soft rock layer practically throughout the entire thickness is in limiting and superlimiting states, and its curve σ is identical to that for variant No. 2, therefore, the mechanical characteristics of the second (harder) layer do not influence on a state of the overlying layer;

– the curve σ for the third (also soft) layer is mainly similar to the curve σ for variant No. 2 and its state similar to limiting state or is already limiting state does not depend on the overlying sandstone; these results state that there is high degree of independence of the lithological variety state under study on a state of an adjacent rock layer;

– mentioned above is confirmed – mainly the prelimiting state of the second layer and particularly stable state of the fourth layer; in the second layer, the local areas of a limiting state of sandstone are observed only on its surfaces and have very limited areas of distribution.

In general, it is possible to make a conclusion that heaving caused by weakening of the soft first and third layers is considerably limited by harder (generally holistic) sandstone and is rather moderately manifested in mine working.

In the structure No. 2 (see Table 15.2), the odd layers have reduced thickness, and even-numbered – the increased thickness (in comparison with the structure No. 1) and how much this change has influenced on the field of the specific stresses in the bottom is of interest (Fig. 15.8).

The variant No. 2 (see Table 15.3) of bedding exceptionally soft rocks in all four layers has a curve σ (see Fig. 15.8, *a*), with some exceptions, very similar to the previous structure. So, the first thin rock layer from argillite is unloaded across the entire width of mine working, and weakening occurs only beyond its limits in the sides. In the remaining rock layers, the qualitative picture of the field σ is similar to the previous structure, but the zone of a prelimiting state (or similar to limiting) extends almost to the sizes of mine working width, and the rock weakening also occurs only in the lateral areas and the volumes of the weakened rocks decrease. Obviously, such peculiarities may decrease the displacement of the bottom rocks into the cavity of mine working and are quite regular: the thickness of the softest rock layer is low and, therefore, the volume of weakened rocks is less, and the second (harder) rock layer, due to its increased thickness, acts as the limiter of heaving. Thus, despite the soft bottom as a whole, the thickness of the least hard first rock layer make a certain impact on the curve of the specific stresses.

The variant No. 3 of bedding in the soil of rocks with an increased hardness (see Fig. 15.8, *b*) converts a curve σ practically to an elastic form, where very limited plastic areas are located only under bearings of the prop stays; the curve σ itself is insignificantly different from that for the previous structure. Therefore, in the conditions of bedding of more hard siltstone along with sandstone, the thickness, both the first, and underlying rock layers influences insignificantly on the specific stresses distribution.

The conclusion made is not applicable to variant No. 6 of alternating soft (odd) layers with sandstone (see Fig. 15.8, *c*). In the first and third soft layers, the curve σ is similar to the previous structure and variants where soft layers occur; a difference is only in sizes of the areas of prelimiting and limiting state. On the contrary, the second rock layer (sandstone) due to an increased thickness completely eliminates the zones of the limiting state in it, but also it limits those in the third (soft) rock layer. This factor induces sharp increase in bottom stability and insignificant manifestations of heaving in mine working.

The structure No. 3 (see Table 15.2) is characterized by the reduced thickness of even-numbered layers and the increased thickness of odd layers; this one, being the direct opposite of the previous structure, will allow to expand the concepts of influence patterns on the curve σ of thickness of adjacent bottom rocks at various variants of their mechanical characteristics distribution (Fig. 15.9).

In variant No. 2 of bedding only soft rocks (see Fig. 15.9, *a*), the first rock layer through the entire thickness is in the superlimiting state that in total with its increased thickness (4 m) causes an intensive heaving development. The zone of relative unloading in underlying layers corresponds approximately to structure No. 1, and in the lateral areas (with respect to the vertical axis of mine working) the zone of the limiting state is slightly increased that also strengthens the heaving manifestations.

In variant No. 3 of bedding everywhere the rocks with an increased hardness (see Fig. 15.9, *b*), the bottom structure does not significantly influence on the curve σ, with the exception of the increased unloading zone sizes in the immediate bottom across the width of mine working.

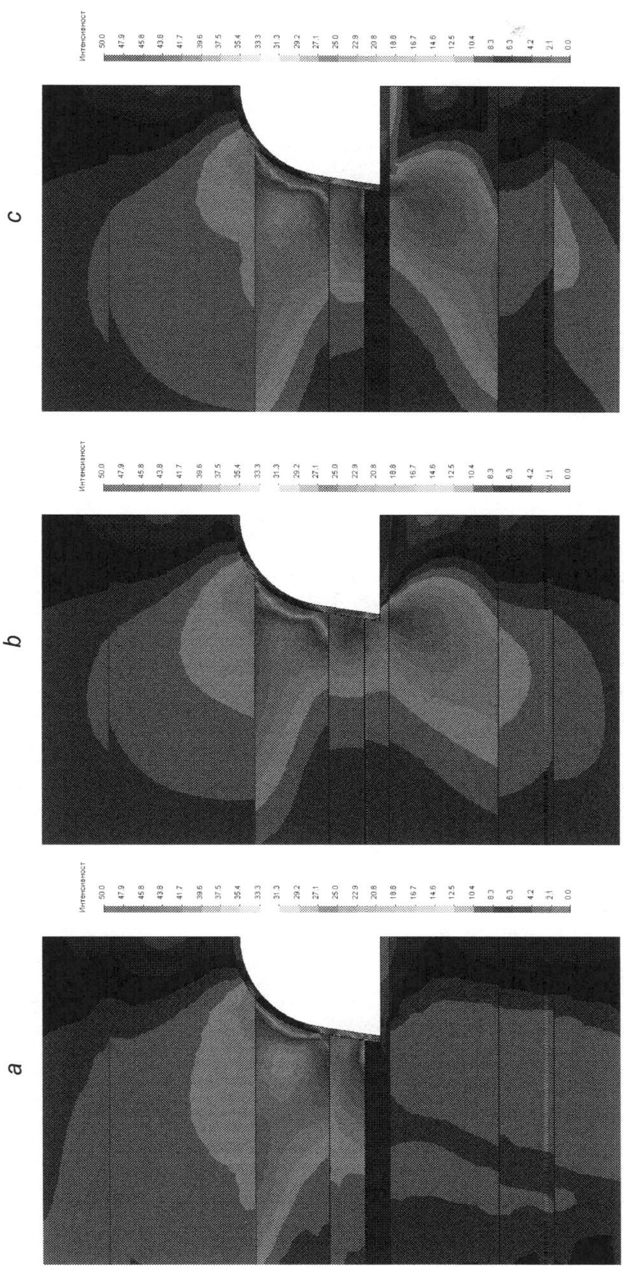

Fig. 15.8. Curves of the stresses σ intensity at variant No. 2 of the bottom structure (from Table 15.2) and variants of mechanical characteristics distribution (from Table 15.3): *a* – No. 2; *b* – No. 3; *c* – No. 6

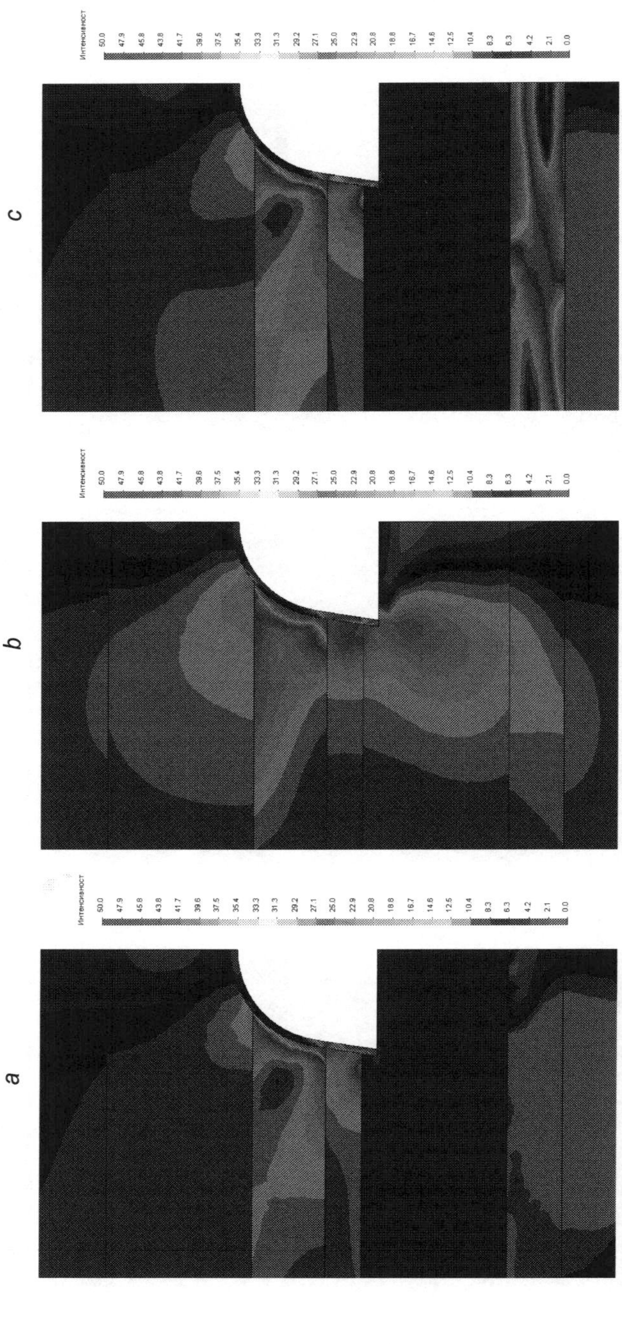

Fig. 15.9. Curves of the stresses σ intensity at variant No. 3 of the bottom structure (from Table 15.2) and variants of mechanical characteristics distribution (from Table 15.3): *a* – No. 2; *b* – No. 3; *c* – No. 6

As it was before, in variant No. 6 of bedding substantially non-uniform adjacent layers (see Fig. 15.9, *c*), the most significant changes of the field σ are observed connected with change in thicknesses of bottom rock layers:

– the first softest layer is characterized by superlimiting state everywhere with almost uniform field of specific stresses σ that intensifies the heaving process of the bottom;

– the third rock layer is also characterized by wider area of the limiting state that increases the load on the second rock layer from sandstone;

– the second, harder rock layer, has the reduced thickness (1.5 m), and the load on it is increased as from the first, so from the third rock layers; it leads to a considerable bending and weakening of sandstone that intensifies an unstable state of the bottom.

Thus, in case of bedding the sandstone in the main bottom, the development of the heaving process is essentially dependent on thicknesses of adjacent rock layers.

Other structure variants (see Table 15.2) and ratios of mechanical characteristics (see Table 15.3) of rock layers of in-seam working bottom confirm and supplement the revealed patterns which should be substantiated in terms of studying the mechanism of the heaving development depending on the structure and properties of rocks of the adjacent laminal bottom.

CONCLUSIONS

The study of the patterns of the bottom heaving manifestations in the preparatory mine workings of the Western Donbas mines has made it possible to generalize the research results, to reveal the mechanism of geomechanical processes occurrence in the bottom and to create a basis for their predictive assessment.

1. The geomechanical model "massif – in-seam working" system has been substantiated comprehensively as part of performance of the main condition – the most reliable reflection:

– the structure and properties of adjacent rocks of the roof and the bottom of coal seams in the full range of their fluctuations with the complex use of these geological surveys, predictive assessment of changes in parameters, analysis of existing results of studies of physical and mechanical properties of the coal-bearing rocks;

– accurate reproduction of the passport of conducting and fastening of mine working with modeling of design peculiarities of the support elements;

– modeling of mechanical properties of all "massif – in-seam working" system elements according to the complete diagram (limiting and superlimiting state of rock, plastic flow of steel) of their materials deformations.

2. The technique has been developed of a computing experiment including seven consecutive stages of the SSS calculation of a system and its analysis for the most objective reflection of the system behavior in the range of changing its

parameters, which are typical for mining-and-geological and mining-engineering conditions of the preparatory mine workings maintenance in the Western Donbas. For this purpose, a set of variants has been developed in a tabular form for combining the geomechanical parameters of the system where the principle of dividing a limited number of basic variants (each of them) into a group of sub-variants was used. This made it possible, on the one hand, to consider an extensive variation of conditions, and on the other hand, to save the computing resource and the time required for calculations as much as possible.

3. The analysis of the SSS calculation results of the system has revealed the ambiguity of influence on the heaving process of the structure and properties of adjacent rocks of the bottom: on the one hand, the well-known tendency to reduce the bottom displacements U_d with increasing resistance of rocks to compression is confirmed; on the other hand, the process is influenced by considerable inhomogeneity of properties (in certain areas of mine fields) of adjacent rock layers (for example, the flooded siltstone of the immediate bottom and much harder sandstone of the main bottom); this effect is enhanced with a wide range of fluctuations (along the length of mine working) in thickness of adjacent layers and a possibility of destruction of low-thickness sandstone. Different tendencies of the heaving development have been established with increasing volumes of weakening rocks of the bottom, where one of the main factors is the peculiarities of its structure, which should be substantiated in terms of revealing this geomechanical process development.

16. PECULIARITIES OF THE HEAVING DEVELOPMENT MECHANISM OF THE BOTTOM ROCKS OF IN-SEAM WORKING

Based on the analysis of fields of the stresses component distribution, some patterns have been revealed of influence on the type of state of laminal rocks of the bottom, its structure and mechanical properties. Changing the rock state parameters generates development of its displacements in the direction of free surfaces, that is, into the cavity of mine working, which is of most interest in terms of assessing its stability. Therefore, the main attention is focused on manifestations of heaving as a result of displacements development in depth and across the width of adjacent rock layers of in-seam working bottom.

The concepts that have developed in geomechanics about the mechanism of occurrence and development of bottom rocks heaving state that the main factor is the increased rock pressure in the sides of mine working, acting like a stamp onto the bottom rocks. The process of pressing-out the soft rocks in mine working bottom is very clearly shown in [73], where by means of vectors, the change is illustrated in the direction of complete displacements of the border rocks from vertical in the roof, oblique in the sides, almost horizontal near the bearings of the support prop stays to oblique and vertical in the bottom (with the opposite sign) closer to the vertical axis of mine working. A similar vector representation has been also obtained in the present work, since the physical mine rock model was used, which reflects the complete diagram of its deformation, namely, the stages of weakening and a loosening are decisive in the process of a plastic flow development in the bottom rocks.

The ambiguity of the bottom rocks structure and properties influence of on its SSS predetermines the ambiguity of heaving manifestations in the bottom of in-seam working. An extremely integrated approach is required here for interpretation of patterns; the analysis of the bottom displacement curves is supported by the reasons identified from the analysis of peculiarities of the field of each stresses component distribution. Such methodology of the analysis has made possible to reveal and explain a number of basic diagrams of the heaving development of the bottom in the preparatory mine workings. It should be noted that the determined patterns of heaving development, connected with structure and properties of adjacent bottom rocks of the coal layer, do not contradict but supplement and concretize the developed concepts [102] about the mechanism of this process development. The SSS analysis of the "massif – in-seam working" system has revealed three common factors that generate manifestations of heaving in mine workings:

– influence of vertical tensile stresses σ_y forming a zone (in the form of "overturned" dome of natural equilibrium) of rocks stratification of the immediate bottom and the upper part of the main bottom;

– pressing-out the very soft rocks which are at a stage of weakening and a loosening into the mine working cavity under the influence of harder coal layer forming similarity to the zone of bearing pressure in the sides of mine working – so-called "stamp effect";

– formation in harder siltstone and sandstones of relatively small thickness of quasi-plastic hinges (under the influence of tensile stresses and increased horizontal compressive stresses σ_x), which increase mobility of this layer and can intensify the heaving process.

All listed factors are present constantly in a varying degree, however the prevailing action of one of them is put into the principle of dividing the heaving process of the bottom rocks into four possible variants of development, reflecting the real mining-and-geological conditions for maintaining the in-seam workings in the Western Donbas mines.

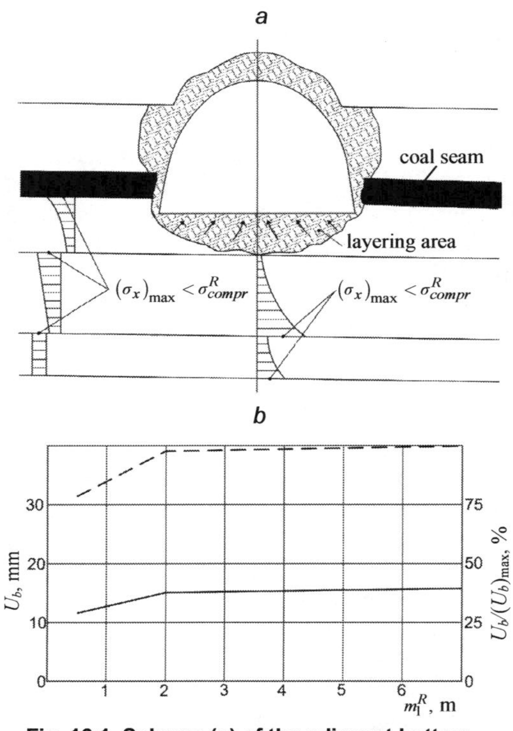

Fig. 16.1. Scheme (*a*) of the adjacent bottom rocks deformation with an increased compressive resistance $\sigma^R_{compr_i}$ and the patterns (*b*) of displacements development U_b with an increase in thickness m_1^R of the first layer: —— U_b; – – $U_b/(U_b)_{max}$

The first group of mining-and-geological conditions is characterized by rather stable state of the bottom of in-seam working, when its adjacent layers are presented by lithological varieties, different in thickness and mechanical properties, but all of them have an increased (in mining-and-geological conditions of the Western Donbas) calculated compressive resistance $\sigma^R_{comp_i} \geq 20$ MPa taking into account the action of factors weakening the rock. Here, the main reason for the manifestation of heaving is in the action of vertical tensile stresses σ_y, which form an area of partially weakened rocks, determined by penetration depth of tensile stresses σ_y, and the width corresponds to the width of mine working in a driving; the shape of the area is alike the dome of natural equilibrium, rotated 180° relative to the vertical axis (Fig. 16.1). Dimensions of the area are enough stable regardless of the bottom rocks structure, there is a stratification in it of rock in the vertical direction (action of tensile stresses σ_y), which is not amplified by other factors. Indeed, the lateral areas of the bottom rocks (outside the width of mine working) are in a stable state: the tensile horizontal stresses σ_x do not arise in the lower part of each layer thickness; in the upper part of each layer, both compressive stresses σ_x and specific stresses σ are far from the value of resistance of rocks to compression $\sigma^R_{compr_i}$. Thus, in these areas, the bottom rocks actively resist the bearing pressure from the coal seam, hence, the "stamp effect" is not observed. On the other hand, the rock layers, underlying the stratification area (as a rule, the maximum depth near the area of vertical axis of mine working does not exceed 2 – 2.5 m) under the mine working, are also stable for the above reasons: in the upper part of each layer, the tensile stresses σ_x do not appear, and in the lower part, the compressive stresses σ_x and specific stresses σ do not exceed the value of $\sigma^R_{compr_i}$ (see Fig. 16.1). Therefore, there are no quasi-plastic hinges in the rock layers, and their state is far from the limiting one. As a result, the heaving of the bottom is generated by exclusively limited area of action of tensile vertical components σ_y, which are not capable of creating any significant displacements of the bottom into the mine working cavity, and in different variants of the bottom structure it does not lead to an excess of $U_b = 20 - 30$ mm.

The peculiarities of influence of the bottom rocks structure are as follows. The area of tensile stresses σ_y distribution is rather constant: with a thin first layer, the tensile stresses σ_y penetrate into the second thicker layer of bottom; with a thick first layer, the tensile stresses σ_y are localized in it only; lithological varieties of the Western Donbas rocks resist equally weakly to tension, so that the process of stratification and displacement of rocks is changed not so significantly. Siltstone and argillite as softer rocks (in comparison with sandstone of the second and third layers) with the increased deformation properties are the main source of the bottom rocks deformations, however, their stable state does not allow developing of displacements outside the area of tensile stresses σ_y, which can be clearly seen on the diagrams (see Fig. 16.1, *b*). With an increase in siltstone thickness from 0.5 to 2.0 m, there is an increase in the bottom displacements U_b by only a few millimeters – this means that the main contribution is made by the zone of stratification from tensile stresses σ_y. But, since the absolute values U_b are low, their relative increase (in relation to the maximum $(U_b)_{\max}$ at $m_1^R = 7$ m) grows more significantly by 25 – 35%. With a further increase in thickness of siltstone, the intensity in the growth of the bottom displacements decreases even more and almost attenuates at $m_1^R \geq 4$ m.

In general, it can be concluded that with the calculated compressive resistance of at least 20 MPa (that is, taking into account the weakening factors) of the adjacent rock layers of the bottom, there are no significant manifestations of heaving in the areas of in-seam working, which are located outside the zone of the stope works influence. Therefore, the first group of mining-and-geological conditions is not of practical interest in terms of planning the actions to counteract the heaving of the bottom rocks; based on this factor, the operational state of mine working is predicted as quite satisfactory, and the analyzed group of mining-and-geological conditions is not studied in further research.

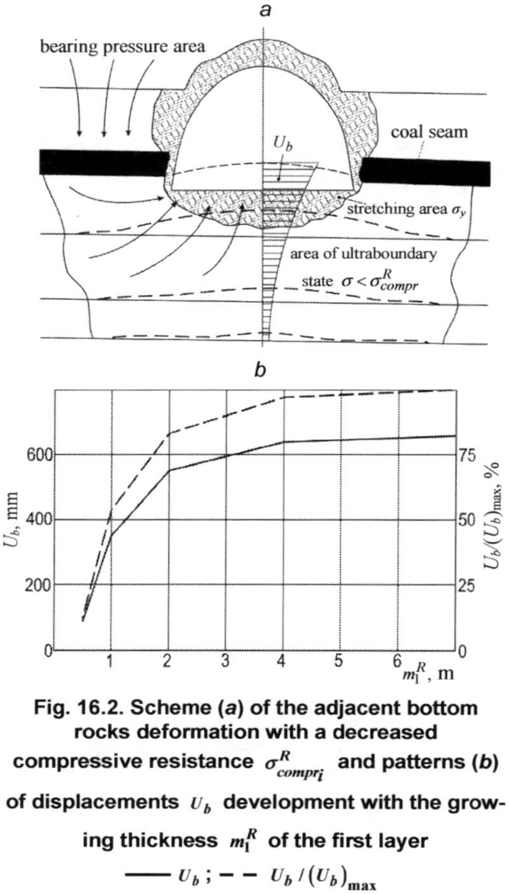

Fig. 16.2. Scheme (*a*) of the adjacent bottom rocks deformation with a decreased compressive resistance $\sigma^R_{compr_i}$ and patterns (*b*) of displacements U_b development with the growing thickness m_1^R of the first layer

—— U_b ; – – $U_b/(U_b)_{max}$

The second group of mining-and-geological conditions is the diametric opposite of the first one, since it reflects the mining-and-geological situation when the adjacent rock layers of the bottom are represented by argillites and siltstones with small hardness which, moreover, are quite often water-flooded. Potentially, this is the most dangerous (from the point of view of the bottom rocks stability) group of mining-and-geological conditions; the corresponding mechanism of the bottom rocks deformation is represented in Fig. 16.2, and to systematize further research, the formed concepts were called as "Scheme I of heaving development".

The roof rocks create an increased load in the sides of mine working, which the coal seam perceives as the hardest element, and transfers this load to the softer rocks of the bottom, which leads to the occurrence of extensive areas of limiting and their superlimiting state. The rock layers experience bending into the mine working cavity, during which they are destroyed, firstly, in the areas of the maximum bending moments action (see Fig. 16.2, *a* – in the central sections of each layer and in the lateral ones – in the zones of bearing pressure), and then throughout the span width of the rock plates. Their unstable state is facilitated by the increased horizontal compressive stresses σ_x in the bottom under the bearing pressure zones, which, taken together, create the additional overturning moment relative to the neutral axis of each rock layer. The condition for occurrence of the increased compressive stresses σ_x is the changeover of rock into the limiting and superlimiting states, caused by their low resistance to compression. Thus, the low hardness of rocks (including that caused by action of the weakening factors, such as water-cut, the fracturing, etc.) at first creates the conditions of an intensive rock layers deflection, then intensifies a bending due to an increased horizontal stresses, and, as a result, the entire rock volume from one zone of an increased rock pressure (in the left side of mine working) to another (in the right side) changes into the limiting and superlimiting states, which are characterized not only by the weakening, but also the loosening of rock, which, increasing in volume, moves in the direction of the lowest resistance, that is, into the mine working cavity. This conclusion about the formation of extensive zones with the limiting (superlimiting) state of bottom rocks is convincingly confirmed by a series of the SSS calculations (see §15.2) for different structure variants, when the value of heaving U_b consistently exceeded 300 – 400 mm (for the thickness of the first layer $m_1^R = 1.5 - 2$ m and more) in the absence of sandstone in the lower layers.

At the same time, if the first softest rock layer has insignificant thickness ($m_1^R = 0.5 - 0.8$ m), and a stronger siltstone or argillite is behind it, then the heaving is relatively small (up to 150 – 200 mm). But, with an increase in m_1^R, there is a sharp intensification of heaving (see Fig. 16.2, *b*), caused by an increase in the volume of the loosening rocks. Nevertheless, the gradient of the function $U_b\left(m_1^R\right)$ growth tends to attenuation already at $m_1^R \geq 2$ m (80 – 85% of displacement from the maximum value of $(U_b)_{\max}$ at $m_1^R = 7$ m, and at $m_1^R \geq 4$ m, the heaving is almost stabilized. This phenomenon is conditioned, in our opinion, by two reasons: firstly, with an increase in the distance from mine working, the value of the massif SSS perturbations decreases; secondly, the increasing volume of the loosened rocks is partially compensated by compression of more remote rocks, which are less weakened, but, nevertheless, are still significantly deformable.

In these conditions (with the calculated compressive resistance $\sigma^R_{compr_i} \leq 10$ MPa), the plastic flow of the bottom rocks is evidently manifested in classical understanding of the heaving phenomenon, which influences on not only adjacent, but also on more remote rock layers.

The mechanism of heaving development is significantly different at rather widespread mining-and-geological situation of bedding the siltstone and argillite with low thickness and underlying harder sandstone in the adjacent bottom of in-seam working, where the key influencing parameters are: thickness m_1^R of the first layer, thickness m_2^R of the second sandstone layer, resistance to compression $\sigma^R_{compr_3}$ of the third layer presented by argillite or siltstone; also the variant of sandstone bedding is possible in the third layer after argillite or siltstone of the second layer.

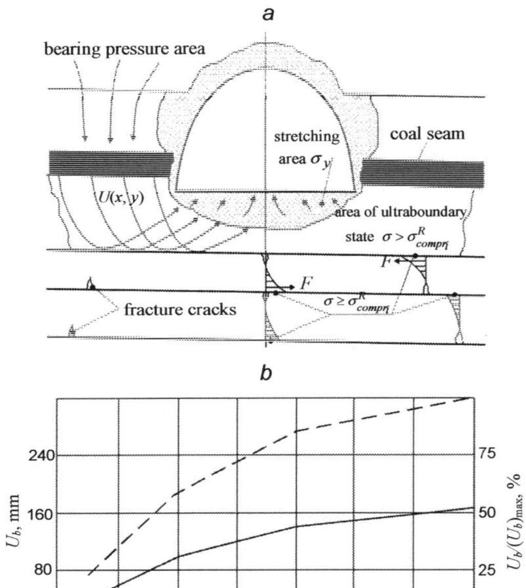

Fig. 16.3. Scheme (*a*) of substantially non-homogenous adjacent bottom rocks deformation and the patterns (*b*) of displacements U_b development with an increase in thickness m_1^R of the first layer while maintaining the stability of the underlying sandstone: —— U_b ; – – · $U_b/(U_b)_{max}$

Fig. 16.3, *a* shows the basic diagram of the most characteristic structure of the bottom rocks, when the weakened argillite or siltstone occurs in the first layer, and it is followed by sandstone and the harder (than the rocks of the first layer) siltstone or argillite. This group of mining-and-geological conditions is reflected by "Scheme I of heaving development". An increased load in the bearing pressure zone, which is perceived by the harder coal layer is transferred to the immediate bottom from weakened (for example, water-flooded) argillite or siltstone. Low calculated compressive resistance of the first layer ($\sigma^R_{compr_1} < 10$ MPa) induces to development in it of an extensive area of the limiting and superlimiting state which due to the rocks loosening generates the significant bottom heaving. For the same reasons, the resistance to rock pressure in the first rock layer decreases and the most load part is transferred to sandstone, the thickness of which is such that it keeps its stable state. In the sandstone rock plate under the influence of an external load, the concentrations of horizontal stresses σ_x (characterizing the plate bending) are formed near the vertical axis of mine working and in its sides: in the center of the rock plate span, the tensile stresses σ_x are locate in the upper part of a layer, and the increased compressive stresses σ_x – in lower part; in the lateral areas of the plate (under the zones of bearing pressure), the curvature of its bending line changes a sign and tensile stresses appear already in the lower part of a plate. As it is known, the Western Donbas rocks have very low tensile strength and in the areas of acting tensile stresses σ_x, the tension cracks are formed, as shown in Fig. 16.3, *a*. Nevertheless, due to the compressive stresses, a pair of resultant forces F is formed, creating the restoring moment, and the rock plate is in a stable state provided that the specific stresses σ (in the areas of acting compressive stresses σ_x) do not exceed the sandstone resistance to compression $\sigma^R_{compr_2}$. With a sufficient strength of the underlying rock layers (third and fourth), there are similar processes of occurring the tension cracks in them only in the areas of acting tensile stresses σ_x, which do not entail the loss of stability of the main bottom.

In these conditions, the heaving development mechanism is significantly transformed because of the location of a weak weakened layer of the immediate bottom between two strong layers: from above – coal layer, from below – rigid sandstone. The weakened argillite or siltstone has considerable mobility (similarity of a plastic flow) and under the influence of “a rigid stamp” (coal layer) are deformed in the vertical and horizontal directions that is prevented (besides stable rocks in the remote areas of mine working sides) by holistic and rigid sandstone in the main bottom. Then, the siltstone displacements do not extend deep into the soil, and their vector changes the direction from oblique to the depth of the bottom to oblique towards the mine working: there happens a kind of ‘reflection’ of the direction of the weakened and the loosened siltstone displacements from a rigid obstruction in the form of sandstone with sufficient thickness. That is, rigid sandstone in the second rock layer strengthens the heaving manifestations due to creating a directed flow of the disturbed rock into the cavity of mine working.

Such an idea about the mechanism of the heaving process development is confirmed by a series of calculations for different adjacent bottom structure (see Fig. 16.3, *b*). Here, the thickness of the first layer m_1^R plays more essential role in comparison with the previous schemes, since the excess volume of the weakened and loosened rock directed by the sandstone towards the mine working, depends on parameter m_1^R. Thus, unlike the previous scheme (see Fig. 16.2, where at $m_1^R = 2$ m, 80 – 85% of U_b displacements from the maximum value are realized, and at $m_1^R = 4$ m, the growth of U_b practically stops), sandstone induces further growth of U_d: at $m_1^R = 2$ m only 55 – 60% are realized, and at $m_1^R = 4$ m – about 85% of the maximum value are realized. Nevertheless, the function $U_b\left(m_1^R\right)$ tends to flattening at $m_1^R > 4$ m. In general, it is possible to conclude that the heaving process (in these mining-and-geological conditions) is mainly influenced by two factors: activation of displacements of soft argillite or siltstone due to the rigid footing presented by sandstone; restriction by rigid sandstone of displacements of underlying bottom rock layers.

In the presented patterns, a significant role is also played by the value of sandstone removal from the mine working contour in terms of its placement, for example, in the third layer. And the layer of harder (water-free) argillite or siltstone occurs between the disturbed layer and sandstone. Here the general tendency is shown of decreasing the "rigid footing effect" with an increase in the depth of sandstone bedding, which predetermines the decrease in the intensity of pressing-out the rocks of the first layer. The influence of a second more stable argillite layer, which partially absorbs deformations of the first layer, is added to this process that in total leads to the reduced heaving, on the one hand, and to an increased heaving 'realization' at the thicknesses of first layer up to 2 m, on the other hand, that is, the mechanism of heaving development is approaching Scheme I.

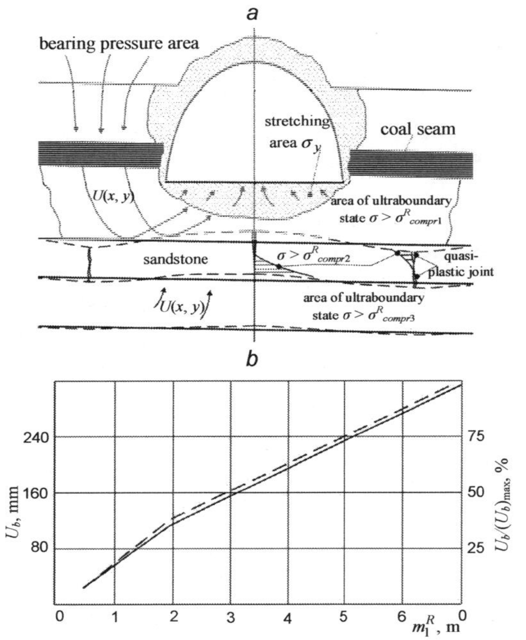

Fig. 16.4. Scheme (*a*) of substantially non-homogenous adjacent bottom rocks deformation and the patterns (*b*) of displacements U_b development with an increase in thickness m_1^R of the first layer when forming the quasi-plastic hinges in the sandstone:
—— U_b ; – – · $U_b/(U_b)_{max}$

One more variant of heaving development of the bottom rocks is significantly different from considered and characterized by the following structure: the rocks with low hardness occur in the immediate bottom, the second rock layer is presented by sandstone with thickness up to 1.5 – 2 m, and the soft siltstone or argillite occurs in the third layer. The behavior of such a structure (which is actually occurs in the coal seams bottom) with alternating layers of reduced and increased strength is characterized by a relatively independent [73, 111] transition of each rock layer to the limiting and superlimiting states that have also been confirmed by the performed calculations. With very soft rocks of the first and third layers, sandstone is a rigid plate loaded both from the side of its roof and from the side of the bottom, and the load distribution is quite unfavorable from the point of view of sandstone stability. Indeed, in the central part of the plate span, an increased vertical load acts (in the direction of mine working) from the weakened third layer, which bends the plate and reveals fractures in the area of acting tensile stresses σ_x. In the lateral parts of the sandstone, a predominantly vertical load acts from the plastic flow of the first rock layer (Fig. 16.4, *a*), which forms quasi-plastic hinges in plate bearings, pressing them into the weakened third layer. Thus, there are significant bending moments in the sandstone (that is evidently observed on the curves σ_x in §15.2), which are capable of destroying it into sections where the bending moment reaches a maximum – in the plate bearings and in the central part of the span. This occurs at a relatively small thickness of sandstone $m_2^R \leq 1.5 - 2$ m and the similarity of thrust system is created from its blocks, which, nevertheless, has some reaction of resistance to rocks displacement towards the bottom of mine working. The occurrence of a thrust system in sandstone depends not only on its thickness m_2^R, but also on the thicknesses of the first m_1^R and third m_3^R soft rock layers. The graphs show the influence of first layer thickness m_1^R on the value of heaving U_b and this pattern characterized by the slow growth of function $U_b\left(m_1^R\right)$, the form of which is approaching linear. The explanation of this pattern can be explained, in our opinion, by two factors: increased mobility of the thrust system in the sandstone, which enhances the intensity of the first layer displacements towards the mine working; increased movement of the thrust system in sandstone due to the plastic flow in the third rock layer. Thus, when both factors act in the same direction, the value of heaving is directly related to the excess volume of rocks from loosening, which depends, first of all, on the thickness of the first layer. Such behavior of the bottom rocks is singled out in a separate "Scheme III of heaving development".

Overall, the SSS analysis results of "massif – in-seam working" system have been logically explained in the proposed heaving development mechanism (in the preparatory mine workings) which is reflected in three basic diagrams in view of the fact that the uniform behavior of diverse and non-uniform structure of the bottom cannot exist conditioned by the nature of geomechanical processes itself. Therefore, the following necessary stage of research is the expansion of the calculation base in terms of involvement of really existing diverse structures and properties of adjacent bottom rocks for a more complete disclosure and specification both the mechanism and the patterns of heaving development.

CONCLUSIONS

1. The ambiguous tendencies in the bottom heaving development of the preparatory mine workings have been explained on the basis of an integrated approach, where the analysis of the curves of the bottom displacements is supported by the reasons revealed when studying the peculiarities of the field of each stresses component distribution. It is this methodology has allowed to establish three general factors generating the heaving manifestations, which are constantly present, but the prevailing action of one of them is put into the principle of dividing

the heaving process into three possible variants of development:

– influence of the vertical tensile stresses σ_y forming a stratification zone of rocks of the immediate bottom and upper part of the main bottom;

– pressing-out the very soft rocks, which are at a stage of weakening and loosening, into the cavity of mine working under the influence of harder coal seam, forming a similarity of a bearing pressure zone in the sides of mine working – the so-called "stamp effect";

– formation in sandstones and harder siltstones of a relatively small thicknesses of quasi-plastic hinges (under the influence of tensile stresses and increased horizontal compressive stresses σ_x), which increase the mobility of this layer and can intensify the heaving process.

2. The noted factors have been logically explained in the proposed heaving development mechanism, which is reflected in three basic diagrams that reveal the nature of geomechanical processes in a very non-uniform and diverse structure of the bottom rocks.

17. STUDY OF PATTERNS RELIABILITY OF RELATION OF THE BOTTOM ROCKS STRUCTURE AND PROPERTIES WITH ITS HEAVING

17.1. DEVELOPMENT OF A COMPUTATIONAL EXPERIMENT TECHNIQUE

Being guided by the general research methodology and the results of the previous calculations, it is advisable to develop the technique of a computational experiment in terms of detailing and systematization of the initial geomechanical parameters for obtaining the reliable quantitative patterns of the heaving manifestation in the bottom of the preparatory mine workings. The development of a technique is implemented in two main directions: detailing the structure of bottom rocks adjacent to the coal seam and specifying the range of changes in the calculated compressive resistance of rocks layers, depending on the intensity of action of their weakening factors (water-cut, fracturing, rheology, etc.). Both directions are conditioned by the revealed three diagrams of the heaving development depending on the structure and properties of rocks of the immediate bottom and adjacent main bottom.

The first direction is related to the set tendencies in the thicknesses m_i^b of rock layers influence on the manifestation of heaving. Moreover, a noticeable influence is made by layers occurring at a depth to 6 m – this fact is clearly consistent with the normative methodology [94], where lithological varieties with a depth of bedding up to one width of mine working in drivage are taken into consideration (relative to our research area – 5.5 – 6 m). Thus, the main contribution to the heaving process is made by 2 – 3 rock layers, excluding low-thickness (0.1 – 0.3 m) rock interlayers and coal interlayers, the probability of occurrence of which at a depth of 6 m is low. To increase the reliability of setting the patterns of relationship between the value of heaving U^b and thicknesses m_i^b of adjacent rocks layers of the bottom, it is necessary to provide a sufficient number of variants of changing the thickness of each rocks layer. This ensures the basic structural variants (see Table 15.2) which generate a set of sub-variants by combining two-three layers into a single thicker layer by eliminating the boundary condition for the separation of lithological varieties in a specific model.

In chapter 15, the nonlinear link of parameters U^b and m_1^b is set that conditions the need for calculations (with a variable m_1^b) in the number of at least five-six values in the range of $0.5 \text{ m} \le m_1^b \le 5.5$ m. It is easy to perform using basic structures according to Table 15.2:

– $m_1^b = 0.5$ m – structure №2;

– $m_1^b = 1$ m – structure №4;

– $m_1^b = 2$ m – structure №1;

– $m_1^b = 3$ m – structure №5;

– $m_1^b = 4$ m – structure №3;

– $m_1^b = 5.5$ m – structure №3 when combining the first two rocks layers.

The second rock layer, taking into account the sufficient probability of bedding both low-thickness (about 1 m), and thick (up to 5.5 m) sandstone, it is expedient to calculate in an interval of 1 m $\leq m_2^b \leq$ 5 m, which can be realized in the following structures:

– $m_2^b = 1$ m – structure №4;

– $m_2^b = 1.5$ m – structure №3;

– $m_2^b = 2$ m – structure №1;

– $m_2^b = 3$ m – structure №2

– $m_2^b = 4$ m – structure №5;

– $m_2^b = 5$ m – structure №4 when combining the second and third layers of the bottom.

The third rock layer is usually presented by argillite or siltstone with different thickness, and taking into account the depth us to 6 m which is of interest, the interval of changing m_3^b is accepted the following 0.5 m $\leq m_3^b \leq$ 5.5 m; with thick first two layers, the third rock layer is low-thickness, and in the opposite case, the third layer has an increased thickness. The structures realizing the specified interval are taken:

– $m_3^b = 0.5$ m – structure №3 when combining the first two rocks layers;

– $m_3^b = 1$ m – structure №5;

– $m_3^b = 2$ m – structure №1;

– $m_3^b = 3$ m – structure №3;

– $m_3^b = 4$ m – structure №4;

– $m_3^b = 5.5$ m – structure №2 when combining the third and fourth layers.

The listed values of the thickness of the first three rock layers in their combination are limited by the previously substantiated depth of the studied bottom. Therefore, with the low-thickness first (second) layer, the combination is calculated with the more thick second (third) layer and vice versa – the number of the studied structures in such a way is significantly limited, without reducing amount of discrete values m_i^b for creation of correlation dependences. As a result, based on five basic structures, seven variants, which are accepted to the SSS research of the bottom rocks, are selected and are presented in Table 17.1

Table 17.1

VARIANTS FOR THE STRUCTURE OF BOTTOM ROCKS AT THE FINAL STAGE OF CALCULATIONS

Number of variant		1	2	3	4	5	6	7
m_i^b, m	m_1^b	0.5	1	1	2	3	4	5.5
	m_2^b	3	1	5	2	4	1.5	3
	m_3^b	5.5	4	3	2	1	3	0.5

The second direction is connected with a more detailed substantiation of the range of changing the calculated resistance R_i^b of the bottom rocks caused not only by a variation of uniaxial compressive strength $\sigma^b_{compr_i}$ of rock in a sample, but also by an extent of the weakening factors influence: water-cut, fracturing and rheology.

The geological survey data testify that only three lithological varieties occur in the bottom of the seams (with rare exception of coal interlayers occurrence with thickness of 0.1 – 0.3 m): argillite, siltstone and sandstone.

The calculated resistance of the bottom rocks to compression R_i^b according to the guiding normative documents [94] is determined by a formula

$$R_i^b = K_c K_w K_t \sigma^b_{compr_i}, \tag{17.1}$$

into which an additional coefficient K_t is introduced, considering the rheological factor of decreasing the compressive resistance over time, which is very relevant for the soft rocks of the Western Donbas, characterized by significant creeping.

According to the listed normative documents, the ranges of changing have been determined for lithological varieties of the bottom rocks of the coefficient K_c of the structural weakening (considering average distance between the surfaces of weakening the rocks in the massif) and the coefficient K_w, considering the weakening of the water-flooded rocks (Table 17.2).

The range of changing the coefficient K_t considering the rheological factor is determined on the basis of large-scale researches [62] of the Western Donbas rocks creeping and its value are presented in Table 17.3.

Using the data in these Tables 17.2 and 17.3, the intervals of changing the calculated resistance to compression R_i^b of the bottom rocks (Table 17.4) are determined by the formula (17.1).

Table 17.2

VALUES OF COEFFICIENTS FOR STRUCTURAL WEAKENING K_c AND WATER-FLOODED MASSIF K_w

Type of rock	K_c	K_w
Argillite	0.4 – 0.9	0.5
Siltstone	0.6 – 0.9	0.6
Sandstone	0.8 – 0.9	0.8

Table 17.3

VALUES OF THE COEFFICIENT K_t CONSIDERING RHEOLOGICAL FACTOR

Type of rock	Rheological factor K / β	K_t
Argillite	0.2 – 0.5	0.71 – 0.89
Siltstone	0.2 – 0.5	0.71 – 0.89
Sandstone	0.1 – 0.3	0.84 – 0.95

Table 17.4

INTERVALS OF THE CALCULATED RESISTANCE TO COMPRESSION OF THE BOTTOM ROCKS OF IN-SEAM WORKINGS

Type of rock	Calculated resistance to compression (water-free / water-flooded) R_i^b, MPa
Argillite	$\frac{3.1 - 22.4}{1.6 - 11.2}$
Siltstone	$\frac{6.0 - 36.0}{3.6 - 21.6}$
Sandstone	$\frac{26.9 - 51.3}{21.5 - 41.0}$

The analysis of these data in Table 17.4 indicates a significantly wide interval of changing the calculated resistance to compression R_i^b of each lithological variety:

– argillite – $R_i^b = 1.6 - 22.4$ MPa;

– siltstone – $R_i^b = 3.6 - 36.0$ MPa;

– sandstones – $R_i^b = 21.5 - 51.3$ MPa.

To limit these intervals, the results of modeling were used in terms of delimitating the mechanism of development of the bottom heaving according to three schemes and mining-and-geological conditions, when the value of heaving the bottom rocks requires additional measures to ensure its stability. Therefore, firstly, the variants for combining the calculated resistance to compression R_i^b across the three rock layers of the bottom should be distinguished separately for each of the three schemes of heaving development; secondly, variants of combinations R_i^b should be considered only for cases when the value of heaving exceeds the permissible norms for safe mine working operation. Being guided by the above considerations, the following variants have been developed for combinations of the calculated resistance to compression R_i^b of adjacent three rock layers of the bottom.

The "Scheme I of heaving development" involves the occurrence of relatively uniform in mechanical characteristics rocks with small hardness (soft argillite and siltstone) in three adjacent rock layers of the bottom, since the modeling results and mine observations indicate sufficient stability of the bottom at the calculated resistance to compression $R_i^b > 10$ MPa. Therefore, we are interested in the state of the bottom rocks at $R_i^b < 10$ MPa, for which the following variants for combining the calculated resistance to compression of three adjacent rock layers (Table 17.5) are developed.

Table 17.5

VARIANTS OF VALUES OF CALCULATED RESISTANCE TO COMPRESSION R_i^b OF THE BOTTOM ROCKS OF MINE WORKING AT "SCHEME I OF HEAVING DEVELOPMEN"

Variants		A_I	B_I	C_I	D_I	E_I	F_I
R_i^b, MPa	R_1^b	3	3	5	5	10	10
	R_2^b	5	10	3	10	3	5
	R_3^b	10	5	10	3	5	3

The "Schemes II and III of heaving development" provide for the occurrence of sandstone in the second rock layer, which either maintains continuity, or a thrust system of rock blocks is formed in it. Therefore, the second rock layer has an increased value of the calculated resistance to compression, which is most expedient to reflect with two values of 30 and 50 MPa. It was also taken into account that the heaving of dangerous value is developec with very soft rocks of the first and third layers, which is limited by the above condition of

$R_{1,3}^b < 10$ MPa. According to these provisions, six variants of the parameter R_i^b combinations have been developed, which should be studied (Table 17.6).

In conclusion should be noted, that the remaining mechanical characteristics of the complete rock deformation diagram correspond to the modelled lithological varieties and were substantiated earlier in Chapter 15.

Table 17.6

VARIANTS OF VALUES OF CALCULATED RESISTANCE TO COMPRESSION R_i^b OF THE BOTTOM ROCKS OF MINE WORKINGAT THE "SCHEMES II AND III OF HEAVING DEVELOPMENT"

Variants		$A_{II,III}$	$B_{II,III}$	$C_{II,III}$	$D_{II,III}$	$E_{II,III}$	$F_{II,III}$
R_i^b, MPa	R_1^b	3	3	5	5	10	10
	R_2^b	30	50	50	30	30	50
	R_3^b	3	10	3	10	3	10

17.2. ESTABLISHING THE PATTERNS OF STRUCTURE AND PROPERTIES INFLUENCE OF BOTTOM ROCKS OF THE COAL SEAM ON HEAVING MANIFESTATIONS IN IN-SEAM WORKINGS

17.2.1. RELATIVELY UNIFORM MECHANICAL PROPERTIES OF ADJACENT ROCK LAYERS OF THE BOTTOM

In accordance with the provisions of the methodology, the computational experiment has been conducted by seven structures (see Table 17.1) of the bottom rocks and by six variants of the ratios of calculated resistance to compression R_i^b of the rocks for the "Scheme I of heaving development" (see Table 17.5).

The results of a computational experiment according to the "Scheme I of heaving development" of the bottom have been analyzed when it is represented by relatively uniform in mechanical characteristics soft lithological varieties with a resistance to compression $R_i^b \le 10$ MPa.

The patterns of changing the value of heaving U^b with an increase in thickness m_1^b of the first rock layer are presented in Fig. 17.1. First of all, different tendencies come under notice of U^b and m_1^b relation, depending on the calculated resistance to compression R_i^b of the first rock layer. So, with the softest first rock layer ($R_1^b = 3$ MPa), with an increase in its thickness, there is a steady increase in the value of heaving for both variants A_I and B_I (see Table 17.5) of the ratio R_i^b over the adjacent layers. This growth tendency is explained by the fact that the softest first rock layer, being in the superlimiting state, makes the main contribution to the heaving development and with an increase in m_1^b, the volume of rocks increases at the stage of loosening, and, consequently, the intensity of their displacement into the cavity of mine working increases. As can be seen from the graphs, the pattern $U^b\left(m_1^b\right)$ is of attenuated nature: with an increase in m_1^b, the growth gradient U^b decreases and at $m_1^b \geq 4$ m the further increase of U^b is low and is stabilized around 940 – 960 mm. This means that at $m_1^b \geq 4$ m, the strength characteristics of the underlying rock layers have already an insignificant influence on the value of heaving, as evidenced by the convergence of the graphs by variants A_I and B_I of distributions over the layers of calculated resistance to compression R_i^b. At $m_1^b < 4$ m, the influence of R_2^b and R_3^b is more significant – a difference in variants A_I and B_I is: 30.6% at $m_1^b = 0.5$ m, 19.6% at 1 m, $m_1^b = 5.1\%$ at $m_1^b = 2$ m, 9.4% at $m_1^b = 3$ m and –1.1% at 4 m. The specified difference is conditioned by a ratio of $R_2^b = 5$ MPa, $R_3^b = 10$ MPa for variant A_I and $R_2^b = 10$ MPa, $R_3^b = 5$ MPa for variant B_I. It is even more amplified by the different thickness of the second and third layers – structure No. 2 is not involved in the graphs plotting ($m_2^b = 1$ m, $m_3^b = 4$ m – Table 17.1), for which, relative to the structure No. 3 ($m_2^b = 5$ m, $m_3^b = 3$ m), the difference of displacements is – 23.6%, that is, in total (19.6% + 23.6%), the influence of parameters $R_{2,3}^b$ and $m_{2,3}^b$ has made 43.2% at $m_1^b = 1$ m.

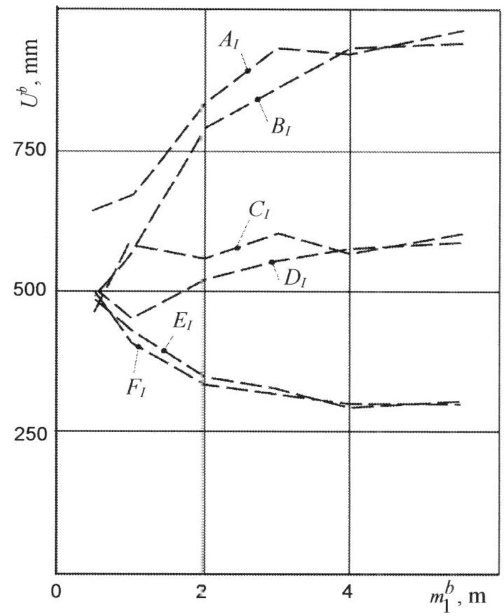

Fig. 17.1. Patterns of changing the value of heaving (according to "Scheme I of heaving development") depending on the thickness m_1^b of the first rock layer for variants of combinations of calculated compressive resistance R_i^b of adjacent rock layers according to Table 17.5

The pattern $U^b\left(m_1^b\right)$ is significantly changed at $R_1^b=5$ MPa (variants C_I and D_I of a ratio according to Table 17.5) and here the main reason is that the calculated resistance to compression R_1^b of the first layer is no longer the smallest of three adjacent rocks layers – it takes an average value in relation to $R_{2,3}^b=3$ MPa and 10 MPa. With an increase in m_1^b, there is some increase of U^b from 460 mm ($m_1^b=0.5$ m) up to 600 mm ($m_1^b=5.5$ m) for variant B_I and from 500 to 585 mm for variant D_I. In percentage terms, the growth of U^b is – 30.4% for variant C_I and 17% for variant D_I, which is significantly less than the corresponding growth of U^b (50% for the variant A_I and 91.8% for the variant B_I) at $R_i^b=3$ MPa. The pattern $U^b\left(m_1^b\right)$ is also exposed to the influence of parameters $R_{2,3}^b$ (a deviation of 28.9% at $m_1^b=1$ m) and $m_{2,3}^b$ (deviations – 31.1% with the opposite sign). Thus, even at $R_i^b=5$ MPa, there is a significant influence of the structure parameters and resistance to compression of the underlying rock layers at $m_1^b=4$ m despite a low gradient of $U^b\left(m_1^b\right)$ change.

Another type of pattern $U^b\left(m_1^b\right)$ is observed when the first rock layer is composed by the hardest rock ($R_1^b=10$ MPa) with respect to the underlying layers ($R_{2,3}^b=3$ and 5 MPa). Here, the variants E_I and F_I of the ratio of the calculated resistance to compression R_1^b from Table 17.5 are used, the difference U^b between which in value U^b is already less significant than that of the above variants; obviously, the predominant influence of the harder first rocks layer is affected. The maximum of heaving about 485 mm (variant E_I) and 500 mm (variant F_I) is observed at a minimum thickness of $m_1^b=0.5$ m and this is quite natural, since there is a minimal influence of a harder rock layer takes place, and the underlying softer rock layers make the main contribution to the heaving development. The fact of the minimum difference in displacements U^b at $m_1^b=0.5$ m for all variants of R_1^b distribution is also natural, regardless of the value of the calculated resistance to compression R_1^b of the first rock layer, which indicates the relationship of its influence intensity with thickness m_1^b. Further, with an increase in m_1^b, the influence of the first rock layer increases, that is realized in decrease in the value of heaving, which at $m_1^b \geq 4$ m tends to a stable value of $U^b = 295 -$ 305 mm.

Based on the results of the revealed patterns according to the "Scheme I of heaving development" of bedding the soft and rather uniform rocks in adjacent bottom layers, it is possible to make the following conclusions:

– different extent of parameters R_1^b and m_1^b influence, depending the thickness m_1^b of the first rocks layer is observed;

– with a low thickness $m_1^b = 0.5$ m, the influence of R_1^b is minimal, and the value of heaving is determined by the parameters of the underlying layers;

– with an increased thickness of the first rocks layer ($m_1^b > 4$ m), the influence of underlying layers is minimized and the defining factor in the heaving development is only resistance to compression R_1^b of the first rocks layer;

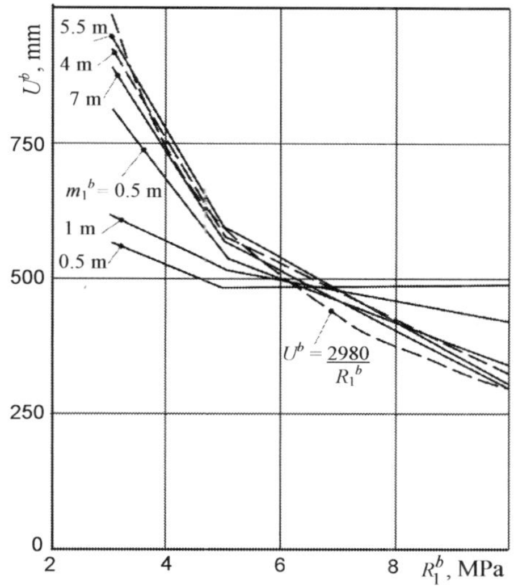

Fig. 17.2. Patterns of changing the value of heaving U^b from the calculated compressive resistance R_1^b of rock of the first layer at varying its thickness m_1^b

– in the range of 0.5 m $\leq m_1^b \leq 4$ m, the value of heaving is influenced by all the studied parameters R_i^b and m_i^b but with a different specific contribution to this process.

Based on the conclusions formulated, the tendencies of the parameters R_i^b and m_1^b influence on the heaving value U^b are analyzed, which form the basis for predicting this phenomenon by constructing correlation relationships. At $m_1^b > 4$ m, only one parameter (besides depth H of the mine working location) – the calculated resistance of the first rock layer to compression R_1^b has a decisive influence. Fig. 17.2 clearly shows an inversely proportional link of U^b and R_1^b at different values of m_1^b: at $m_1^b < 2$ m, an inversely proportional link is expressed less obviously as there is a significant influence of parameters $R_{2,3}^b$ and $m_{2,3}^b$ of the underlying rock layers (see Fig. 17.1); with an increase in m_1^b, this influence is weakened and at $m_1^b > 4$ m, it practically disappears. In this area, other parameters do not introduce significant distortions into the pattern of U^b and R_1^b linking, which causes the reliability of its approximation by an equation of the form (a dashed curve in Fig. 17.2)

$$U^b\left(R_1^b\right) = \frac{2980}{R_1^b}, \text{ mm.} \tag{17.2}$$

To substantiate the link function U^b and m_1^b, the graphs in Fig. 17.1 are also used, which indicate the necessity to choose the approximating equation providing stabilization of value U^b at $m_1^b > 4$ m regardless of other parameters values. Here, the function with one of the terms in the form of an exponential with argument m_1^b is the most suitable, which gives an asymptotic approximation of the equation $U^b\left(m_1^b\right)$ with high m_1^b to a certain constant value

$$U^b\left(m_1^b\right) = \frac{a_1}{1 + a_2 \exp\left(-a_3 m_1^b\right)}, \tag{17.3}$$

where a_1, a_2, a_3 – approximation coefficients.

Now let us assess the influence of parameters $R_{2,3}^b$ and $m_{2,3}^b$ of the underlying rock layers in the range of $m_1^b \le 4$ m. In the guideline normative documents [71, 94], the averaged calculated compressive resistance of rocks of the non-uniform bottom is calculated as the "weight-average" value by summing the product of the resistance to compression R_i^R by the coefficient of each layer influence, which is determined depending on its thickness m_i^b and the distance of location of the mine working bottom from the surface. That is, there is an assumption of necessity to consider the extent of the underlying rock layers' influence by the combined parameter

$$P = \varphi\left(R_{2,3}^b, m_{2,3}^b\right), \tag{17.4}$$

the expediency of using which is verified by the results of a computational experiment.

The influence of the "weight-average" calculated resistance of the second and third layers is assessed in relation to the calculated compressive resistance of the first layer R_1^R; thus, a certain generalized parameter P is formed

$$P=\frac{K_2 m_2^b R_2^b + K_3 m_3^b R_3^b}{\left(K_2 m_2^b + K_3 m_3^b\right) R_1^b}, \tag{17.5}$$

indicating the specific weight of resistance to compression $R_{2,3}^b$ of underlying layers in relation to the first layer R_1^b (here K_2 and K_3 – coefficients of the thicknesses $m_{2,3}^b$ influence of the second and third rock layers; changes in the range of $0 \le K_{2,3} \le 1$). Also, it has been determined that the influence of the listed parameters on the heaving value substantially depends on the thickness m_1^b of the first rock layer in the range of $m_1^b < 4$ m; therefore, the patterns of link are observed between the heaving U^b and the generalized parameter P for different discrete values of m_1^b according to the studied variants of the bottom structures (see Table 17.1). Moreover, not the absolute, but the relative value of change in heaving will be the most informative value exposed to influence of parameter P

$$\Delta=\frac{U_s^b - U^b}{U_s^b}\cdot 100\ \%, \tag{17.6}$$

where U^b – an instantaneous value of heaving for a specific value of thickness m_1^b of the first rock layer; U_s^b – the steady-state value of heaving at $m_1^b = 5.5$ m (variant No. 6 of structure according to Table 17.1).

The patterns of change $\Delta(P)$ are shown in Fig. 17.3 (at $K_2 = K_3 = 1$) for five values of thickness m_1^b of the first layer, which confirm earlier revealed tendency of decreasing the intensity of the underlying rock layers influence with an increase in the thickness of the first layer. It can be seen from the graphs that at $m_1^b = 4$ m the influence of parameter P can be assessed as insignificant – deviations in the value of heaving are

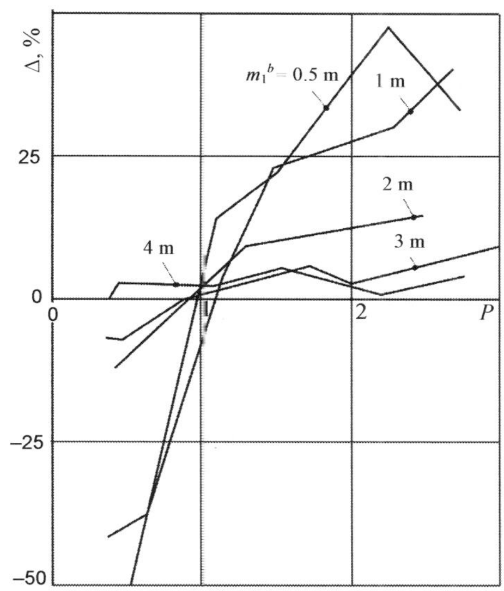

Fig. 17.3. Relative change Δ in the value of heaving depending on the generalized parameter P of the underlying rock layers influence at varying thickness m_1^b of the first layer

0 – 5.8% for all variants of the ratio of the calculated compressive resistance R_i^b from Table 17.5. At $m_1^b = 3$ m, certain increase of Δ is already shown (with an increase in P) from – 6.7% (variant F_I) to 9.6% (variant B_I, see Table 17.5). A further decrease in the thickness of the first layer leads to an increase in the gradient of increasing the function $\Delta(P)$ with simultaneous expanding the range of a deviation Δ fluctuations: thus, at $m_1^b = 1$ m the value of Δ varies in the range from 41% (variant E_I) to 40.4% (variant B_I); at $m_1^b = 0.5$ m, the range of Δ increases from – 66.7% (variant F_I) to 47.9% (variant B_I).

At low values of the parameter P (is determined by the lowered values of $R_{2,3}^b$ in comparison with R_1^b), the influence of underlying rock layers tends towards an increase in heaving, which is quite explainable and consistent with the existing expectations and regulatory techniques of prediction; with the increased values of P, the influence of underlying layers tends towards a decrease in heaving that also does not contradict well-known studies. Such general features in patterns of $\Delta(P)$ at different thicknesses of m_1^b draw attention:

– firstly, more intensive growth of function $\Delta(P)$ in the area $P \le 1.1 - 1.5$ and some flattening of $\Delta(P)$ in the area $P > 1.1 - 1.5$;

– transition from negative values of Δ (increase in heaving) to positive values (decrease in heaving) occurs in the range of $0.9 \le P \le 1.1$ (except for the graph for $m_1^b = 4$ m, where the influence of underlying layers is practically absent); the value of $P = 1$ characterizes the same value of calculated compressive resistance of all three layers, that is, the homogeneity of the bottom by factor R_i^b and in this case, a change in the thickness of some layer in favor of another should not influence on the value of heaving.

The fairly constant nature of the patterns $\Delta(P)$ at different thicknesses of the first layer indicates that the generalized parameter P according to formula (17.5) quite objectively reflects the influence of parameters $m_{2,3}^b$ and $R_{2,3}^b$ of the underlying layers, which was used to derive regression equations for predicting the heaving of the bottom rocks of in-seam working according to the "Scheme I" of this process development.

17.2.2. SUBSTANTIALLY NON-UNIFORM MECHANICAL PROPERTIES OF ADJACENT ROCK LAYERS OF THE BOTTOM

The "Schemes II and III of heaving development" in the bottom of in-seam working are characterized by placement of sandstone with different thickness directly under soft argillite or siltstone.

The presence of sandstone in the second bottom layer significantly changes the patterns of development of the heaving value U^b with increasing the thickness m_1^b of the immediate bottom, represented by soft argillite (Fig. 17.4). Here, unlike the "Scheme I", there is a stable pattern of increase in U^b with increasing thickness m_1^b of the immediate bottom, regardless of its calculated compressive resistance R_1^b, the lower the growth gradient U^b, which is caused by a decrease in the volume of loosened rocks of the first layer "pressed-out" by rigid coal seam and sandstone into the mine working cavity. An increase in function $U^b\left(m_1^b\right)$ is explained by the fact that, with the soft immediate bottom an increase in its thickness m_1^b contributes to an increase in the volume of the most mobile weakened rocks, which, when moving deeper into the bottom (from rigid coal seam), change the direction of displacements thanks to rigid sandstone and an excess of their volume (from weakening and a loosening) has one free direction of movement – the cavity of the mine working. The influence of thicknesses m_1^b with its increase is weakened and, just as in case of the "Scheme I of heaving development", it can be assessed as insignificant increase of U^b at $m_1^b > 4$ m – the explanation for this phenomenon was offered earlier in §17.2.1.

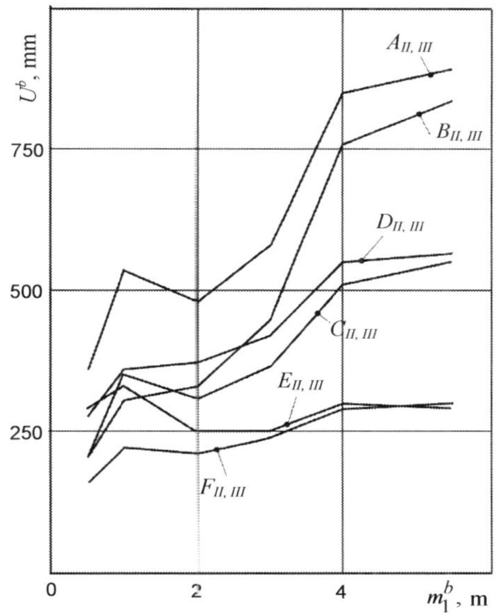

Fig. 17.4. Patterns of changing the value of heaving (according to schemes II and III of its development) depending on the thickness m_1^b of immediate bottom for variants of combinations of calculated compressive resistance R_i^b of adjacent layers according to Table 17.5

In the general sufficiently stable pattern of U^b growth with increasing m_1^b, two areas of the function $U^b\left(m_1^b\right)$ perturbation at $m_1^b = 0.5 - 1$ m and $m_1^b = 3 - 4$ m are observed regardless of R_1^b distribution over the adjacent rock layers. In our opinion, these disturbances (a sharp increase in heaving U^b) are conditioned by the low thickness of sandstone ($m_2^b = 1$ m at $m_1^b = 1$ m and $m_2^R = 1.5$ m at $m_1^R = 4$ m), which, under the influence of the overlying and underlying soft weakening layers, loses continuity and, to a lesser extent, resists to bottom heaving. This process of loss of continuity by the sandstone of the main bottom under the influence of rock pressure manifestations requires additional study in terms of determining the conditions for the occurrence of this phenomenon.

It is noteworthy that with some combinations of the bottom structure m_i^b and calculated compressive resistance R_i^b of adjacent rock layers, including soft argillite, the manifestation of heaving are not so significant and are within the admissible value U_{adm}^b, which is assessed by experts at the level of

$$U_{adm}^b \le 200 - 300 \text{ mm}. \qquad (17.7)$$

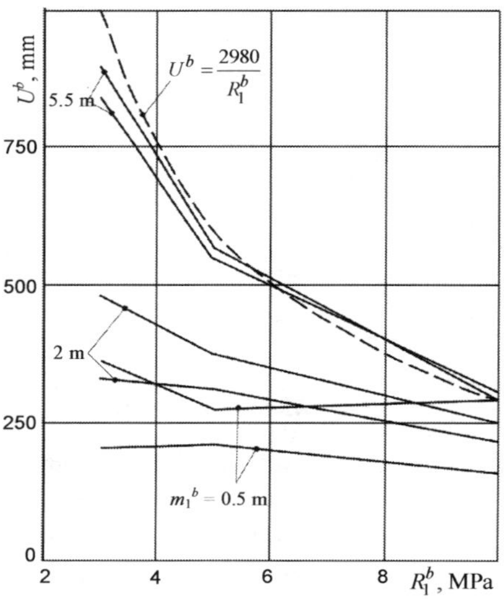

Fig. 17.5. Patterns of changing the value of heaving U^b depending on the calculated compressive resistance R_1^b of rocks of the immediate bottom at its varying thickness m_1^b, structure and properties of the main bottom (see Table 17.1 and 17.6)

It has been set that all the studied parameters R_i^b and m_1^b, the specific contribution of which should be studied in more detail, make an influence on the restriction of heaving.

In Fig. 17.5, the main attention is focused on the impact of the calculated compressive resistance R_1^b and thickness m_1^b of the immediate bottom on the value U^b. Similarly to the "Scheme I of heaving development" at $m_1^b > 4$ m, the influence of structure and properties of the underlying layers becomes insignificant: thus, at $m_1^b = 5.5$ m, an increase in R_2^b from 30 to 50 MPa and in R_3^b from 3 to 10 MPa leads to a change in the value of heaving within $0 - 6.1\%$. It is also noteworthy that with an insignificant influence of underlying layers of the bottom, the pattern $U^b\left(R_1^b, m_1^b\right)$ is described by the same equation (17.2), as by the "Scheme I of heaving development" (see Fig. 17.2 and 17.5, a dashed line). At $m_1^b < 4$ m, the influence of the underlying layers increases, and the value of heaving decreases to an admissible value U_{adm}^b due to the harder sandstone occurring in the main bottom (the second layer): with an increase in R_2^b from 30 to 50 MPa, the heaving decreases to 43% at $m_1^b = 0.5$ m and to 31.3% at $m_1^b = 2$ m.

Here, the most representative is a comparison of variants No. 2 and No. 3 of the structure of the bottom according to Table 17.1, for which the thickness of the immediate bottom is constant ($m_1^b = 1$ m), and the thicknesses of the underlying layers are different, as well as their resistance to compression according to the variants from Table 17.6.

The patterns of influence (Fig. 17.6) of the generalized parameter P are not so explicitly expressed, as in the "Scheme I of heaving development", since in the schemes II and III (with a small thickness of the immediate bottom), a significant impact is made by sandstone of the main bottom, moreover, both the value of its calculated compressive resistance R_2^b and the thickness m_2^b. This influence is superimposed also by variable parameters R_3^b and m_3^b in total with different value of R_1^b (according to Table 17.6).

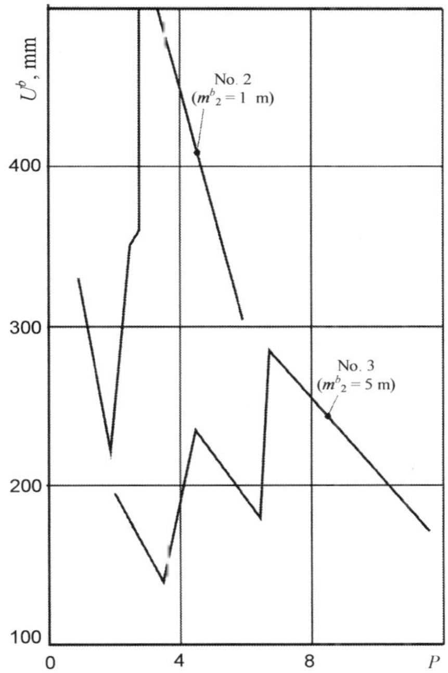

Fig. 17.6. On an assessment of influence on the value of heaving U^b of the structure and resistance to compression of the rocks of the main bottom at a reduced thickness ($m_1^b = 1$ m) of the immediate bottom (structures №2 and №3 of Table 17.1)

Nevertheless, the pattern of the considerable restriction of heaving is clearly observed, when the sandstone occurring with a thickness of $m_2^b = 5$ m in the main bottom: at any ratios of other parameters, U^b decreases from 1.53 times (variant $D_{II,III}$) to 1.94 times (variant $C_{II,III}$), and the value of this decrease is relatively stable (in the specified range) for all variants of the ratio R_i^b according to Table 17.6.

Moreover, in case of $R_2^b = 50$ MPa, the value of heaving is stably less than 200 mm for all variants of the ratio R_i^b; it should be recalled here that according to geological prediction, on separate areas of mine fields, the thickness of the sandstone is $m_2^b \geq 5$ m, so that this study has a real reflection in geological structures.

The maxima of U^b, according to graphs in Fig. 17.6, correspond, as a rule, to the reduced resistance to compression of sandstone of $R_2^b = 30$ MPa along with the reduced strength of other layers – considerable ranges of fluctuations U^b in different variants of combinations R_i^b have appeared from here.

In addition, a preservation of the sandstone continuity for variant of structure No. 3 ($m_2^b = 5$ m) and mainly the discontinuity of sandstone in variant No. 2 ($m_2^b = 1$ m) can be clearly observed on the stresses components curves, which ultimately results in increase in the bottom heaving (see Fig. 17.6).

Based on the analysis of the modelling results according to the three schemes of heaving development, a number of immediate tasks arise to ensure the reliability and accuracy of predicting the manifestations of the bottom heaving in in-seam working:

– conducting the correlation and dispersive analysis with the construction of regression equations for calculating the value of heaving;

– identifying the areas the bottom with satisfactory operational condition, where special measures are not required;

– delimitating the areas according to Schemes I, II and III of heaving development to increase the reliability and accuracy of predicting heaving.

18. ASSESSING THE STATE OF BOTTOM IN IN-SEAM WORKINGS

18.1. CONSTRUCTION OF REGRESSION EQUATIONS FOR CALCULATING THE VALUE OF HEAVING IN THE BOTTOM ROCKS

To implement the above tasks, first of all, it is necessary to construct the regression equations for calculating the value of heaving U^b as functions of parameters R_i^b and m_i^b. Such an operation was performed on the basis of calculation data for modelling the geomechanical processes according to variants of the bottom structure (Table 17.1) and by the variants for combining the rocks strength properties R_i^b for the Scheme I (Table 17.5), Schemes II and III (Table 17.6) of heaving development.

It has been previously determined that at $m_1^b > 4$ m the influence of the parameters $R_{2,3}^b$ and $m_{2,3}^b$ of the main bottom is insignificant, and the pattern of changing the heaving depending on parameters R_1^b and m_1^b of the immediate bottom is most reliably described by the functions (17.2) and (17.3). This fact is as indirect confirmation of the modelling accuracy, since it reflects the unified nature of the process, when, regardless of Schemes I, II and III of the heaving development, with the thick immediate bottom, the differences in the structure and properties of bottom rocks are leveled and combined into a single geomechanical process of heaving of the bottom rocks in in-seam working.

With a reduced thickness of the immediate bottom, a significant influence is revealed of parameters $R_{2,3}^b$ and $m_{2,3}^b$ of the main bottom on the value of heaving U^b as a characteristic of which a generalized parameter P is offered according to the formula (17.5). This parameter can be presented as an analogue of the relative 'weight-average' calculated compressive resistance of three adjacent rock layers of the bottom, and the coefficients K_2 and K_3 reflect the thickness influence of the main bottom layers and their removal from the surface of mine working bottom. The correlation-dispersion analysis of results of the SSS calculations of the bottom rocks has revealed the most accurate values of these coefficients (with a reliability of 0.82), at which the generalized parameter takes the form

$$P = \frac{0.6m_2^b + 0.4m_3^b R_3^b}{\left(0.6m_2^b + 0.4m_3^b\right) R_1^b}. \tag{18.1}$$

The analysis of patterns of the parameters R_i^b and m_i^b influence on the value of heaving U^b has led to the main conclusion taking into account (18.1): not only in qualitative, but also partially in the quantitative terms, these patterns are of the same type, which reflects the uniform nature of the heaving phenomenon. The differences between themselves of the revealed Schemes I, II and III of heaving development are in the peculiarities of influence of the immediate bottom thickness m_1^b, which is most expedient to present as a factor to a constant function $U^b\left(R_i^b, m_i^b\right)$.

More clearly the peculiarities of m_1^R influence with different schemes of heaving development are shown in Fig. 18.1, where, with identical parameters $R_{1,3}^b$ and $m_{1,3}^b$, the dependences $U^b\left(m_1^b\right)$ for all three schemes are given. It can be seen from the graphs that sandstone ($R_2^b = 40$ MPa) in the main bottom restricts the development of heaving, and this process is more intensively restricted when the sandstone retains continuity (Scheme II) as a result of its increased thickness m_2^b. The dashed lines show a hypothetical situation of lack of loosening of the immediate bottom, then the effect of "reflecting" by sandstone of displacements of heaving first layer disappears, and the very value of U^b becomes even less that quite reflects the nature of heaving phenomenon. Thus, there is a double influence of the sandstone of the main bottom: on the one hand, it restricts heaving due to its increased hardness, on the other hand, it intensifies heaving of the immediate bottom due to the effect of "reflection" of its displacements in the direction of mine working cavity; but, the first kind of sandstone influence is prevailing. The uniformity or similarity of patterns $U^b\left(m_1^b\right)$ for all three schemes of heaving development also draw attention.

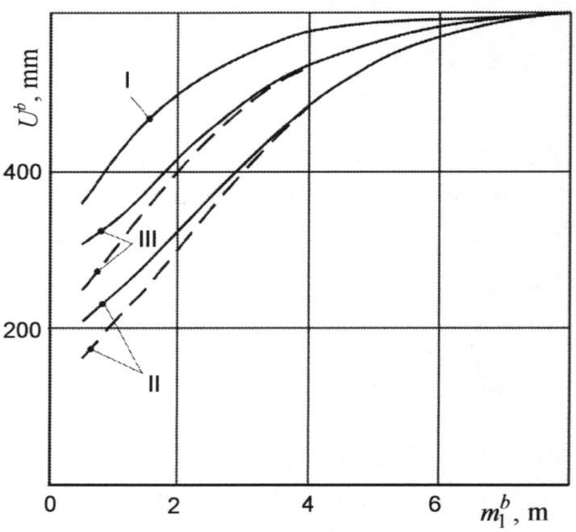

Fig. 18.1. To the comparative analysis of dependences $U^b\left(m_1^b\right)$ for three schemes I, II and III of the heaving development in in-seam workings: —— taking into account effect of "reflection" at loosening of the immediate bottom; – – · lack of loosening of the immediate bottom

As a result, the following regression equations with a sufficiently high coefficient of multiple correlations (0.69 – 0.76) are received that characterizes the reliability of the revealed dependences for predicting the value of heaving of the bottom rocks in in-seam working:

$$U^b = \frac{2980\Phi_{I,II,III} K_H K_S}{R_1^b + \left(\dfrac{0.6 m_2^b R_2^b + 0.4 m_3^b R_3^b}{0.6 m_2^b + 0.4 m_3^b} - R_1^b \right) \exp\left(-0.8 m_1^b\right)}, \tag{18.2}$$

where $\Phi_I = 1$ – for Scheme I; $\Phi_{II} = 1 + 0.5\, exp\left(-1.2 m_1^R\right)$ – for Scheme II; $\Phi_{III} = 1 + 0.5 \exp\left(-1.5 m_1^b\right)$ – for Scheme III; K_H – coefficient considering the influence of depth H of mine working location; K_S – coefficient of a standard size influence of the mine working section, determined according to work [68] depending on the cross-sectional area S_{fin} in the clear to subsidence; for the types TSYS-9.5 – 17.7 and MYS-A3 – 9.2 – 18.3 of supports, wh ch are the most widely applied in the Western Donbas; values of coefficient are changed within the range of $K_S = 0.74 - 1.23$ and are determined according to table [68].

18.2. THE LINK OF VALUE OF THE BOTTOM ROCKS HEAVING WITH THE DEPTH OF MINE WORKING LOCATION

Almost all known studies of the heaving processes of the mine workings bottom and existing techniques for predicting this phenomenon, including normative documents, indicate a close link of parameters U^b and H, for example:

– in the works [67, 68], a parabolic function $U^b(H)$ is recommended;

– in normative documents [94], – non-linear combined function $U^b(H)$ is recommended with an increase in a growth gradient U^b in case of increasing H, which is qualitatively similar to the previous one;

– in the modern guideline document [71], the linear relation of U^b and H is used.

In these studies the set dependence (18.2) has been obtained at a depth $H = 400$ m of the mine working location. But, earlier, a peculiarity of modelling the heaving processes according to the complete diagram of rock deformation (taking into account its weakening and a loosening) was noted, which is characterized by the fact that when calculating, the depth H of the mine working location is a variable parameter and it was enough to fix the value U^b for specific values of the current parameter H within one and the same design variant. This was realized during the research. Here, the well-known methodological technique was used for introducing a correction coefficient for different depth H of the mine working location: in case of $H = 400$ m $K_H = 1.0$; in case of $H \neq 400$ m $K_H \neq 1$.

An analysis of plans for mining operations in the Western Donbas mines today and for the near future has allowed to substantiate the most objective interval of changing the mine working depth of 200 m$\leq H \leq$ 600 m for which the corresponding calculations have been performed. The calculation results are summarized in Table 18.1.

From the Table 18.1 it is evident that there is some nonlinearity of the function $U^b(H)$, which is intermediate between the linear relation [71] and nonlinear dependences [67, 68, 94], which indirectly testifies the accuracy of the determined patterns.

Table 18.1

VALUES OF THE COEFFICIENT K_H FOR ACCOUNTING THE DEPTH H OF MINE WORKING LOCATION

Average depth of mine working location, H, m	200	250	300	350	400	450	500	550	600
K_H	0.43	0.56	0.7	0.85	1.0	1.14	1.3	1.47	1.65

Thus, using the data from Table 18.1, according to the formula (18.2), the predicted value is calculated of the bottom heaving for the corresponding range of the depth H of in-seam workings location at the Western Donbas mines.

18.3. SEPARATION OF MINING-AND-GEOLOGICAL CONDITIONS ACCORDING TO THE NATURE OF HEAVING DEVELOPMENT AND DEGREE OF THE BOTTOM ROCKS STABILITY

An equally important task is to differentiate the mining-and-geological conditions of in-seam workings operation by the factor of necessity (or its lack) to hold measures for resisting the heaving of bottom rocks. The determination of the areas of satisfactory bottom state of mine working is based on the obtained regression equations (18.2), and the criterion is the upper boundary of the condition (17.7), that is $U^b_{adm} \leq 300$ mm according to modern studies and existing recommendations for flat-lying Donbas seams.

As a result of the joint solution of the equations (18.2) and a condition (17.7), the following dependence has been obtained that determines the conditions of satisfactory state of the bottom

$$R_1^b \geq \left[9.93\exp\left(0.8m_1^R\right)\Phi_{I,II,III}K_H K_S - \frac{0.6m_2^b R_2^b + 0.4m_3^b R_3^b}{0.6m_2^b + 0.4m_3^b} \right]\left[\exp\left(0.8m_1^b\right) - 1\right]^{-1}. \quad (18.3)$$

According to the formula (18.3) for the analysis of the determined patterns, an example of calculating the minimum value of calculated compressive resistance R_1^R of the immediate bottom depending on its thickness for the following initial data has been performed:

– Scheme I: $R_{2,3}^b = 5$ MPa, $m_{2,3}^b = 3$ m;

– Scheme II: $R_2^b = 40$ MPa, $R_3^b = 5$ MPa, $m_2^b = 3$ m, $m_3^b = 3$ m;

– Scheme III: $R_2^b = 40$ MPa, $R_3^b = 5$ MPa, $m_2^b = 1$ m, $m_3^b = 3$ m.

For the "Scheme I of heaving development", an intensive increase in the required value R_1^b at $m_1^b < 2$ m is observed (Fig. 18.2) to maintain the bottom in a satisfactory state. This is conditioned by two reasons: firstly, the main bottom is represented by soft rocks ($R_{2,3}^b = 5$ MPa), for restriction of their heaving it is required a rather stable immediate bottom, and, with decreasing thickness, the stresses during bending increase and a higher rock hardness is required; secondly, with a decrease in m_1^b, the soft rocks of the main bottom influence more significantly on heaving manifestations in mine working, as they approach to its contour. With a significant thickness of the immediate bottom ($m_1^b > 3$ m), the value R_1^b is stabilized at the level of $R_1^b = 10$ MPa, which provides a satisfactory state of the mine working bottom.

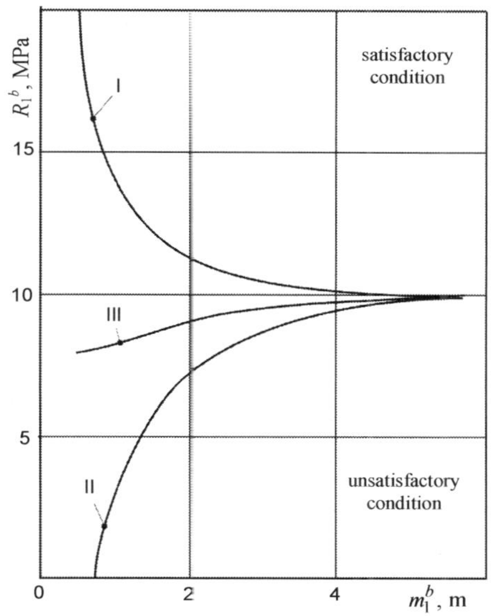

Fig. 18.2. Example of predicting (in terms of parameters R_1^b and m_1^b) the areas of satisfactory state of the bottom rocks of in-seam working for three schemes I, II and III of heaving development

A different picture is observed with the "Scheme II of heaving development" – here, at $m_1^b \leq 1$ m for almost any hardness of the immediate bottom, sufficiently thick sandstone ($m_2^b = 3$ m) provides a satisfactory state of the mine working bottom; with an increase in the thickness of the immediate bottom, the calculated compressive resistance of its rock increases, the value of which asymptotically approaches to $R_1^b = 10$ MPa increases.

The “Scheme III of heaving development” takes an intermediate position – here, low-thickness ($m_2^b = 1$ m) sandstone restricts the heaving, but even at $R_1^b = 0.5$ m to maintain the mine working bottom in a satisfactory state, its R_1^b reaches 7.9 MPa; with an increase in the thickness of the immediate bottom, its necessary resistance to compression increases with a low gradient up to a stable value of $R_1^b = 10$ MPa.

In general, the formula (18.3) for predicting the state of the in-seam working bottom quite logically explains different situations of heaving manifestation depending on the structure and properties of the rocks of the immediate and main bottom. The inequality (18.3) has a simple recording form, which enables making prompt calculations in an explicit form and deciding the need to hold measures for resisting the heaving of bottom in the preparatory mine workings.

The final task in this subchapter is to determine which of these schemes (I, II or III) of the process of the bottom heaving will take place in specific conditions of mine working operation with the purpose of predicting this phenomenon and developing the measures for its restriction. The Scheme I does not need any criteria of its identification, since it is characterized by lack of sandstone in the bottom, which is composed of soft rocks, relatively uniform in mechanical properties (argillite, siltstone). A different situation arises when separating the Schemes II and III of heaving development, which differ from each other only by a condition of continuity or disturbance of sandstone in the main bottom. Therefore, the condition for maintaining the continuity of the sandstone by determining its minimum thickness $\left(m_2^b\right)_{\min}$, at which the disturbances in the sandstone do not occur was accepted as the criterion for identifying the Scheme II or Scheme III, according to which the process of heaving development can be predicted.

The task set has been solved on the basis of correlation-dispersion analysis of modelling results in a detailed study of all stresses components curves in sandstone. It has been determined that the closest link of minimum permissible sandstone thickness is observed with the following parameters: the calculated compressive resistance of sandstone R_2^b, the ratio $\frac{m_1^b}{R_1^b}$ for the immediate bottom and the ratio $\frac{m_3^b}{R_3^b}$ for the layer of the main bottom located under the sandstone. The analysis results are presented in Fig. 18.3, where the dependency graphs are provided, dividing the areas of maintaining the sandstone continuity (Scheme II) and its disturbance (Scheme III). The revealed patterns are such that with an increase in the sandstone strength R_2^b and a decrease in the parameters $\frac{m_{1,3}^b}{R_{1,3}^b}$ (increase in $R_{1,3}^b$ or decrease in $m_{1,3}^b$), the minimum sandstone thickness $\left(m_2^b\right)_{min}$ decreases at which it maintains the continuity.

These patterns are quite logical and explainable from the point of view of the geomechanics of the heaving development: the higher the hardness of the rocks, the less intense heaving; the lower the thicknesses $m_{1,3}^b$ of the weakest rock layers, the smaller the volume of the weakened rocks is involved in the heaving process, and it is easier to restrict it with harder sandstone. Thus, at $\frac{m_{1,3}^b}{R_{1,3}^b}=0.1$ m/MPa in the range of $R_2^b=30-50$ MP, the sandstone continuity is maintained at its thickness of $\left(m_2^b\right)_{min}=0.85-1.1$ m; with an increase in thickness $m_{1,3}^b$ and a decrease in compressive resistance $R_{1,3}^b$ of the softest layers, the load on the sandstone increases and with an increase in the minimum sandstone thickness up to $\left(m_2^b\right)_{min}=2.69-3.48$ m, its destruction is prevented.

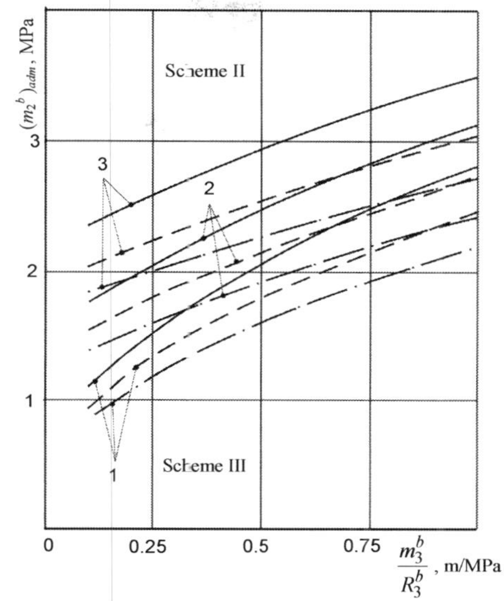

Fig. 18.3. Separation of conditions for heaving development in mine working according to Schemes II or III:

1 – $\frac{m_1^b}{R_1^b}=0.1$ m/MPa; 2 – $\frac{m_1^b}{R_1^b}=0.5$ m/MPa; 3 – $\frac{m_1^b}{R_1^b}=1$ m/MPa; —— $R_2^b=30$ MPa, – – $R_2^b=40$ MPa, – · · $R_2^b=50$ MPa

In addition to a graphical illustration of separation criterion for Schemes II or III of the heaving development, a regression equation is obtained on the basis of correlation-dispersion analysis that determines the minimum value of the sandstone thickness at which it maintains its continuity (Scheme II)

$$\left(m_2^b\right)_{min} \geq 3.6\sqrt{\frac{1}{R_2^b}\left(11\frac{m_1^b}{R_1^b}+17\frac{m_3^b}{R_3^b}\right)},\ \text{m} \tag{18.4}$$

Thus, all the basic patterns for creating a methodology of the prediction of the heaving manifestations in the bottom rocks in the preparatory mine workings have been determined.

19. TECHNIQUE FOR PREDICTING AND MONITORING OF BOTTOM IN IN-SEAM WORKINGS OF THE WESTERN DONBAS MINES

19.1. GENERAL INFORMATION AND CALCULATIONS ALGORITHM

Monitoring the state of bottom along the entire length of in-seam working based on the predicting the value of its heaving is an actual practical task for the timely planning of measures to ensure proper standards and the implementation of the rules for the safe mining operations. The mining and geological situation of the mine working maintenance, which changes along its length, requires (among other conditions) a preliminary assessment of the bottom stability and identification of areas that are problematic from this point of view. Here, in addition to the traditional bottom ripping, the preliminary planning is possible with the subsequent implementation of measures to increase the bottom stability, including during the period of mine working operation. For example, with a soft immediate bottom with a thickness of up to 1.5 – 2 m, an increase in the depth of ripping is possible to reduce the volume of unstable rocks, if other technical standards permit this for its operation. Also, it is necessary to study the expediency of restricting water inflow in areas with water-flooded bottom, where more intensive heaving is predicted. It is not necessary to ignore the measures to strengthen the bottom, strengthen the prop stays of the frame support and the side rocks with roof-bolts, erecting an arched floor and the like. All these technical solutions are provided on the basis of performing the calculation complex of the value U^b of the bottom heaving in each of the mine working area, where the mining and geological conditions of its maintenance change significantly. The calculation of the value of the bottom rocks heaving and its assessment is carried out in the following sequence.

1. Based on the mining and geological prediction, the mine working is divided into sections with essentially different conditions of its maintenance: a change in the structure of rocks of the main and immediate bottom (lithotype, thickness of the layers, the degree of water-cut, the intensity of fracturing), and variation in the strength characteristics of adjacent rock layers. For each selected area, the initial data on the parameters are compiled: the thickness of the layer m_i^b, the resistance to compression of its rock $\sigma^b_{compr_i}$, the degree of water-cut, and the intensity of fracturing.

2. The calculated compressive resistance of the rocks of the adjacent three bottom layers is determined using the formula $R^b_{1,2,3}$ (17.1), Tables 17.2 and 17.3.

3. A preliminary assessment of the bottom state is conducted according to the

condition: when the calculated compressive resistance of the rocks of all three adjacent layers is not less than 10 MPa, the bottom is sufficiently stable and does not require any measures to resist heaving; if the condition of $R_i^b \geq 10$ MPa is not fulfilled, the state of at least one of the three adjacent bottom layers should be further assessed.

4. Based on the data on the structure of the main bottom, the scheme of heaving development is determined:

– Scheme I – the absence of sandstone in the adjacent layers of the main bottom;

– Scheme II, III – the presence of sandstone in the adjacent layers of the main bottom.

The choice of Scheme II or Scheme III is made according to condition (18.4) by comparing the boundary value $\left(m_2^b\right)_{adm}$ with the actual value of the sandstone thickness.

5. The final assessment of the degree of bottom stability is made according to the expression (18.3): at a value R_1^b not less than calculated according to the formula (18.3), the state of the bottom rocks is satisfactory; if the condition (18.3) is not fulfilled, appropriate measures should be taken to ensure the stability of the bottom rocks in in-seam working.

6. The predicted value U^b of the bottom heaving is calculated by the formula (18.2) to specify the measures to resist heaving; for example, the scope of works for the bottom ripping is determined.

The developed algorithm for assessing the state of bottom rocks is recommended to be used in combination with the current normative document [94] on the choice of the mine working support for the adoption of complex technical solutions on their effective and safe maintenance.

19.2. FORMATION OF THE INITIAL DATA BASE FOR PERFORMING CALCULATIONS

The following geomechanical and mining-engineering parameters are used in the developed technique for predicting the heaving of the bottom rocks in in-seam workings:

– calculated depth H of mine working location;

– thickness $m_{1,2,3}^b$ and calculated compressive resistance $R_{1,2,3}^b$ of three adjacent layers of the immediate and main bottom of the coal seam;

– the cross-sectional area S_{fin} of mine working in the clear to subsidence.

Formation of the initial data for predicting the state of the bottom rocks of in-seam working is performed on the basis of information on geological survey (mining and geological prediction) for a specific mine working, technical documentation for its construction using the main provisions of the normative document [71] and the results of studies [68].

The calculated depth H of mine working location is determined by the formula

$$H = H_{des} \cdot k \text{ , m,} \tag{19.1}$$

where H – is the design actual depth of mine working location or its section from the daylight area; k – a coefficient that takes into account the difference in the stressed state of the rock massif from the initial one, formed by the deadweight of the coal-overlaying formation to the daylight area; for ordinary conditions outside the zone of the stope works influence and tectonic disturbances $k = 1$; while maintaining mine working in disjunctive breaks or curves of folds $k = 1.5$; in the zones of increased rock pressure, the value k is determined according to the conclusion of Ukrainian Research Development Surveying Institute.

The thickness $m_{1,2,3}^{b}$ of the three adjacent rock layers of the immediate and main bottom of the seam is determined from the mining data and geological prediction of the conditions for maintaining a specific mine working. Since the task of prognostic monitoring is to assess the state of the bottom rocks over the entire length of mine working, its length is divided into areas with a relatively consistent structure of the bottom rocks. When replacing the lithological varieties (changing the structure of bottom rocks) in at least one of the three layers $m_{1,2,3}^{b}$, this area is allocated as an independent one for a separate calculation of the expected value of heaving of the bottom rocks.

The calculated compressive resistance $R_{1,2,3}^{b}$ of the rocks of three adjacent layers of the immediate and main bottom is determined by the formula (17.1) using Tables 17.2 and 17.3. In formula (17.1), the uniaxial compressive strength of rocks in a sample $\sigma_{compr_{1,2,3}}^{b}$ is selected according to the data of mining-geological prediction. From the point of view of fulfilling the tasks of monitoring the state of the bottom rocks of mine working, it is divided into additional sections (to those already available in terms of parameters $m_{1,2,3}^{b}$), where a sharp (more than 30%) change in the intensity of fracturing or water-cut of the bottom rocks is predicted.

The cross-sectional area S_{fin} of mine working in the clear to subsidence is determined from the technical documentation for the construction of a specific mine working. When changing S_{fin} along its length, these areas are subject to additional calculation in the course of performing prognostic monitoring of the bottom rocks state. The coefficient k_S of the standard size influence of the mine working cross-section is determined according to the studies [68] by the Table 19.1.

Table 19.1

VALUES OF COEFFICIENT K_S OF THE STANDARD SIZE INFLUENCE OF MINE WORKING CROSS-SECTION

Type of support	Cross-sectional area in the clear to subsidence S_{fin}, m^2					
MYS - A3	9.2	11.2	13.8	15.5	18.3	
k_S	0.74	0.86	1.00	1.09	1.23	
TSYS	9.5	10.5	11.7	12.1	14.4	17.7
k_S	0.76	0.82	0.89	0.91	1.03	1.20

19.3. ASSESSMENT OF THE STATE OF BOTTOM ROCKS AND PREDICTION OF THE VALUE OF ITS HEAVING

In accordance with the calculation algorithm according to §19.1, the calculations are performed in the following sequence.

1. Mine working is divided into sections with significantly different structure and properties of the rocks of the immediate and main bottom; a preliminary selection of sites is performed on which the unsatisfying state of the bottom rocks is possible, regardless of its structure: with a calculated compressive resistance of less than 10 MPa of at least one of the three adjacent rock layers of the immediate and main bottom, a calculation procedure should be performed to predict the value of heaving.

2. A scheme is selected of heaving development of the bottom rocks, depending on its structure according to §19.1. In the absence of sandstone in the three adjacent rock layers of the bottom, the peculiarities of heaving development are reflected by the Scheme I. In the presence of sandstone in the three adjacent rock layers of the bottom, its behavior is described by the Schemes II and III of heaving development. To separate the schemes II and III, criterion (18.4) is used, which determines the boundary value of the thickness $\left(m_2^b\right)_{\min}$ of the second layer represented by a sandstone. The thickness of sandstone m_2^b in the specified section of mine working length is compared with the value $\left(m_2^b\right)_{\min}$ calculated from formula (18.4). If $m_2^b \geq \left(m_2^b\right)_{\min}$, the sandstone retains its continuity and the heaving process is developed according to the Scheme II of heaving development; if $m_2^b < \left(m_2^b\right)_{\min}$, the sandstone is divided into blocks with the formation of a thrust system, – the development of heaving is developed according to the Scheme III.

3. In the studied mine working area, the state of mine working bottom is tested

for the suitability for operation under the condition that the predicted value of heaving U^b does not exceed the permissible value of 300 mm. The condition (18.3) is the criterion for assessment: if the value of calculated compressive resistance of the immediate bottom R_1^b exceeds that calculated by the formula (18.3), then the predicted value of heaving does not exceed the permissible standards of mine working operation; otherwise, it is necessary to take measures to restrict the value of heaving or to eliminate its consequences.

4. To substantiate the measures against the bottom rocks heaving, its predicted value U^b is calculated using the formula (18.2). The calculations are performed for each of the sections of the mine working length where the unsatisfactory state of the bottom rocks is predicted according to point 3 of this subchapter. The final monitoring result is presented in the form of a list of mine working areas with an unsatisfactory bottom state, in which the length of the area and the predicted value U^b of heaving are recorded. According to these indicators, the types are determined and the volumes of maintenance and repair works for ensuring the stability of the bottom rocks in in-seam working are calculated.

19.4. AN EXAMPLE OF A PREDICTIVE CALCULATING THE BOTTOM ROCKS HEAVING OF IN-SEAM WORKING

1. *Initial data.* It is planned to construct a belt entry with a cross section in the clear to subsidence of $S_{fin} = 11.7$ m^2 under the frame support TSYS-11.7. The drift is carried out to the rise of the coal seam (with an average incidence angle of 3.5°) outside the zones of tectonic disturbances at a depth of $H_{des} = 340 - 490$ m. The structure and properties of the adjacent rock layers of the belt entry bottom are characterized by the following parameters.

The immediate bottom is represented mainly by argillite with a thickness of $m_1^b = 0.5 - 2.7$ m with an average compressive resistance in the sample of $\sigma^b_{compr_1} = 18.2$ MPa; the intensity of fracturing – 5 – 8 cracks per running meter; in the areas between PK0 - PK12, PK48 - PK71, PK106 - PK112, the rocks of the immediate bottom are water-flooded. In the areas PK72 – PK105, argillite is predicted to be replaced by siltstone with a thickness of $m_1^b = 1.8 - 3.3$ m, $\sigma^b_{compr_1} = 20.6$ MPa; the rocks are non-fractured, water-free.

Main bottom. When calculating, the parameters of the structure and properties of the two rock layers adjacent to mine working with a thickness of $m_{2,3}^b$ and compressive resistance in the sample $\sigma^{b}_{compr_{2,3}}$ are taken into account. The second rock layer (relative to the bottom of mine working) is represented by siltstone with a thickness of $m_2^b = 2.1 - 5.7$ m, $\sigma^{b}_{compr_2} = 20.6$ MPa with replacement within the area PK48 – PK112 with sandstone with a thickness of $m_2^b = 0.9 - 4.3$ m, $\sigma^{b}_{compr_2} = 41.0$ MPa. Both lithotypes are not fractured, and their water-cut is predicted in the areas between PK0 – PK12, PK48 – PK71, PK106 – PK112. The third rock layer is represented by non-fractured argillite with a thickness of $m_3^b = 2.6 - 6.7$ m, $\sigma^{b}_{compr_3} = 14.5$ MPa. In the areas between PK0 – PK12, PK48 – PK71 and PK106 – PK112, argillite is water-flooded.

We determine the calculated depth of the mine working location in accordance with the formula (19.1). Since, according to the conditions of the example, the belt entry is located outside the zone of tectonic disturbances, then we take $k = 1$ and the calculated depth H coincides with the design depth H_{des}

$$H = H_{des} \cdot 1 = 340 - 490 \text{ m}.$$

We calculate the calculated value of the compressive resistance $R^R_{1,2,3}$ of three adjacent layers of the immediate and main bottom, taking into account the action of factors weakening the rock in accordance with formula (17.1), Tables 17.2 and 17.3.

$$R_i = K_c K_w K_t \sigma^{b}_{compr_i} \ (i = 1, 2, 3).$$

For the convenience of calculations and the systematization of the subsequent prediction, the results are summarized in Table 19.2 for selected areas of the length of mine working. The areas are divided according to two factors: replacement of lithological varieties and degree of water-cut. In total 6 areas have been identified, for each of which, for each lithological variety, the values of the coefficients K_c, K_w, K_t of weakening are determined.

The depth of mine working location ($H = H_{des}$) is determined according to the plan of mining operations for each of the identified areas of mine working. The average value of the coefficient K_H (see Table 18.1) is determined from the depth influence values; the results are summarized in Table 19.3.

Table 19.2

EXAMPLE OF DIVIDING THE MINE WORKING INTO AREAS FOR CALCULATING HEAVING OF BOTTOM ROCKS

No.	Picket number	Number of the bottom layer											
		1				2				3			
		Weakening factors			R_1^b, MPa	Weakening factors			R_2^b, MPa	Weakening factors			R_3^b, MPa
		K_c	K_w	K_t		K_c	K_w	K_t		K_c	K_w	K_t	
1	PK0 – PK12	argillite			2.91	siltstone			8.90	argillite			5.22
		0.4	0.5	0.8		0.9	0.6	0.8		0.9	0.5	0.8	
2	PK13 – PK47	argillite			5.82	siltstone			14.83	argillite			10.44
		0.4	1.0	0.8		0.9	1.0	0.8		0.9	1.0	0.8	
3	PK48 – PK71	argillite			2.91	sandstone			26.57	argillite			5.22
		0.4	0.5	0.8		0.9	0.8	0.9		0.9	0.5	0.8	
4	PK72 – PK105	argillite			14.83	sandstone			33.21	argillite			10.44
		0.9	1.0	0.8		0.9	1.0	0.9		0.9	1.0	0.8	
5	PK106 – PK112	argillite			2.91	sandstone			26.57	argillite			5.22
		0.4	0.5	0.8		0.9	0.8	0.9		0.9	0.5	0.8	
6	PK113 – PK123	argillite			5.82	siltstone			14.83	argillite			10.44
		0.4	1.0	0.8		0.9	1.0	0.8		0.9	1.0	0.8	

Table 19.3

AVERAGE VALUES OF THE COEFFICIENT K_H FOR AREAS

Picket number	PK0 – PK12	PK13 – PK47	PK48 – PK71	PK72 – PK105	PK106 – PK112	PK113 – PK123
K_H	1.24	1.15	1.05	0.95	0.87	0.84

The coefficient k_S of the standard size influence of the mine working cross-section is determined from Table 19.1. For the section under the TSYS-11.7 support, the value is $k_S = 0.89$.

Thus, the initial database has been completely formed for performing the predictive calculation of heaving of the bottom rocks in the belt entry.

2. We make a preliminary assessment of the bottom rocks stability in terms of the calculated compressive resistance $R_{1,2,3}^b$ of the three adjacent layers: at $R_{1,2,3}^b \geq 10$ MPa, the bottom rocks are in a stable state. Only one area of the six selected – from PK72 to PK105, satisfies the preliminary condition of stability; the remaining five areas require further assessment.

3. We determine a scheme for development of heaving of the bottom rocks in mine working in each of the five areas, depending on the structure of the immediate and main bottom.

In the three areas between PK0 – PK12, PK13 – PK47 and PK113 – PK123, the three rock layers adjacent to mine working are represented by argillite and siltstone with sufficiently similar mechanical properties, that is, the structure of the bottom can be characterized as relatively homogeneous. Therefore, these areas belong to the "Scheme I of heaving development" of the bottom rocks.

In the remaining two areas (PK48 – PK71 and PK106 – PK112), sandstone is present in the second rock layer, the mechanical properties of which are significantly different compared to argillite and siltstone. Here, the development of heaving of the bottom rocks occurs according to the Scheme II or III. To select a particular scheme, a condition with the initial data according to the Tables 19.2 and 19.3 is used. The intervals of change in the thicknesses m_i^b of the rock layers for each area of the mine working length are determined from the data of mining and geological prediction and are given in Table 19.4. Here is a line with the calculated values of the parameters m_i^b (used in all calculations), which are selected from the above interval of change in m_i^b according to the following rules. Studies have revealed that the most intensive heaving of the bottom rocks is observed at a combination of the minimum thickness of the hardest layer and the maximum thicknesses of the less harder layers. Therefore, for a harder sandstone, the minimum thickness value in each area of the mine working length is chosen as the calculated thickness m_2^b; the maximum values within the specific ranges of changes are used for argillite and siltstone $m_{1,3}^b$.

According to the condition (18.4), we determine a scheme for development of heaving of the bottom rocks in the areas of the mine working length.

PK48 – PK71:

$$\left(m_2^b\right)_{\min} \geq 3.6\sqrt{\frac{1}{R_2^b}\left(11\frac{m_1^b}{R_1^b}+17\frac{m_3^b}{R_3^b}\right)};$$

$$3.6\text{ m} \geq 3.6\sqrt{\frac{1}{26.57}\left(11\frac{2.5}{2.91}+17\frac{3.9}{5.22}\right)}=3.29\text{ m};$$

3.6 m > 3.29 m – the condition of the sandstone continuity is fulfilled; consequently, the development of heaving of the bottom rocks takes place according to the Scheme II.

Table 19.4

DATA ON THE RANGES OF CHANGES AND THE VALUES OF CALCULATED THICKNESS $m^R_{1,2,3}$ OF THE BOTTOM ROCK LAYERS IN THE AREAS OF MINE WORKING

Picket number		PK0 – PK-12	PK13 – PK47	PK48 – PK71	PK106 – PK112	PK113 – PK123
Thickness interval, m	m_1^R	2.3 – 2.7	0.6 – 2.3	0.6 – 2.5	0.5 – 0.7	0.7 – 1.1
	m_2^R	2.5 – 3.2	3.2 – 5.7	3.6 – 4.3	0.9 – 1.5	2.1 – 2.8
	m_3^R	3.1 – 4.4	2.6 – 4.4	2.6 – 3.9	5.0 – 6.7	4.9 – 6.7
Calculated thickness, m	m_1^R	2.7	2.3	2.5	0.7	1.1
	m_2^R	2.5	3.2	3.6	0.9	2.1
	m_3^R	4.4	4.4	3.9	6.7	6.7

PK106 – PK112:

$$0.9 \text{ m} \geq 3.6\sqrt{\frac{1}{26.57}\left(11\frac{0.7}{2.91}+17\frac{6.7}{5.22}\right)}=3.45 \text{ m};$$

$0.9 \text{ m} < 3.45 \text{ m}$ – the condition of the sandstone continuity is not fulfilled, the development of heaving of the bottom rocks takes place according to the Scheme III.

As a result, for each of the five "problematic" areas of the mine working length, a specific scheme of development of heaving of the bottom rocks has been determined.

4. We test the state of the bottom rocks of mine working for the suitability for operation under the condition (18.3) of limited heaving, not exceeding the permissible value of 300 mm.

PK0 – PK12:

$$R_1^b \geq \left[9.93\exp\left(0.8m_1^b\right)\Phi_{I,II,III}K_H K_S - \frac{0.6m_2^b R_2^b + 0.4m_3^b R_3^b}{0.6m_2^b + 0.4m_3^b}\right]\left[\exp\left(0.8m_1^b\right)-1\right]^{-1};$$

$\Phi_I = 1$ for the Scheme I, $\Phi_{II} = 1+0,5\,exp\left(-1.2m_1^R\right)$ for the Scheme II and $\Phi_{III} = 1+0.5\,exp\left(-1.5m_1^R\right)$ for the Scheme III of development of the bottom rocks heaving.

$$2.91 \geq \left[9.93\,exp(0.8\cdot 2.7)1\cdot 1.24\cdot 0.89 - \frac{0.6\cdot 2.5\cdot 8.90+0.4\cdot 4.4\cdot 5.22}{0.6\cdot 2.5+0.4\cdot 4.4}\right]\times$$

$$\times\left[\exp(0.8\cdot 2.7)-1\right]^{-1} = 11.49;$$

2.91 MPa $<$ 11.49 MPa – the condition of the permissible value of compressive resistance of the bottom rocks is not fulfilled, – it is necessary to make a predictive assessment in this area of the intensity of the bottom rocks heaving, and, based on the results, provide for appropriate measures to restrict it.

According to the above example, the following results have been obtained for the remaining areas of the mine working length:

– PK13 – PK47, PK48 –PK71, PK106 – PK112 – the condition of the permissible value of heaving is not fulfilled and it is necessary to plan measures to resist heaving of the bottom rocks in these areas;

– PK113 – PK123 – the condition of the permissible value of heaving is fulfilled ($U^b = 264$ mm); hence, mine working remains suitability for operation without any additional measures to restrict heaving of the bottom rocks.

5. Determine the value of heaving U^b by the formula

$$U^b = \frac{2980\Phi_{I,II,III} K_H K_S}{R_1^b + \left(\dfrac{0.6 m_2^b R_2^R + 0.4 m_3^b R_3^b}{0.6 m_2^b + 0.4 m_3^b} - R_1^b \right) \exp\left(-0.8 m_1^b\right)}.$$

In the area PK0 – PK12:

$$U^b = \frac{2980 \cdot 1 \cdot 1.24 \cdot 0.89}{2.91 + \left(\dfrac{0.6 \cdot 2.5 \cdot 8.9 + 0.4 \cdot 4.4 \cdot 5.22}{0.6 \cdot 2.5 + 0.4 \cdot 4.4} - 2.91 \right) \exp(-0.8 \cdot 2.7)} = 967 \text{ mm}.$$

Similar to the above calculation, the following results were obtained in other areas:

PK13 – PK47: $U^b = 435$ mm;

PK48 – PK71: $U^b = 583$ mm;

PK106 – PK112: $U^b = 430$ mm.

Thus, according to the results of monitoring the state of the bottom rocks of the belt entry, four areas have been identified where the predicted value of heaving exceeds the permissible value. Particular attention should be paid to the area between PK0 – PK12 with a predicted value of heaving of the bottom rocks of about 1 m.

Summing up the research results, it is possible to state the fulfillment of the main purpose of the work – the creation of a scientifically grounded base for advance planning of measures to limit the negative consequences of heaving of the bottom rocks in in-seam workings.

CONCLUSIONS

As a result of the implementation of the research tasks, a technique for predicting heaving of the bottom rocks in in-seam workings has been developed, reflecting the specific character of the deformation of the adjacent coal-bearing stratum of the Western Donbas. The main conclusions are as follows.

1. The continuity of the previous studies results made it possible to develop a methodology for conducting a computational experiment to a level that takes into account the action of the main weakening factors of the massif (water-cut, fracturing, rheology), while maintaining the multivariate geomechanical calculations for creating a database to determine the multiparameter dependences, characterizing the manifestations of rock pressure in the bottom of in-seam workings.

2. A comprehensive substantiation has been performed for the patterns of the main geomechanical factors influence (rock structure and properties, depth of location) on the manifestations of heaving of the bottom rocks of mine workings. For each of the three determined schemes for the development of rock heaving process, the correlation dependences have been obtained for calculating the value of heaving with a sufficiently high coefficient of multiple correlation. These dependences take into account the distribution of thickness and compressive resistance of the rock layers over the depth of the immediate and main bottom, but are distinguished by the simplicity and availability of use when predicting bottom heaving under different scheme variants of its development. Division of the mining and geological conditions according to the nature of heaving development in the rocks has been brought to the criterion equation, which enables to numerically determine the process scheme.

3. It has been proved that with a value of the calculated compressive resistance $R_i^b \geq 10$ MPa of the rocks of all three adjacent bottom layers at a depth of at least 6 m, the rocks are sufficiently stable, and the limited value of heaving (up to 200 – 300 mm) does not significantly influence on the operational characteristics of mine working as a whole. The boundaries of areas with satisfactory and unstable state of the bottom rocks have been determined and a criterion equation for separating these areas has been obtained, which makes possible to identify the problem areas along the length of mine working at the design stage and substantiate the measures to restrict or eliminate the negative influence of heaving in the bottom rocks.

4. Based on the research results, a technique and an algorithm for predicting heaving of the bottom rocks in in-seam workings of the Western Donbas mines have been developed, which make it possible to quickly assess its state and develop technical solutions to ensure the proper operational parameters of mine working as a whole. Therefore, the proposed predicting technique is recommended to be used as an addition to the existing normative documents on maintaining mine workings and, on its basis, to carry out prognostic monitoring of the geomechanical phenomenon of heaving of the bottom rocks, which is extremely relevant for the conditions of the Western Donbas mines.

LIST OF REFERENCES

1. Баришніков А.С. Закономірності деформування системи «кріплення – масив» у слабких породах у зоні впливу очисних робіт: автореф. дис. на здобуття наук. ступеня канд. техн. наук: спец. 05.15.01 «Маркшейдерія» / А.С. Баришніков. – Дніпропетровськ, 2016. – 20 с.

2. Estimate of efficiency of two-level rock bolting at junctions in coal mines / P. Grechishkin, A. Pozolotin, D. Zayatdinov [et al.] // Gornyi Zhurnal. – 2015. – No. 8. – P. 48 – 52.

3. Analysis of yielding steel arch support with rock bolts / T. Majcherczyk, Z. Niedbalski, P. Malkowski [et al.] // Journal of Mining Science. – 2014. – No. 59(3). – P. 641 – 654.

4. Stress evolution in roadway rock bolts during mining in a fully mechanized longwall face, and an evaluation of rock bolt support design / K. Zhang, G. Zhang, R. Hou [et al.] // Rock mechanics and rock engineering. – 2015. – No. 48(6). – P. 2421 – 2433.

5. Maghous S. Three-dimensional numerical simulation of rock deformation in bolt-supported tunnels: a homogenization approach / S. Maghous, D. Bernaud // Tunn. Undergr. Space Technol. – 2015. – No. 31. – P. 68 – 79.

6. Gao F.Q. Effect of pre-tensioned rock bolts on stress redistribution around a roadway – insight from numerical modeling / F.Q. Gao, H.P. Kang // J China Univ. Min. Technol. – 2008. – No. 18(4). – P. 509 – 515.

7. Fengshan Han. Study on Rock Bolt Support of Roadway of Coal Mine using Neural network / Han Fengshan, Wu Xinli // Applied Mechanics and Materials. – 2014. – Vol. 448 – 453. – P. 3799 – 3802.

8. Gale W.J. Computer Simulation of ground behavior and rock bolt interaction at Emerald Mine / W. Gale, C. Mark, D. Oyler // In Proceedings of the 21st International Conference on Ground Control in Mining, 3 – 5 August 2004. – West Virginia University, 2004. – P. 27 – 34.

9. Systematic monitoring of the performance of anchor systems in fractured rock masses / HY. Sun, LNY Wong [et al.] // Int. J Rock Mech. Min. Sci. – 2010. – 47(6). – P. 1038 – 1045.

10. Jayanthu S. Behavior in Development and Depillaring workings supported with Roof Bolt and Cable Bolts / S. Jayanthu, V. Venkateswarlu // Strata, Short Term Course on – Trends in Strata Control Techniques & Instrumentation for Enhancing Safety in Coal Mines, 28 – 31 July 2008. – Mining Dept. NIT Rourkela. – 2008. – P. 145 – 153.

11. Chen Y. Discussion on bolt and cable supporting technique of the complex roof / Y. Chen, X. Yu, J. Bai // Energy Technology and Management. – 2005. – №6. – 6 – 7.

12. Kang H. Rock bolting theory and complete technology for coal roadways / H. Kang, J. Wang // China Coal Industry Publishing House. – 2007. – P. 13 – 28.

13. Composite element model of the fully grouted rock bolt / S. Chen, S. Qiang [et al.] // Rock Mech. Rock Eng. – 2004. – 37(3). – P. 193 – 212.

14. Xiu Z. Rockbolting in Gateroads of Top Coal Caving Longwall Faces / Z. Xiu, J. Jiang // Aachen International Mining Symposium. – 2004. – P. 195 – 210.

15. Дарков А.В. Строительная механика: монография / А.В. Дарков, В.И. Кузнецов. – М.: Высшая школа, 1962. – 744 с.

16. Сопротивление материалов: учебник / под общ. ред. Г.С. Писаренко. – К.: Вища школа, 1979. – 696 с.

17. Борисов А.А. Новые методы расчета штанговой крепи: монография / А.А. Борисов. – М.: Госгортехиздат, 1962. – 64 с.

18. Семевский В.Н. Штанговая крепь: монография / В.Н. Семевский. – М.: Недра, 1965. – 328 с.

19. Гелескул М.Н. Справочник по креплению капитальных и подготовительных горных выработок: справочник / М.Н. Гелескул, В.Н. Каретников. – М.: Недра, 1982. – 479 с.

20. Геомеханика нагружения и расчет параметров крепежной и охранной систем подготовительных выработок шахт Западного Донбасса: монография / [Бондаренко В.И., Ковалевская И.А., Симанович Г.А. и др.]. – Д.: ЛізуновПрес, 2014. – 228 с.

21. Ковалевская И.А. Экспериментальные исследования перемещений опор стоек крепи пластовой выработки / И.А. Ковалевская, В.Н. Почепов, А.Р. Мамайкин // Школа подземной разработки: материалы III междунар. науч.-практ. конф. – Д.: ЛізуновПрес, 2009. – С. 392 – 397.

22. Краев Ю.К. Оперативный способ увеличения несущей способности крепи, взаимодействующей со слабыми породами / Ю.К. Краев, М.В. Корнилов // Строительство шахт, рудников и подземных сооружений. – Свердловск: СГИ. – 1988. – С. 39 – 43.

23. Байкенжин М.А. Разработка способа поддержания пластовых подготовительных выработок глубоких горизонтов шахт податливой рамно-анкерной крепью: автореф. дисс. на соискание учен. степени канд. техн. наук: спец. 05.15.02 «Подземная разработка месторождений полезных ископаемых» / М.А. Байкенжин. – Алма-Ата, 1987. – 19 с.

24. Егоров С.И. Обобщение опыта крепления и охраны подготовительных выработок / С.И. Егоров, Ю.М. Халимендик, Э.П. Курченко и др. // Уголь Украины. – 2002. – №5. – С. 19 – 21.

25. Касьян Н.Н. О перспективах применения анкерной крепи на угольных шахтах Донбасса / Н.Н. Касьян, Ю.А. Петренко, А.О. Новиков // Научные труды ДонНТУ. – 2009. – № 10. – С. 109 – 115.

26. Кулешов Е.С. Оценка состояния выработок с анкерной крепью на шахтах Российского Донбасса (проблемы и решения) / Е.С. Кулешов // Горный информационно-аналитический бюллетень. – 2005. – № 2. – С. 174 – 177.

27. Якоби О. Практика управления горным давлением: монография / О. Якоби. – М.: Недра, 1987. – 566 с.

28. Черняк И.Л. Управление горным давлением в подготовительных выработках глубоких шахт: учебник / И.Л. Черняк. – М.: Недра, 1984. – 304 с.

29. Kravets V. Increasing stability of underground mine workings by forming new geotechnical properties of adjoining layers via roller compaction / V. Kravets, S. Zaychenko, G. Gayko // Mining of Mineral Deposits. – 2016. – 10(1), 44 – 49.

30. Исследование скорости подвигания на работу лав / Н.И. Лобков, А.К. Носач, Л.Л. Бачурин [и др.] // Проблемы механики горно-металлургического комплекса. – Д.: НГУ, 2002. – С. 178 – 180.

31. Оценка влияния скорости подвигания очистного забоя на вмещающие боковые породы / М.А. Ильяшов, Н.И. Лобков, Е.Н. Халимендиков [и др.] // Уголь Украины. – 2008. – №8. – С. 11 – 13.

32. Лобков Н.И. Геодинамическое обоснование характера обрушения кровли в очистных забоях / Н.И. Лобков, В.В. Порубай // Физико-технические проблемы горного производства. – Донецк: ИФГП НАНУ, 2004. – Вып. 7. – С. 181 – 187.

33. Progressive method of anchor-concrete support of capital mine workings and tunnels / V.G. Kravets, G.I. Gaiko, S.V. Zaichenko [et al.] // Mining of Mineral Deposits. – 2015. – 9(3). – 307 – 312.

34. Комиссаров С.Н. Управление массивом горных пород вокруг очистных выработок / С.Н. Комиссаров. – М.: Недра, 1983. – 237 с.

35. Свистун Р.Н. Обоснование и расчет параметров крепежной системы выемочных выработок с учетом асимметрии ее нагружения: дисс. ... кандидата техн. наук: 05.15.02 / Свистун Руслан Николаевич. – Д., 2014. – 171 с.

36. Ryabtsev O. Studies of stationary supporting zone sizes varied in the course of mining operations in deep horizons / O. Ryabtsev, O. Voloshyn // Mining of Mineral Deposits. – 2013. – P. 71 – 77.

37. Фисенко Г.Л. Предельные состояния горных пород вокруг выработок / Г.Л. Фисенко. – М.: Недра, 1976. – 272 с.

38. Зборщик М.П. Охрана выработок глубоких шахт в зоне разгрузки / М.П. Зборщик, В.В. Назимко. – К.: Техніка, 1991. – 248 с.

39. Свистун Р.Н. Прогноз и оценка достоверности расчета перемещений контура пластовой выработки / [Бондаренко В.И., Ковалевская И.А., Симанович Г.А., Свистун Р.Н.] // Геотехническая механика: сб. науч. трудов ИГТМ НАН Украины. – Д., 2009. – Вып. 83. – С. 35 – 48.

40. Булат А.Ф. Опорно-анкерное крепление горных выработок угольных шахт / А.Ф. Булат, В.В. Виноградов. – Д.: ИГТМ НАН Украины, 2002. – 372 с.

41. Бесцеликовая отработка пластов / [Худин Ю.Л., Устинов М.И., Брайцев и А.В. др.]. – М.: Недра, 1983. – 280 с.

42. Повышение устойчивости подготовительных выработок угольных шахт / [Заславский И.Ю., Компанец В.Ф., Файвишенко А.Г., Клещенков В.Н.]. – М.: Недра, 1991. – 235 с.

43. Борзых А.Ф. Поэтапное проведение и поддержание прилегающей к лаве выработки / А.Ф. Борзых, В.И. Липкович, А.И. Рачев // Уголь Украины. – 2002. – №7. – С. 20 – 21.

44. Заславский И.Ю. Проявления горного давления в выемочных выработках, охраняемых жесткими литыми полосами / И.Ю. Заславский, А.Г. Файви-

шенко // Напряжения и деформации в массивах горных пород при разработке месторождений полезных ископаемых. – М.: ИПКОН АН СССР, 1988.– С. 122 – 128.

45. Зборщик М.П. Сохранение устойчивости подготовительных выработок высоконагруженных лав при отработке пологих пластов / М.П. Зборщик, А.В. Костенко // Уголь Украины. – 2003. – №4. – С. 7 – 11.

46. Медяник В.Ю. Обоснование параметров способа охраны подготовительных выработок при комбинированной системе разработки пологих пластов на больших глубинах: автореф. дисс. на соискание учен. степени канд. техн. наук: спец. 05.15.02 «Подземная разработка месторождений полезных ископаемых» / В.Ю. Медяник. – Д.: НГУ, 2005. – 17 с.

47. Лозовский С.П. Обоснование параметров способа повышения устойчивости подготовительных выработок с пучащей почвой в зоне влияния очистных работ. автореф. дисс. на соискание учен. степени канд. техн. наук: спец. 05.15.02 «Подземная разработка месторождений полезных ископаемых» / С.П. Лозовский. – Д.: НГУ, 2004. – 16 с.

48. Слесарев В.Д. Управление горным давлением при разработке угольных пластов Донецкого бассейна / В.Д. Слесарев. – М.: Углетехиздат, 1952. – 379 с.

49. Расчет напряжений в зонах влияния целиков и краевых частей смежных пластов свиты / В.В. Зубков, И.А. Зубкова, А.М. Линьков [и др.] // Разработка месторождений полезных ископаемых. – К., 1982. – №63. – С. 48 – 54.

50. Булычев Н. С. Механика подземных сооружений / Н. С. Булычев. – М.: Недра, 1989. – 270 с.

51. Симанович Г. А. Управление режимами взаимодействия породного массива с крепью горных выработок на основе регулирования ее деформационно-силовой характеристикой: дисс. … доктора техн. наук: 05.15.02; 05.15.11 / Симанович Геннадий Анатольевич. – Днепропетровск, 1993. – 408 с.

52. Ковалевская И. А. Геомеханика взаимодействия элементов системы «горный массив – упрочненные породы – крепь подземных выработок» / И.А. Ковалевская. – Днепропетровск: НГА Украины, 2003. – 234 с.

53. Підготовчі виробки на пологих пластах. Вибір кріплення, способів охорони: СОУ 10.1.00185790.011:2007. – Донецьк: ДонВУГІ, 2008. – 114 с. – (Нормативний документ Мінвуглепрому України. Стандарт).

54. Рябцев О.В. Численное моделирование процесса разрушения горных пород / О.В. Рябцев, А.И. Волошин, Ю.А. Морозов // Уголь Украины. – 2016. – №8. – С. 39 – 43.

55. Экспериментальные исследования устойчивости повторно используемых выемочных выработок на пологих пластах Донбасса / [Бондаренко В.И., Ковалевская И.А., Симанович Г.А. и др.]. – Д.: ЛізуновПрес, 2012. – 426 с.

56. Геомеханика нагружения крепи очистных и подготовительных выработок в слоистом массиве слабых пород / [В.И. Бондаренко, И.А. Ковалевская, Г.А. Симанович, В.Г. Черватюк]. – Д.: ЛізуновПрес, 2012. – 236 с.

57. Экспериментальные исследования пучения пород почвы подготовительных выработок на пологих пластах Донбасса / [В.И. Бондаренко, И.А. Ковалевская, Г.А. Симанович, В.Г. Снигур]. – Д.: ЛізуновПрес, 2012. – 224 с.

58. Взаимодействие грузонесущих элементов крепежной системы выемочных выработок «массив – рама – анкер» / [Бондаренко В.И., Ковалевская И.А., Симанович Г.А. и др.]. – Д.: Літограф, 2015. – 214 с.

59. Баклашов И.В. Деформирование и разрушение породных массивов / И.В. Баклашов. – М.: Недра. 1988. – 270 с.

60. Булычев Н.С. Механика подземных сооружений / Н.С. Булычев. – М.: Недра, 1982. – 272 с.

61. Борисов А.А. Механика горных пород и массивов / А.А. Борисов. – М.: Недра, 1980. – 360 с.

62. Усаченко Б.М. Свойства пород и устойчивость горных выработок / Б.М. Усаченко. – К.: Наукова думка, 1979. – 136 с.

63. Усаченко Б.М. Геомеханика охраны выработок в слабометаморфизированных породах / Б.М. Усаченко, В.Н. Чередниченко, И.Е. Головчанский. – К.: Наукова думка, 1990. – 144 с.

64. Ержанов Ж.С. Метод конечных элементов в задачах механики горных пород / Ж.С. Ержанов, Т.Д. Каримбаев. – Алма-Ата: Наука, 1975. – 238 с.

65. Амусин Б.З. Метод конечных элементов при решении задач горной геомеханики / Б.З. Амусин, А.Б. Фадеев. – М.: Недра, 1975. – 144 с.

66. Фадеев А.Б. Метод конечных элементов в геомеханике / А.Б. Фадеев. – М.: Недра, 1987. – 224 с.

67. Развитие научных основ повышения устойчивости горных выработок шахт Западного Донбасса / [Бондаренко В.И., Мартовицкий А.В., Ковалевская И.А., Черватюк В.Г.]. – Д.: ЛізуновПрес, 2010. – 340 с.

68. Методы расчета перемещений и упрочнения приконтурных пород горных выработок шахт Западного Донбасса / [Бондаренко В.И., Ковалевская И.А., Симанович Г.А. и др.]. – Д.: Дріант, 2010. – 328 с.

69. Барон Л.И. Определение свойств горных пород / Барон Л.И., Логунцов Б.М., Позин Е.З. – М.: Госгортехиздат, 1962. – 232 с.

70. Свойства горных пород и методы их определения / [Ильницкая Е.И., Тедер Р.И., Ватолин Е.С., Кунтыш М.Ф.]. – М.: Недра, 1969. – 232 с.

71. СОУ 10.1.00185790.011:2007. Підготовчі виробки на пологих пластах. Вибір кріплення, способів і засобів охорони. – К: Мінвуглепром України, 2007. – 113 с.

72. Инструкция по поддержанию горных выработок на шахтах Западного Донбасса. – СПб. – Павлоград: ВНИМИ – ЗДНПЦ «Геомеханика», 1994. – 95 с.

73. Компьютерное моделирование напряженно-деформированного состояния мелкослоистого породного массива вокруг пластовой выработки. Книга 2. Предельное и запредельное состояние системы «порода – крепь» / [Бондаренко В.И., Ковалевская И.А., Симанович Г.А., Фомичев В.В.]. – Д.: Системные технологии, 2007. – 198 с.

74. Виноградов В.В. Геомеханика управления состоянием массива вблизи горных выработок / В.В. Виноградов. – К.: Наукова думка, 1989. – 192 с.

75. Геомеханика взаимодействия анкерной и рамной крепей горных выработок в единой грузонесущей системе / [Бондаренко В.И., Ковалевская И.А., Симанович Г.А. и др.]. – Д.: ЛізуновПрес, 2010. – 174 с.

76. Мельников Н.И. Анкерная крепь / Н.И. Мельников. – М.: Недра, 1980. – 252 с.

77. Ковалевская И.А. Геомеханика управления устойчивостью пространственной системы «массив – упрочненные породы – крепь подземных выработок»: дисс. ... доктора техн. наук: 05.15.02 / Ковалевская Ирина Анатольевна. – Д., 2004. – 349 с.

78. Совершенствование способов охраны подготовительных выработок: монография / [Симанович А.М., Сребный М.А., Малов В.И., Белинский И.Л.]. – Донецк: Донбасс, 1973. – 121 с.

79. Симанович А.М. Охрана выработок на глубоких горизонтах: монография / А.М. Симанович, М.А. Сребный. – М.: Недра, 1976. – 144 с.

80. Скипочка С.И. Геомеханика охраны выемочных штреков в неустойчивых породах: монография / Скипочка С.И., Мухин А.В., Черватюк В.Г. – Д.: НГА Украины, 2002. – 125 с.

81. Ковалевская И.А. Взаимодействие анкерной и комбинированной крепей с породным массивом и разработка метода расчета их рациональных параметров: дисс. ... кандидата техн. наук: 05.15.02; 05.15.11 / Ковалевская Ирина Анатольевна. – Д.: ГГАУ, 1995. – 200 с.

82. Круковський О.П. Наукові основи технології опорно-анкерного кріплення гірничих виробок вугільних шахт: автореф. дисс. ... доктора техн. наук: спец. 05.15.02 «Підземна розробка родовищ корисних копалин» / О.П. Круковський. – Д., 2012. – 36 с.

83. Симанович Г.А. Определение нагрузок на крепь выемочного штрека в зоне влияния очистных работ / Г.А. Симанович, П.П. Небатов // Науковий вісник НГА України. – 1999. – № 5. – С. 43 – 46.

84. Байсаров Л.В. Геомеханика и технология поддержания повторно используемых горных выработок: монография / Байсаров Л.В., Ильяшов М.А., Демченко А.И. – Д.: ЧП «Лира ЛТД», 2005. – 240 с.

85. Охрана штреков литыми полосами при разработке пологих пластов средней мощности / М.А. Ильяшов, М.П. Зборщик, А.А. Яйцов [и др.] // Уголь Украины. – 2001. – № 9. – С. 3 – 6.

86. Formation of limit equilibrium zone in marginal rocks of excavation mine working / I. Kovalevska I., G. Symanovych, O. Gusiev [etc.] // Materialy Szkoly Eksploatacij Podzemnej-2015. – [Electronic resource] 1 electronic optical drive (2,25 MB; CD-ROM). – Krakow: Katedra Gornictwa Podziemnego.

87. КД 12.01.01.501 – 98. Система забезпечення надійного та безпечного функціонування гірничих виробок із анкерним кріпленням. Загальні технічні вимоги. – Д.: ІГТМ НАНУ, 1999. – 42 с.

88. КД 12.01.01. 502 – 98. Система забезпечення надійного та безпечного функціонування гірничих виробок із анкерним кріпленням. Порядок та організація. – Д.: ІГТМ НАНУ, 1999. – 14 с.

89. Протодьяконов М.М. Давление горных пород и рудничное крепление. Ч. 1. Давление горных пород / М.М. Протодъяконов. – М.-Л.: ОГИЗ, 1931. – 104 с.

90. Цимбаревич П.М. Механика горных пород: учеб. / П.М. Цимбаревич. – М.: Углетехиздат, 1948. – 247 с.

91. Kovalevska I. Research into stress-strain state of reinforced marginal massif of extraction mine working by combined anchoring system / I. Kovalevska, M. Barabash, O. Gusiev // Mining of Mineral Deposits. – 2016. – 10(1). – P. 31 – 36.

92. СНиП II-94-80. Подземные горные выработки. Нормы проектирования. – М.: Стройиздат, 1982. – 272 с.

93. Руководство по размещению, выбору и расчету крепи и защите подземных горных выработок. – М.: Стройиздат, 1980. – 374 с.

94. КД 12.01.01.201-98. Расположение, охрана и поддержание горных выработок при отработке угольных пластов на шахтах. Руководящий нормативный документ Министерства угольной промышленности Украины. Методические указания. – К.: УкрНИМИ, 1998. – 149.

95. Аналитико-экспериментальные исследования устойчивости выемочных выработок и расчет параметров крепежной системы / [Бондаренко В.И., Ковалевская И.А., Симанович Г.А. и др.]. – Д.: ЛізуновПрес, 2013. – 178 с.

96. Широков А.П. Расчет анкерной крепи для различных условий применения / Широков А.П., Лидер В.А., Писляков Б.Г. – М.: Недра, 1976. – 208 с.

97. Мероприятия по снижению и предотвращению пучения пород почвы в горных выработках: Угольный портал: [Электронный ресурс]. – режим доступа: http: //www.coal.dp.ua/index.php?option-com_content&view – article/-02.04.2013.

98. Татаринов А.А. Прогнозування та підвищення стійкості виробок в умовах слабометаморфізованих порід Західного Донбасу: автореф. дис. на здобуття наук. ступеня канд. техн. наук: спец. 05.15.04 «Шахтне і підземне будівництво» / А.А. Татаринов. – Дніпропетровськ, 1999. – 15 с.

99. Гапеев С.Н. Исследование процесса пучения почвы протяженной горной выработки лабораторными и численными методами / С.Н. Гапеев // Сб. научн. тр. НГУ. – 2003. – Т.1. – № 17. – С. 350 – 355.

100. Литвинский Г.Г. Прогноз пучения пород почвы горных выработок – вероятностный аспект / Г.Г. Литвинский, Э.В. Фесенко // Сб. научн. тр. ДонГТУ. – 2010. – № 31. – С. 168 – 177.

101. Заславский Ю.З. Расчет смещений почвы выработки, расположенной в слоистом горном массиве / Ю.З. Заславский, В.П. Киндур // Проектирование и строительство угольных предприятий. – 1970. – № 2. – С. 51 – 53.

102. Черняк И.Л. Предотвращение пучения почвы горных выработок / И.Л. Черняк – М.: Недра, 1978. – 237 с.

103. Кириченко В.Я. О механике пучения почвы выработок, сооружаемых в слабых породах / В.Я. Кириченко, А.В. Шлиголь, В.Н. Рева // Шахтное

строительство. – 1988. – № 11. – С. 3 – 5.

104. Пирскиий А.А. Шахтные исследования пучения почвы в выработках Западного Донбасса / А.А. Пирский, С.Н. Стовпник // Уголь Украины. – 1989. – № 11. – С. 2 – 3.

105. Скворцов Г.Г. Инженерно-геологическая характеристика пучения пород на угольных месторождениях СССР / Г.Г. Скворцов, А.К. Кориковская // Советская геология. – 1961. – № 11. – С. 158 – 164.

106. Ростовцев Д.С. Управление кровлей при разработке пологопадающих пластов каменного угля / Д.С. Ростовцев. – Харьков – К.: Гос. науч.-техн. изд-во Украины, 1935. – 378 с.

107. Савостьянов А.В. Управление состоянием массива горных пород: учеб. пособие / А.В. Савостьянов, В.Г. Клочков. – К.: УМК ВО, 1992. – 276 с.

108. Бондаренко В.И. Теория управления состоянием массива горных пород: учеб. / В.И. Бондаренко, М.А. Ильяшов, Н.К. Руденко. – Д.: ЛізуновПрес, 2012. – 320 с.

109. Шашенко А.Н. Управление устойчивостью протяженных выработок глубоких шахт / А.Н. Шашенко, А.В. Солодянкин, А.В. Мартовицкий. – Д.: ЛізуновПрес, 2012. – 384 с.

110. Усаченко Б.М. Охрана подготовительных выработок глубоких горизонтов шахт Западного Донбасса / Б.М. Усаченко, В.Я. Кириченко, А.В. Шмиголь. – М.: ЦНИЭИуголь, 1992. – 168 с.

111. Бондаренко В.И. Прогноз перемещений контура пластовой выработки в слоистом массиве слабых пород / В.И. Бондаренко, М.А. Ильяшов, И.А. Ковалевская, Г.А. Симанович, В.В. Фомичев. – Д.: Системные технологии, 2008. – 193 с.

112. Прогнозный каталог шахтопластов Донецкого угольного бассейна с характеристикой горно-геологических факторов и явлений. – М.: ИГД им. А.А. Скочинского, 1982. – 267 с.

113. Кадастр угольных шахтопластов, предусмотренных к отработке шахтами и разрезами Госуглепрома Украины. – Донецк: Изд-во Минуглепрома Украины, 2001. – 125 с.

114. Ставрогин А.Н. Прочность горных пород и устойчивость выработок на больших глубинах / А.Н. Ставрогин, А.Г. Протосеня. – М.: Недра, 1985. – 271 с.

115. Компьютерное моделирование напряженно-деформированного состояния мелкослоистого породного массива вокруг пластовой выработки. Книга 1. Допредельное деформирование системы «порода-крепь»: монография / [В.И. Бондаренко, И.А. Ковалевская, Г.А. Симанович, В.В. Фомичев]. – Д.: Системные технологии, 2006. – 172 с.

116. Ферстер Э. Методы корреляционного и регрессионного анализа / Э. Ферстер, Б. Ренц. – М.: Финансы и статистика, 1983. – 302 с.

CONTENTS

INTRODUCTION 3

Chapter I. RESEARCH INTO STABILITY OF REUSABLE IN-SEAM WORKINGS WHEN MINING THIN COAL SEAMS 5

1 ANALYSIS OF ROCK PRESSURE MANIFESTATION IN THE PREPARATORY MINE WORKINGS AT THE STRUCTURAL TRANSFORMATION OF ENCLOSING ROCKS BY THE ROOF-BOLT STRENGTHENING 5

1.1 Experience of combined roof-bolting systems application for strengthening the roof rocks of mine workings 5

1.2 A mechanism for increasing the stability of mine workings by combined roof-bolting systems 7

1.3 The interaction of support in mine workings with border rocks under the conditions of areas formation of their limiting state 11

Conclusions 16

2 RESEARCH AND ANALYSIS OF THE EXPERIENCE OF THE PREPARATORY MINE WORKINGS REUSE AT YUBILEINA MINE 17

2.1 Mining research methodology 17

2.2 Analysis of mining research results 19

Conclusions 35

3 THE COMPUTATIONAL EXPERIMENT SUBSTANTIATION AND STRESS-STRAIN STATE ANALYSIS OF BORDER ROCKS, AS WELL AS OF THE BASIC FASTENING SYSTEM OF PREPARATORY MINE WORKINGS IN THE ZONE OF STOPE WORKS INFLUENCE 37

3.1 Construction of geomechanical models in conducting the computational experiments 37

3.2 Analysis of stresses intensity in the border rocks of the preparatory mine working in the bearing pressure zone ahead of the longwall face 46

3.3 Research and analysis of the stresses intensity of the basic fastening system in the bearing pressure zone ahead of the longwall face 47

3.4	The stress intensity of border rocks in the zones of beginning of protective strip resistance and stabilization of the rock pressure manifestation	49
3.5	The stress intensity analysis in basic fastening and protection systems behind the longwall face	50
	Conclusions	53
4	RESEARCH AND ANALYSIS OF THE STRESS STATE OF THE LOAD-BEARING ELEMENTS IN THE PREPARATORY MINE WORKING "MASSIVE – FRAME – COMBINED ROOF-BOLTING SYSTEM"	54
4.1	Stress state of rocks enclosing mine working in the area of stabilization of rock pressure manifestations	54
4.2	The stress state of fastening system elements in the preparatory mine working and comparison with the basic variant of its fastening	59
	Conclusions	66
5	ESTABLISHING AND ANALYZING THE PATTERNS OF GEOMECHANICAL FACTORS INFLUENCE ON THE RESISTANCE OF THE COMBINED ROOF-BOLTING SYSTEM TO ROCK PRESSURE	67
5.1	Substantiation of the criterion for assessing the level of the roof-bolts resistance to the rock pressure manifestations	67
5.2	The influence of the depth of mine working location	68
5.3	Influence of the average calculated rocks resistance to compression of the adjacent massif	70
5.4	Influence of the ratio of the average calculated resistance to compression of the immediate roof rocks to its thickness	71
	Conclusion	73
6	DEVELOPMENT AND SUBSTANTIATION OF THE PARAMETERS CALCULATION OF THE ROOF ROCKS STRENGTHENING BY A COMBINED ROOF-BOLTING SYSTEM IN THE PREPARATORY MINE WORKINGS	74
6.1	Substantiation and calculation of the parameters of the resin-grouted roof bolts location in the rocks of mine working arch	74

6.2 Substantiation and calculation of strengthening the rocks in the arch of mine working with rope bolts 80

6.3 Algorithm for calculating the parameters of the combined roof-bolting system for strengthening the rocks of the arch in the preparatory mine workings 86

Conclusions 88

7 THE INFLUENCE OF THE CENTRAL PROP STAYS OF THE STRENGTHENING SUPPORT ON THE FASTENING SYSTEM STRESS-STRAIN STATE OF MINE WORKING 90

7.1 Analysis of the operation modes of the central prop stays of the strengthening support in the preparatory mine workings 90

7.2 The research of the fastening system stress-strain state of mine working with the yielding prop stay in the strengthening support 96

7.2.1 Analysis of vertical stresses components 97

7.2.2 The study of the curve of the horizontal stresses components 99

7.2.3 Analysis of stresses intensity 101

Conclusions 103

Synthesis of the research results to Chapter I 104

Chapter II. STRUCTURAL TRANSFORMATIONS OF THE ROCK MASSIF AND STABILITY OF MINE WORKINGS UNDER THE DOWNWARD AND JOINT MINING OF C_8^l AND C_9 COAL SEAMS 106

8 GEOMECHANICAL AND METHODICAL ASPECTS OF RESEARCH INTO STABILITY OF THE REUSABLE MINE WORKINGS WHEN OVERWORKING AND UNDERWORKING A PARTING IN SEAMS C_8^l AND C_9 106

8.1 Geomechanical peculiarities of C_8^l seam mining 106

8.2 Development and substantiation of the geomechanical model of the 861 belt entry 110

9	RESEARCH AND STRESS-STRAIN STATE ANALYSIS OF THE ROCK MASSIF ENCLOSING THE 861 BELT ENTRY AT THE BASIC AND RECOMMENDED SCHEMES OF MAINTAINING	114
9.1	Anomalies of the rock pressure manifestations in the rock massif in the vicinity of the 861 belt entry at the basic scheme of maintaining	114
9.1.1	Analysis of vertical stresses components distribution	114
9.1.2	Distribution of horizontal stresses components	116
9.1.3	Research of stresses intensity	119
9.2	Research and stress-strain state analysis of coal-bearing massif around the 861 belt entry at the recommended scheme of its maintenance	120
9.2.1	Analysis of distribution of vertical stresses components	121
9.2.2	Distribution of horizontal stresses components	123
9.2.3	Research of stresses intensity	126
9.2.4	Comparative stress-strain state analysis of the enclosing rock massive with different schemes of maintaining the 861 belt entry	128
9.2.5	Stress-strain state analysis of a parting rocks of seams C_8^l and C_9 in the area of the 861 belt entry location	130
10	RESEARCH AND ANALYSIS OF THE STRESS-STRAIN STATE OF FASTENING AND PROTECTION BASIC SYSTEMS ELEMENTS OF THE 861 BELT ENTRY OF THE SEAM C_8^l	136
10.1	Analysis of vertical stresses components	136
10.2	Determination of horizontal stresses components	138
11	RESEARCH AND ANALYSIS OF THE STRESS-STRAIN STATE OF FASTENING AND PROTECTION RECOMMENDED SYSTEMS ELEMENTS FOT MAINTAINING THE 861 BELT ENTRY OF THE SEAM C_8^l	143
11.1	Analysis of vertical stresses component	143
11.2	Horizontal stresses components distribution	145
11.3	Stresses intensity research	146
11.4	Recommendations for selecting the parameters of the fastening and protection systems of reused mine workings through the example of mining the 861 longwall face	150

Summarizing the research results to Chapter II 152

Chapter III. RESEARCH INTO HEAVING OF MINE WORKINGS BOTTOM WHEN MINING THE FLAT-LYING SEAMS OF THE WESTERN DONBAS 154

12 ANALYSIS OF THE PROBLEM OF PREDICTING THE BOTTOM ROCKS HEAVING OF MINE WORKINGS 154

12.1 Technical and economical aspects of mine workings maintenance during rocks heaving 154

12.2 Analysis of concepts of the mechanism of occurring the bottom rocks heaving and factors affecting its development 155

12.3 Analysis of prediction techniques of the bottom rocks heaving in mine workings 157

12.4 Structure, purpose and objectives of research in the geological and industrial Western Donbas region 166

13 CONSTRUCTION OF GEOMECHANICAL MODEL OF THE “MASSIF – IN-SEAM WORKING” SYSTEM 169

13.1 Structure of adjacent coal-bearing rock stratum of the mined-out coal seams of the Western Donbas 169

13.1.1 General provisions 169

13.1.2 Lithotype and thickness of adjacent roof rock layers of coal seams 170

13.1.3 Structure and thickness of mined coal seams 171

13.1.4 Lithotype and thickness of adjacent rock layers of the coal seam bottom 174

13.2 Modeling the soft mine rocks deformation in the vicinity of in-seam working 177

13.3 Substantiating parameters of in-seam workings and their support 183

Conclusions 185

14 TESTING THE MODEL OF THE GEOMECHANICAL SYSTEM “MASSIVE – IN-SEAM WORKING” 187

14.1 The substantiation of parameters of the coal-bearing rock massif 187

14.2 Parameters substantiation of the in-seam working and support 192

	Conclusions	195
15	RESEARCH AND ANALYSIS OF INFLUENCE OF STRUCTURE AND PROPERTIES OF THE COAL-BEARING ROCK STRATUM ON THE STRESS-STRAIN STATE OF THE "MASSIF – IN-SEAM WORKING" SYSTEM	197
15.1	Substantiation of variants of stress-strain state calculation of "massif – in-seam working" systems and the methodology of conducting the computational experiment	197
15.2	Stress-strain state analysis of the bottom rocks in in-seam working	203
15.2.1	Analysis of vertical stresses components	203
15.2.2	Analysis of horizontal stresses components	208
15.2.3	Analysis of the specific stresses components	216
	Conclusions	221
16	PECULIARITIES OF THE HEAVING DEVELOPMENT MECHANISM OF THE BOTTOM ROCKS OF IN-SEAM WORKING	223
	Conclusions	231
17	STUDY OF PATTERNS RELIABILITY OF RELATION OF THE BOTTOM ROCKS STRUCTURE AND PROPERTIES WITH ITS HEAVING	233
17.1	Development of a computational experiment technique	233
17.2	Establishing the patterns of structure and properties influence of bottom rocks of the coal seam on heaving manifestations in in-seam workings	238
17.2.1	Relatively uniform mechanical properties of adjacent rock layers of the bottom	238
17.2.2	Substantially non-uniform mechanical properties of adjacent rock layers of the bottom	245
18	ASSESSING THE STATE OF BOTTOM IN IN-SEAM WORKINGS	249
18.1	Construction of regression equations for calculating the value of heaving in the bottom rocks	249
18.2	The link of value of the bottom rocks heaving with the depth of mine working location	251
18.3	Separation of mining-and-geological conditions according to the nature of heaving development and degree of the bottom rocks stability	252

19 TECHNIQUE FOR PREDICTING AND MONITORING OF BOTTOM IN IN-SEAM WORKINGS OF THE WESTERN DONBAS MINES 256
19.1 General information and calculations algorithm 256
19.2 Formation of the initial data base for performing calculations 257
19.3 Assessment of the state of bottom rocks and prediction of the value of its heaving 259
19.4 An example of a predictive calculating the bottom rocks heaving of in-seam working 260
Conclusions 266
LIST OF REFERENCES 267